Global Dialogue EXPO 2000

Springer

*Berlin
Heidelberg
New York
Barcelona
Hong Kong
London
Milan
Paris
Singapore
Tokyo*

Detlef Virchow • Joachim von Braun (Eds.)

Villages in the Future

Crops, Jobs and Livelihood

With a CD-ROM

Springer

EDITORS:
Dr. Detlef Virchow
Center for Development Research
at the University of Bonn
Walter-Flex-Straße 3
53113 Bonn
Germany
E-mail: *d.virchow@uni-bonn.de*

Prof. Dr. Joachim von Braun
Center for Development Research
at the University of Bonn
Walter-Flex-Straße 3
53113 Bonn
Germany
E-mail: *jvonbraun@uni-bonn.de*

ISBN 3-540-42467-9 Springer-Verlag Berlin Heidelberg New York

Library of Congress Cataloging-in-Publication Data
Villages in the future: crops, jobs, and livelihood / Detlef Virchow, Joachim von Braun (eds.). p.cm.– (Global dialogue EXPO 2000)
Papers presented at the Global Dialogue in Hannover at the World Exposition in August 2000.
Includes bibliographical references (p.).
ISBN 3540424679 1. Rural development–Congresses. 2. Villages–Congresses. 3. Economic policy–Congresses. I. Virchow, D. (Detlef), 1959- II. Von Braun, Joachim, 1950- III. Series.
HN49.C6 V53 2001 307.1'412–dc21 2001049392

This work is subject to copyright. All rights are reserved, whether the whole or part of the material is concerned, specifically the rights of translation, reprinting, reuse of illustrations, recitations, broadcasting, reproduction on microfilm or in any other way, and storage in data banks. Duplication of this publication or parts thereof is permitted only under the provisions of the German Copyright Law of September 9, 1965, in its current version, and permission for use must always be obtained from Springer-Verlag. Violations are liable for prosecution under the German Copyright Law.

Springer-Verlag Berlin Heidelberg New York
a member of BertelsmannSpringer Science+Business Media GmbH
http://www.springer.de
© Springer-Verlag Berlin Heidelberg 2001
Printed in Germany

The use of general descriptive names, registered names, trademarks, etc. in this publication does not imply, even in the absence of a specific statement, that such names are exempt from the relevant protective laws and regulations and therefore free general use.

Cover Design:
Typesetting: Camera-ready by the editors

SPIN: 10767701   30/3130 – 5 4 3 2 1 0 – Printed on acid free paper

Preface

This book is not the usual volume about the proceedings of an academic conference. In some respects, it is a more modest undertaking. We are not aiming for a grand, state-of-the-art synthesis of rural development but are merely offering selective contributions to solving an immense problem. In other respects, this book is more ambitious. Our goal is to contribute to a process and an international debate about the future of the village in the 21st century.

This volume is part of ZEF's contribution to the discussion about rural change under the umbrella of the *Global Dialogues* of the World Exposition EXPO 2000 in Hannover (Germany). It was initiated in 1998 and included conference dialogues on information technologies for the poor, on women farmers, and on biotechnology for the smallholder sector in developing countries. This volume is based on papers presented at the *Global Dialogue: "The Role of the Village in the 21st Century: Crops, Jobs and Livelihood"* in Hannover at the World Exposition in August 2000. The *Global Dialogue* was organized by the Center for Development Research (ZEF) of the University of Bonn in cooperation with numerous partners. The *Global Dialogue* took place to facilitate learning from worldwide experiences, both the successes and the failures, in rural development. EXPO 2000 was the ideal framework for exploring in a global context visions for the future of rural areas and villages in the 21st century. We intend to follow up the *Global Dialogue* with research and policy debates.

The *Global Dialogue* brought together three hundred participants from fifty-five countries. It involved broad-based and open debate among representatives from the scientific community, politics, NGOs, the private sector and from selected rural areas of the world. The *Dialogue* allowed for an exchange of ideas between participants who would otherwise not have had the chance to meet. The participants noted that rural areas have tremendous experience in innovation and that they may benefit in the next decades particularly from such new technologies as communication technologies and biotechnology.

The publication wants to contribute to the ongoing discussion about the future of the rural areas. The purpose of this book is not to point out the misery in rural areas and those possibly responsible for it. Rather, the focus is on identifying new courses of action and on learning from successful experiences that facilitate improved living conditions in rural areas and involve productive, complementary relationships between cities and villages. The major theme of this publication is the search for fair and efficient institutions and for appropriate modern technologies that could help to promote policies and strategies for the future of the rural areas and to overcome the growing divides between rural and urban areas.

During the three day event, the participants contributed to the dialogue in one of the 13 sessions or by presenting a selected project at the "Red Table" in the demonstration area of the conference venue. Unfortunately, not all of the contributions to the *Global Dialogue* could be integrated into this publication. Nonetheless, this book aspires to depict the diversity of issues related to the promotion of the quality of life in rural areas. Consequently, the selection of articles ranges from scientific papers analyzing particular issues of rural development to short reports from practitioners describing examples of sustainable solutions for the rural populations in different regions of the world. In addition, the quoted declarations from other issue-related conferences in the year 2000 and the illustrations by *Manfred Kern* demonstrate the close linkages of the contributions to other relevant international efforts and to the Agenda 21 issues respectively[1]. The diverse backgrounds of the authors, from science, politics, NGOs and the private sector, illustrate the engagement of the many groups involved in the future of villages and rural areas. The enthusiastic response by the many participants and the media is reassuring and has shown us that the *Global Dialogue* meets a need.

The over-all message from the contributions in this publication is unanimous: there is a promising future for the rural areas worldwide if adequate policies can be enforced and more efficient and fair institutions can be created. Therefore, this publication aims to be a springboard for continuing the dialogue about the future of rural areas and improved urban-rural linkages. A CD-Rom, which can be used to foster the dialogue and to reflect the multi-dimensional approach taken, is attached.

We would especially like to thank everyone who contributed to this publication with their papers. Furthermore, for the technical assistance in processing and editing the contributions as well as in producing the video, we thank Angelika Wilcke, Monika Reule, Margaret Lampe, Annika Dröge and the video production team of "Spectrafilm."

Finally we want to thank our generous sponsors and co-organizers from the private and public sector: AVENTIS CropScience, German Agency for Technical Co-operation (GTZ), Kreditanstalt für Wiederaufbau (KfW), AgriExpo, Schleswig-Holstein Agricultural Work Group, Centre Technique de Coopération Agricole et Rurale (CTA), DETECON GmbH, Deutsche Welle, Fischerwerke (Fischer-Dübel), International Food Policy Research Institute (IFPRI), German Agro Action (DWHH), German Cooperative and Raiffeisen Confederation (DGRV), German Foundation for International Development (DSE), German Foundation for World Population (DSW), Kali & Salz GmbH, United Nations Centre for Human Settlements (UNCHS / Habitat), the German Federal Ministry for Economic Cooperation and Development (BMZ) and Gesellschaft der Freunde – ZEF e.V. They contributed not only financially but also with valuable insights and ideas. Hence, the *Global Dialogue* as well as this resulting publication represent a real public-private partnership.

ZEF-Bonn, May 2001

Detlef Virchow   Joachim von Braun

---

[1] The illustrations are taken from: Manfred Kern (2000): Agenda 21: Programme of Action for Sustainable Development. Aventis CropScience, Frankfurt.

Table of Contents

Preface V
Table of Contents VII
List of Contributing Authors XI
List of Abbreviations... XIX

Introduction... 1

1 Village Futures: Concept, Overview and Policy Implications 3
   Joachim von Braun and Detlef Virchow
2 The Crucial Importance of Urban-Rural Linkages................................21
   Klaus Töpfer
3 The Independence of the Village 25
   M.S. Swaminathan
4 A New Vision of the Rural World's Role in a Globalized
   Environment 27
   Margarita Marino de Botero

I. Political and Institutional Framework................................... 31

Policies to Foster Rural Development...33

Introductory Statement 33
   Michael Bohnet
5 The Need to Reconcile Village and Globalization................................35
   Günter Dresrüsse
6 Policies to Stimulate Agricultural and Rural Growth 43
   Rajul Pandya-Lorch
7 Decentralization and Active Community Participation in India 51
   Anita Das
8 Gandhi and Village Governance 59
   Santosh Kumar Sharma
9 Rural Incomes and Employment in Russia's Transition Phase.............67
   Renata Ianbykh
10 Policies to Manage Local Public Goods in an EU Context...................77
    Martin Scheele
11 German Policy for an Integrated Rural Development...........................85
    Wilhelm Schopen
12 Potsdam Declaration Rural21 91

Grassroots Movements...93

Introductory Statement 93
   Anil Gupta
13 The Active Rural Society in West and Central Africa 97
   Sana F.K. Jatta

14 Akina Mama wa Afrika – a Women's Development Organization 107
　　Sandra Okoed
15 The Darewadi Watershed Development Project in Maharashtra,
　　India 111
　　Robert D'Costa and Abraham Samuel
16 Grassroots Movements as a Means to Overcome Rural Financial
　　Market Failures 115
　　Franz Heidhues
17 The Barefoot Approach 119
　　Bunker Roy
18 IFOAM – the International Organic Grassroots Movement 123
　　Bernward Geier
19 SEKEM - an Egyptian Cultural and Economic Initiative 125
　　Klaus Merckens and Ahmed Shalaby

II. Sustainable Use of Natural Resources 127

Sustainable Energy Systems 129

Introductory Statement 129
　　Peter Hennicke
20 Rural Development and Sustainable Energy Systems 133
　　El Houssine Bartali
21 The Route to Sustainable Energy Systems in Germany 141
　　Erich Deppe
22 Renewable Energy Sources for Community Sustainability in India 145
　　Chaman Lal Gupta
23 Sustainable Energy Systems in Tanzania 149
　　Stanford A.J. Mwakasonda

Sustainable Soil Management 155

Introductory Statement 155
　　Helmut Eger
24 Balanced Fertilization: Integral Part of Sustainable Soil
　　Management 157
　　Adolf Krauss
25 Land Degradation: Causes and Prevention 169
　　Paul L.G. Vlek, Daniel Hillel and Sonya Teimann
26 The United Nations Convention to Combat Desertification 173
　　Hama Arba Diallo
27 Do African Soils only Sustain Subsistence Agriculture? 175
　　Uzo Mokwunye
28 Natural Resource Use in Agriculture in Sub-Saharan Africa: Myths
　　and Realities 185
　　Henk Breman

Enough Water for All..*189*

    Introductory Statement 189
        Winfried Polte
    29 Creating Institutional Arrangements for Managing Water-Scarce
       River Basins..191
        Douglas J. Merrey
    30 More Water for Village-Based Irrigation...203
        Hans P. Aeppli
    31 Access, Participatory Parity and Democracy in User Based
       Irrigation Organizations...215
        Margreet Zwarteveen
    32 Water Conservation and Demand Management in Africa 223
        André Dzikus

III. Broadening the Technological Base 233

Information and Communication Technologies...........................*235*

    Introductory Statement 235
        Subbiah Arunachalam
    33 The Role of Information in Rural Development 237
        Heather E. Hudson
    34 Okinawa Charter on Global Information Society...............................251
    35 Jobs, Livelihood and Information Technology 255
        Srinivasan Ramani
    36 Worldwide Access to ICT: The Digital Divide....................................259
        Christophe Brun and Anton Mangstl
    37 The Phone and the Future: Village Pay Phones in Bangladesh............263
        Abdul Bayes
    38 Local Participation in the Information Society 275
        Arnulf Heuermann
    39 Telecommunications in Sub-Saharan Africa..277
        Miko Rwayitare

The Future of Agriculture *279*

    Introductory Statement 279
        Per Pinstrup-Andersen
    40 What Can We Expect from Agrotechnologies in the Future?..............281
        Manfred Kern
    41 Declaration of Hamburg: Meeting Future Human Needs 295
    42 The Future of Agriculture in the Industrialized World 297
        William Lockeretz
    43 Ensuring the Future of Sustainable Agriculture 305
        Patricia Howard-Borjas and Kees Jansen
    44 Dresden Declaration: Towards a Global System for Agricultural
       Research for Development..321

Bio- and Crop Technologies for Agriculture *323*

Introductory Statement 323
    Barry Thomas and Walter Dannigkeit

45 The Impact of Biotechnology on Plant Breeding 325
    Christian Jung

46 Potential of Biotechnologies to Enhance Animal Production 329
    Ernst Kalm

47 Can India Afford to Ignore Biotechnological Applications? 339
    Usha Barwale Zehr

48 The Conservation of Genetic Resources: A Global Effort 345
    J.M.M. Engels

IV. Urban – Rural Linkages 349

Urban – Rural Linkages *351*

Introductory Statement 351
    Mathias Hundsalz

49 Urban-Rural Linkages in the Context of Decentralized Rural
    Development 355
    Sudhir Wanmali

50 Berlin Declaration on the Urban Future 367

51 Employment and Migration in the Urban Future of Southeast Asia 369
    Yap Kioe Sheng and Radhika Savant Mohit

52 Employment and Migration Perspectives in Africa 377
    John O. Oucho

53 Reunified Germany: Separate Rural Developments 385
    Claudia Neu

54 Large-Scale Creation of Sustainable Livelihoods 391
    Shrashtant Patara and Ashok Khosla

55 Urban World - Rural World 397
    Kirtee Shah

Index 401

List of Contributing Authors

Aeppli, Hans P.
Kreditanstalt für Wiederaufbau (KfW), Palmengartenstraße 5-9
60046 Frankfurt/Main
Germany

Arunachalam, Subbiah
MS Swaminathan Research Foundation, Third Cross Street, Taramani
Chennai 600113
India

Bartali, El Houssine
International Commission of Agricultural Engineering (CIGR)
BP 6202 ANAFID, 10101 Rabat
Morocco

Bayes, Abdul
Jahangirnagar University, Savar Dhaka
Bangladesh

Bohnet, Michael
German Federal Ministry for Economic Cooperation and Development
(BMZ), Friedrich-Ebert-Allee 40, 53113 Bonn
Germany

von Braun, Joachim
Center for Development Research, Walter-Flex-Str. 3, 53113 Bonn
Germany

Breman, Henk
International Fertilizer Development Center – Africa (IFDC-Africa)
BP 4483, Lomé
Togo

Brun, Christophe
Food and Agriculture Organization of the United Nations (FAO)
Via delle Terme di Caracalla, 00100 Rome
Italy

D'Costa, Robert
Watershed Organisation Trust (WOTR), "Paryavaran" Behind Market
Yard, Ahmednagar 414 001 Maharashtra
India

Dannigkeit, Walter
AVENTIS CropScience, 65926 Frankfurt/Main
Germany

Das, Anita
Government of Madhya Pradesh, B-13, Char Imli, Bhopal-462016
Madhya Pradesh
India

de Botero, Margarita
Colombia Institute for Science and Technology
Diagonal 72 No. 2-30 Este, Apt. 101 Santafé de Bogotá
Colombia

Deppe, Erich
Stadtwerke Hannover; Ihmeplatz 2, 30449 Hannover
Germany

Diallo, Hama Arba
Secretariat of the United Nations Convention to Combat Desertification
Martin-Luther-King-Str. 8, 53175 Bonn
Germany

Dresrüsse, Günter
Deutsche Gesellschaft für Technische Zusammenarbeit (GTZ) GmbH
Dag-Hammarskjöld-Weg 1-5, 65726 Eschborn
Germany

Dzikus, André
United Nations Centre for Human Settlements (UNCHS / Habitat)
Box 30030, Nairobi
Kenya

Eger, Helmut
Deutsche Gesellschaft für Technische Zusammenarbeit (GTZ) GmbH
Dag-Hammarskjöld-Weg 1-5, 65726 Eschborn
Germany

Engels, J.M.M.
International Plant Genetic Resources Institute (IPGRI)
Via delle Sette Chiese 142, 00145 Rome
Italy

Geier, Bernward
International Federation of Organic Agriculture (IFOAM)
66636 Tholey-Theley
Germany

Gupta, Anil
Honey Bee Network / Indian Institute of Management
Ahmedabad 380015
India

Gupta, Chaman Lal
Sri Aurobindo Ashram, Pondicherry - 605002
India

Heidhues, Franz
University of Hohenheim, 70593 Stuttgart
Germany

Hennicke, Peter
Wuppertal Institute for Climate, Environment and Energy
P.O.Box 10 04 80, 42004 Wuppertal
Germany

Heuermann, Arnulf
DETECON, Oberkasseler Str. 2, 53227 Bonn
Germany

Hillel, Daniel
Karkur
Israel

Howard-Borjas, Patricia
Wageningen University, Hollandseweg 1, 6706 KN, Wageningen
The Netherlands

Hudson, Heather E.
University of San Francisco, 2130 Fulton Street
San Francisco, CA 94117-1045
USA

Hundsalz, Mathias
United Nations Centre for Human Settlements (UNCHS / Habitat)
P.O.Box 30030, Nairobi
Kenya

Ianbykh, Renata
Agrarian Institute, B. Kharitonievski 21-1, 103064 Moscow
Russia

Jansen, Kees
Wageningen University, Hollandseweg 1, 6706 KN, Wageningen
The Netherlands

Jatta, Sana F.K.
International Fund for Agricultural Development (IFAD)
Via del Serafico, 107, 00142 Rome
Italy

Jung, Christian
Christian-Albrechts-University of Kiel, Olshausenstr. 40, 24108 Kiel
Germany

Kalm, Ernst
   Christian-Albrechts-University of Kiel, Olshausenstr. 40, 24108 Kiel
   Germany

Kern, Manfred
   AVENTIS CropScience, 65926 Frankfurt/Main
   Germany

Khosla, Ashok
   Development Alternatives, B-32 TARA Crescent, 110022 New Delhi
   India

Krauss, Adolf
   International Potash Institute (IPI), Schneidergasse 27, CH-4001 Basel
   Switzerland

Lockeretz, William
   Tufts University, 132 Curtis Street, Medford, MA 02155
   USA

Mangstl, Anton
   Food and Agriculture Organization of the United Nations (FAO)
   Via delle Terme di Caracalla, 00100 Rome
   Italy

Merckens, Klaus
   Society for Cultural Development, P.O.Box 1535 Alf Maskan
   ET 11777 Cairo
   Egypt

Merrey, Douglas J.
   International Water Management Institute (IWMI)
   P.O.Box 2075, Colombo
   Sri Lanka

Mokwunye, Uzo
   United Nations University Institute for Natural Resources in Africa
   Private Mail Bag, Kotoka International Airport Accra
   Ghana

Mwakasonda, Stanford A.J.
   Centre for Energy, Environment, Science and Technology (CEEST),
   P.O.Box 5511, Dar es Salaam
   Tanzania

Neu, Claudia
   University of Rostock, Rostock
   Germany

Okoed, Sandra
Akina Mama wa Afrika, P.O.Box 24130, Kampala
Uganda

Oucho, John O.
University of Botswana, Private Bag 0022, Garborone
Botswana

Pandya-Lorch, Rajul
International Food Policy Research Institute (IFPRI)
2033 K Street, N.W., Washington, DC 20006
USA

Patara, Shrashtant
Development Alternatives, B-32, TARA Crescent Qutub Inst. Area
New Delhi - 110016
India

Pinstrup-Andersen, Per
International Food Policy Research Institute (IFPRI)
2033 K Street, N.W., Washington, DC 20006
USA

Polte, Winfried
Kreditanstalt für Wiederaufbau (KfW), Frankfurt/Main
Palmengartenstraße 5-9, 60046 Frankfurt/Main
Germany

Ramani, Srinivasan
Silverline Technologies Ltd., Flat 74, Nandina, Building No. 6
K.C. Marg, Mumbai 400 050 Dhokali, Thane
India

Roy, Bunker
Barefoot College, Tilona 305 816, Madanganj Ajmer District Rajasthan
India

Rwayitare, Miko
Telecel International, 62 Wierda RD East, Sandton, 2146
South Africa

Samuel, Abraham
Watershed Organisation Trust (WOTR), "Paryavaran" Behind Market
Yard, Ahmednagar 414 001 Maharashtra
India

Savant Mohit, Radhika
Economic and Social Commission for Asia and the Pacific (ESCAP)
Rajdamnern Nok Avenue, Bangkok 10200
Thailand

Scheele, Martin
    EU Commission, Rue de la Loi 130 1/145, B-1040 Brussels
    Belgium

Schopen, Wilhelm
    German Federal Ministry of Agriculture (BML), Postfach, 53107 Bonn
    Germany

Shah, Kirtee
    Ahmedabad Study Action Group (ASAG), P.O. Box 51/A
    Ahmedabad 380-009
    India

Shalaby, Ahmed
    Society for Cultural Development, Alf Maskan, P.O.Box 2834, Cairo
    Egypt

Sharma, Santosh Kumar
    People First, B-32, TARA Crescent, Qutab Institutional Area,
    New Delhi 110016
    India

Swaminathan, M.S.
    MS Swaminathan Research Foundation,, Third Cross Street, Taramani
    Chennai 600113
    India

Teiman, Sonya
    Center for Development Research, Walter-Flex-Str. 3, 53113 Bonn
    Germany

Thomas, Barry
    AVENTIS CropScience, 65926 Frankfurt/Main
    Germany

Töpfer, Klaus
    United Nations Environment Program (UNEP), P.O.Box 30030, Nairobi
    Kenya

Virchow, Detlef
    Center for Development Research, Walter-Flex-Str. 3, 53113 Bonn
    Germany

Vlek, Paul L.G.
    Center for Development Research, Walter-Flex-Str. 3, 53113 Bonn
    Germany

Wanmali, Sudhir
    International Food Policy Research Institute (IFPRI)
    2033 K Street, N.W., Washington, DC 20006
    USA

Yap, Kioe Sheng
> Economic and Social Commission for Asia and the Pacific (ESCAP)
> Rajdamnern Nok Avenue, Bangkok 10200
> Thailand

Zehr, **Usha Barwale**
> Maharashtra Hybrid Seeds Company Limited, Jalna
> India

Zwarteveen, **Margreet**
> Wageningen University, Nieuwe Kanaal 11, Wageningen
> The Netherlands

List of Abbreviations

AI	Artificial insemination
BC	Barefoot College, India
BSE	Bovine Spongiform Encephalopathy, the "mad cow disease"
CAP	Common Agricultural Policy
CBD	Convention on Biological Diversity
CGIAR	Consultative Group on International Agricultural Research
CMA	Catchment Management Agency
CS	Consumer surplus
DCs	Developing countries
DNA	Deoxyribonucleic acid
DWAF	Department of Water Affairs and Forestry, South Africa
ET	Embryo transfer
FSRD	Fund for the Support of Rural Development, Russia
GATT	General Agreement on Tariffs and Trade
GB	Grameen Bank, Bangladesh
GFAR	Global Forum on Agricultural Research
GIS	Geographic Information System
GM food	Genetically modified food
GPS	Global Positioning System
GTZ	German Agency for Technical Co-operation
ICM	Integrated Crop Management
ICs	Industrialized countries
ICT	Information and communication technology
IDRC	International Development Research Center, Canada
IFAD	International Fund for Agricultural Development
IFDC	International Fertilizer Development Center
IFIs	International Financing Institutions
IFOAM	International Federation of Organic Agriculture Movements
IFPRI	International Food Policy Research Institute
IMT	Irrigation management transfer
IPI	International Potash Institute
IPM	Integrated Pest Management
IPNS	Integrated Plant Nutrition Systems
IRM	Integrated Resource Management
IWMI	International Water Management Institute
JFM	Joint Forestry Management, India
LCP	Least-Cost Planning Program

LPG	Landwirtschaftliche Produktionsgenossenschaft (former agricultural cooperatives in the German Democratic Republic)
NGO	Non-governmental organization
PGRFA	Plant genetic resources for food and agriculture
PHP	Private household plot
RUR	Russian Ruble
SA	Sustainable agriculture
SADeF Program	Sahelian Areas Development Fund, Mali
SAP	Structural Adjustment Programs
SC	Scheduled castes
SCD	Egyptian Society for Cultural Development
SHGs	Self-help groups
SRLPP	Sustainable Rural Livelihoods Pilot Project, Russia
ST	Scheduled tribes
UNCCD	United Nations Convention to Combat Desertification
UNDP	United Nations Development Programme
VEG	Volkseigene Güter (former state farms in the German Democratic Republic)
VPPs	Village pay phones
VWC	Village watershed committee
WANA Region	West Asia and North Africa Region
WDM	Water demand management
WOTR	Watershed Organization Trust, India
WTO	World Trade Organization

INTRODUCTION

1 Village Futures: Concept, Overview and Policy Implications

Joachim von Braun and Detlef Virchow

Issues and Questions

In a "virtual global village" scaled to 1,000 inhabitants, 165 people earn less than US $ 1 per day, 130 are malnourished, 206 people above the age of 15 are illiterate, and 306 persons are below the age of 14. The village's richest person owns more than the combined incomes of its 577 poorest people. Of course, despite the evolution of the global information society, which fundamentally bridges the distance between town and countryside, the world as a whole is *not* a village. While globalization impacts deeply on rural areas and their linkages to urban centers in particular, rural communities of tremendous diversity remain the homes of a dominant share of the world population. However, life in these communities is bound to change with historically unprecedented speed in the coming decades. This volume deals with how that change may be guided by local, national and global policies and actions in order to enhance people's livelihood.

What is the future role of the villages and the rural areas in the global context, which seems to be dominated by globalization and urbanization? In the next 25 years, 90% of the population increase in developing countries will be seen in the urban areas. By the year 2025, less than 50% of the population of Africa and of Asia and less than 20% of the American and European populations will live in rural areas.

For many policy-makers, "village" is associated with backwardness, inflexible traditions and an adverse attitude towards technological innovations. At the same time, rural areas have tremendous experience in indigenous innovation.[1] Rural development is exposed to prejudices, but there are some facts about rural areas that underline their continued, if not increasing, significance:

- The demand for food and for (solar) energy-based products will grow disproportionately to the increase in population and will be determined by the rise in purchasing power. This will lead to the need for increased agricultural production.
- The rural people are the custodians of most of the natural resources essential for sustainable development (water, soils, genetic resources, and biodiversity), but simultaneously they are running the risk of wasting them by using the resources in most inefficient ways.

---

[1] As villages are an integral feature of rural areas, the terms will be used interchangeably although it is recognized that the term rural areas encompasses more than villages.

- Of the 6 billion people inhabiting the planet today, more than one billion are living in severe poverty, and the vast majority of these are living in rural areas.
- Rural poverty is endangering political and economic stability; wherever agriculture is unproductive and the food situation is insecure, the risk of national (rural-urban) and international migration as well as of armed conflicts will increase.

People living and working in rural areas and decision makers in governments and in small and large companies are confronted with difficult questions concerning the future of rural areas and rural life on the globe:

- Will the villagers end up in (mega) cities?
- Will the rural areas be left behind in isolation, remoteness and hopelessness by fast development in the cities?
- Will the rich and immensely diverse cultures (including languages) of the world's rural areas vanish?
- Will the emerging technological and institutional changes further increase the political insignificance of villagers?

There is a paradox here: the rural areas today are both places of misery and of beauty. Problems and potentials have moved much closer together, due to technological options and the recognition of the key roles of good policy and governance. Global improvement in living standards and food security in rural areas achieved through appropriate policies, institutions (rights) and technology is also very much in the interest of those living in the urban areas of the developing world and of people in the largely urbanized industrialized countries.

A New Concept for "Rural Development" Is Needed

A general problem in addressing the issue of rural development is the lack of concepts and visions about the future of villages. However, in view of the gross policy failures of the past, one should be cautious of generalized concepts. In many parts of the world, rural areas and village life have been weakened or destroyed by badly designed policies based on misleading concepts and theories, by a lack of rule of law and by deficient governance in the 20th century. The inhumane "industrialization" of agriculture with the Kolkhozes in the former Soviet Union, the inefficient Chinese commune system, the massive expropriation of small farmers in parts of Latin America, urban policy bias and undue taxation of agriculture in many low-income countries are prominent examples. A new and careful consideration of concepts is overdue. Trends toward rule of law, democracy and international competition may be conducive for a fresh approach to rural development. But rural development cannot start over again from scratch. All the above-mentioned aberrations of rural policies created extremely protracted path-dependencies. To address the challenges by top-down planning or by emotional calls for "saving the village cultures" will not solve the multi-faceted problems. Not only plans by experts are needed, but also developmental processes need to be set in motion. A search for concepts may best be managed through institutionalized debate and information sharing. Research-based information that is accessible to all interested groups is essential. Thus, a precondition for conceptualization seems to be structured debates among scientists, policy makers, business leaders and practitioners, i.e., rural people and their leaders. Such debate should be based on relevant, sound theories as well as on ideas from the grassroots level and should consider the problems of rural areas in specific local and regional contexts.

The future of rural areas will be determined by those who are involved in rural activities: politicians, people from the administrations, members of the private sector, scientists, rural development activists, the inhabitants from the cities and, most important, the people from the rural areas themselves. After all, it will be the farmers, pastoralists, rural artisans, the women and men working in rural industry and services and especially the young generation who will be contributing to the future development of rural areas. Not one-shot conferences but rather long-term dialogue about "rural inclusion" with broad-based international and regional participation is needed. Such a dialogue about the future of the rural areas – going far beyond what can be covered within this publication – should focus on four interlinked principal questions:

1. What political frameworks will guarantee both, that cultural and institutional values are enhanced and safeguarded, and that economic growth and improved livelihood are fostered?
2. How can the sustainable use of natural resources – i.e., strategies for the consumption of energy, water use and the management of land – be achieved without conflicting with the goal of improving peoples' livelihood?
3. How can new technologies, especially information technologies and modern biotechnologies, best be utilized for the enhancement of rural areas?
4. How can efficient and fair linkages between urban and rural areas be developed in the key areas of policy coordination, employment, migration, infrastructure and transport?

The interdependency and cross-cutting relevance of these four themes are essential to increasing rural prosperity and thus benefiting the people's overall welfare. However, it would be naive to assume that such debate could be initiated, executed and brought to fruition simply "by invitation" from altruistic, visionary mediators. Key actors would simply not follow the results of such a debate as complex conflicts, power, and group interests are at stake. In the long run, democratic frameworks and a strengthened base of rights must provide the institutionalized framework for the rural development debate at the appropriate levels. Catalyzing the establishment of such frameworks is one intention of this dialogue.

A Political and Institutional Framework to Improve Rural Livelihood

Political and institutional innovation in many of the world's rural areas is out of step with the current and future needs for livelihood improvement, with demographics, with technological opportunities, and with the need for sustainable utilization of natural resources. New political and institutional frameworks are needed. Elements of a conducive framework for the improvement of the rural areas are as manifold as the diversity of rural areas. There are, however, some general features of enabling frameworks. They have to:

- meet basic needs for all, such as food, water, energy, housing, and education;
- safeguard cultural and institutional values and enhance people's participation and empowerment through democracy and rights;
- enable readiness for social and technological innovations;
- provide increasing employment and business opportunities; and
- strengthen productive urban-rural linkages.

The political and institutional frameworks proposed and their potential impact for rural areas will have to be matched against these general objectives.

Political and institutional frameworks for rural development are not only set at a local level. Overall trade and fiscal policies also must be considered, as they very much determine rural growth and employment potentials. Rural economies have the potential to produce a large proportion of their output in terms of tradable goods and services. Market-expanding information technologies will further enhance these possibilities. However, the incentives to actually realize these potentials are undermined by protectionism and taxation in many parts of the world and on the global level. While overcoming these policy distortions will create complex winner and loser patterns, the overall effect will be advantageous for rural development worldwide.

Decentralization Policies

From the perspective of rural development, a fresh look at the opportunities and constraints of decentralization – political, administrative, and fiscal – is needed. In the 1990s, a trend toward rapid decentralization from the central governments to states, districts and even villages was observed, for instance in Latin America, Russia, India, China, and parts of Africa (see *Das* and Dethier, 2000).

Decentralization can improve the effectiveness of national governments and local authorities by increasing transparency and accountability. However, national governments as well as rural authorities cannot always guarantee the supply of basic needs such as health care, education, infrastructure, and electricity to the inhabitants of rural areas. The capacity of the central as well as local governments have to be improved to strengthen economic, social, and political conditions which determine the success of the efforts of other key actors, such as farmers, the private sector and NGOs in rural development (see *Pandya-Lorch*). Besides expanding the capacities of local authorities, arrangements are needed which transcend traditional administrative boundaries between cities and rural areas, which are capable of managing settlements and their economic, social and environmental linkages at the regional level and which are able to facilitate the supply of basic public goods in rural areas.

Administrative decentralization and transparency and accountability by themselves, however, cannot bring about governance that effectively reduces poverty. It seems, for instance, that the more financially dependent a country is on its citizens (and not on other income sources such as mineral resources), the more likely it is that scarce resources will be used to promote human development and thereby reduce poverty (Moore et al., 1999). Political decentralization, e.g., elections at local levels, goes a long way toward inclusion of the rural poor. Still, even with democratic frameworks, the rural elite often manages to capture control of resources unless there is an opportunity for building a vibrant civil society with easy access to relevant information. This accessibility can be facilitated by information and communications technologies. Furthermore, investment in education and training to fuel the process of democratic decentralization is needed (see *Dresrüsse*).

Grassroots Movements

The inter-linkages between the different rural stakeholders are keys to rural innovation. During the 1990s, hundreds of thousands of rural non-governmental organizations were founded around the world. They comprise an important response to the challenges of rural development and of institutional deficiencies, and they can enhance traditional rural structures (see *Jatta* or *Heidhues*). They also will play a key role in fostering the *Global Dialogue* needed for rural development in the future. In many instances, these organizations enforce increased participation in their specific political context (see, e.g., *D'Costa and Samuel* or *Merckens and Shalaby*). Without the participation of the rural population, sustainable development in rural areas is not possible. However, participatory approaches based on interest groups should not be idealized. Along with the benefits, costs are involved for the participants, and the required institution building must take into account the existing organizational weaknesses and path-dependencies (e.g., Cleaver, 1999).

Another challenge is emerging: the growing interdependence between villages and cities increasingly confronts local village-based institutions and organizations. These urban-rural linkages also are influencing traditional values and behavior. Simply investing in current rural organizations and strengthening their institutional bases is not going to facilitate the needed inclusive rural development. Increasingly, organizations will have to bridge rural and urban areas with services such as education, health, banking, and insurance. This poses a challenge to rural grassroots organizations, such as cooperatives, as well as to urban-based organizations, such as small-business groups and trade unions. Numerous organizations have to adjust in order to create an inclusive basis for such development, and rural groups must play their part in it. Grassroots movements can also help to enforce rural peoples' effective access to the judiciary systems in order to claim their rights.

Culture, Knowledge and Education

Modernizing rural change is perhaps in conflict with the rich cultural heritages of rural people. But only by investing in education and knowledge-transfer can the rural population capture the new opportunities. The crucial importance of investment in education is shown, for instance, in China, where government expenditure on education has had the largest impact on poverty reduction and on production growth (Fan et al., 2000). New, non-agricultural jobs can only be created in rural areas if the inhabitants have the required skills. Educational improvement affects economic behavior, family patterns, social identities and cultural norms, but the success of future rural innovations relies not only on the modern educational system and the linkages to the "outside-world" but also on the traditional structure of the rural areas. Traditional values and social identities – to the extent that they are not oppressive – should be kept alive to give the rural population support and security in the evolving changes.

We do not know today if the world's next two generations will be able to live in good health, how their food security can be assured or what kind of energy they will be able to use. Therefore, to face the future, rural areas need innovations. In general opinion, the tradition-based, village-centered processes of innovation hinder standardized education and access to modern research and information systems. But instead of seeing modern and traditional approaches as oppositional, one must look for their complementary elements. Traditional and modern education and knowledge systems can both contribute to varying degrees to creating an atmosphere that enables innovation in rural areas. The information systems, especially the media, as well as the public and private, national and international research systems should share the responsibility for promoting innovations that contribute to enhanced rural livelihood.

Human Actions and Natural Resources

Natural resources are one of the most important rural assets, but they are threatened by human actions both in urban and rural areas. The demand for natural resources is increasing because of population and income growth, soil mining, water wastage, energy inefficiency, and the degradation of genetic resources. These threaten not only the economic basis and livelihood of the rural population but also undermine efforts to sustain global social and economic development. This tendency is increased by "urban footprints", degrading rural natural resources. So, for instance, migration and rapid urban growth may improve the efficiency of the energy supply, but migration also increases the demand for timber from rural areas for house construction, thus threatening the forests with unsustainable utilization (Wells, 2000). Hence, the call for sustainable resource management in the light of growing human needs must address energy, water, soil and genetic resources. Natural resources can only be maintained with intelligent management. However, despite their known value, natural resources are exploited in unsustainable ways. In this area, institutional innovations are of critical importance. The apparent under-valuation of natural resources has led to a dangerous delay in innovations. In our increasingly crowded world, rural communities and entrepreneurs must receive strong incentives for sustainable natural resource management.

Energy

A permanent rural energy supply would greatly impact the improvement of the people's well-being. Hence, it will be necessary to increase the appropriate and efficient use of energy to support the integration of villages and rural areas into modern economies. The challenge facing the rural energy supply is the need to increase human welfare while at the same time reducing the average per capita energy utilization (see *Hennicke*). In general, the rural areas have the greatest diversity of energy resources. However, these resources often are not being utilized efficiently, which leads to their over-exploitation and degradation. Concerning energy efficiency in households and transportation, the choice of which path to take remains an issue of technology research. This could be the most important source of energy savings. What alternative forms of energy supply, both stationary and mobile, are available and necessary, and what are their potentials (see *Mwakasonda*)? Certainly, rural electrification for all remains a primary necessity.

Soil

Our "virtual global village" of 1,000 inhabitants encompasses 2,210 ha of land, and 340 ha show signs of human-induced degradation. This is caused by a combination of factors, which are rooted in poverty, inadequate incentives, lack of knowledge and often crime. The land degradation issues require due consideration, because villages will only survive with environmentally sound land and soil management (see *Vlek et al.*). Besides relying heavily on institutional and political frame conditions, soil management also depends on the technologies used in agriculture (see *Mokwunye* and *Krauss*). Furthermore, villages and cities consume soil for buildings, transport, etc. This aspect of soil consumption needs to be more thoroughly researched, because with increasing urbanization, there will be strong competition for the soil in urban areas between those involved in high-value agriculture and those in infrastructure.

Water

Rural and urban areas are also connected by water usage. Of the 1,000 inhabitants of our "global village", 333 have no access to safe drinking water, and 494 are without adequate sanitation. In the future in rural areas, development in agriculture, improved rural livelihood and economic diversification will lead to increased tensions regarding different stakeholders' demands for and management of water. Within the next 25 years, water shortages may pose serious threats to human health in our global village, and it also will threaten sustainable protection of the environment (see *Merrey*). Some predict that water shortages threaten to reduce global food production by more than 10%, thus increasing the risk of food security (Postel, 1999). Today, however, water is wasted on a large scale, mainly through inefficient irrigation and river management. The solution again is a combination of institutional and technological innovations. A major constraint will be the necessity for people in many areas of the world to accept that water will no longer be free. Reflecting these challenges, it is inevitable that local, proactive approaches to developing institutional and political conditions include adequate incentives for improving environmentally sound, sustainable water utilization (see *Zwarteveen* and *Dzikus*). In addition, national and trans-national supportive frameworks have to be created to enable efficient water distribution und utilization.

Broadening the Technological Base

Technological changes may revolutionize the working conditions and the livelihood of the world's rural population. Fundamental changes are in the making due to modern information and communication technologies (ICTs) as well as biotechnologies. Important technological trends will be the transition from chains to networks in "production–supply–consumption" relations, the redefinition of economies of scale in many industry and services sectors, and transformations through the implementation of bio-, information- and Nano-technologies. But will the new technologies be appropriately designed for rural resource conditions, and will they be accessible to rural people? Or is access to new technologies a privilege reserved for rich countries? Not only the institutional restrictions which have to be overcome to make new technologies more accessible have to be considered, but also debate is needed not just about risks but also about opportunities and the potential loss of future opportunities.

As the case of modern ICTs proves, further inconsistencies in the utilization of new technologies will expand the gap between rich and poor. The inhabitants of rural areas often do not even have access to the basic technologies, which could help them to improve their lives. What opportunities exist to enable the rural areas to "leapfrog" into the application of the most useful modern technologies in order to capture basic comparative advantages (see *Rwayitare*)? The technological divide between rich and poor regions and between rural and urban areas will not be overcome by a few marginal changes. For rural areas in developing countries, the risk of being left behind holds especially true in the cases of the modern ICTs and modern development in agricultural technologies.

Information and Communication Technologies

In the global economy, the new ICTs have the capacity to bridge the gaps between regions and villages and therefore are able to provide substantial benefits through improved market access and more favorable prices. In addition, substantial sociocultural benefits especially to relatively poor households in rural areas can be identified (see *Bayes*). Missing out on these technologies also means hindering a prosperous development of the rural service industries, which are based on these new technologies. Furthermore, ICTs can be powerful tools for education, freedom of information and democratization of rural areas (see *Hudson*).

The speed of diffusion of new information technologies has been increasing. It took 74 years for 50 million people to be connected by telephone, but it only took 38 years for the same number of people to have radios. It then took 11 years for this number of people to have a television and 4 years for 50 million to be connected by Internet. Unfortunately, rural areas always are at risk of receiving services too late or of being left out completely. In our "virtual global village", for instance, only 418 of the 1,000 inhabitants listen to the radio, 247 watch television, 96 read newspapers, and 10 have access to Internet hosts. And in this era of modern ICTs, 500 inhabitants still have to walk two hours or more to the nearest telephone and in reality never use it in a lifetime. Although ICTs may be a powerful tool to close the urban-rural gap, they also may be the fastest accelerator for the marginalization of those already "left-out", thus contributing to the growing divide between informed and non-informed people (see *Brun and Mangstl*), unless appropriate policy actions are taken.

The Future of Agriculture

The utilization of existing technologies and further research on new and adapted technologies have to be promoted in order to increase agricultural production and food safety while simultaneously reducing the degradation of natural resources. When public research and extension capacities are insufficient to serve the innovation needs of the developing world's smallholder sector, agricultural and rural growth will lag behind. A range of technological options needs exploration, such as Integrated Plant Nutrition Systems (IPNS), Integrated Pest Management (IPM), water development and water saving technologies, biotechnology, and precision agriculture (see *Kern*). The critical question here is: do new technologies in agriculture lead to a deeper divide between high-technology agriculture in industrialized countries and low-technology agriculture in developing countries? And which role will the "ecological" approaches to agricultural production play (see *Lockeretz* and *Howard-Borjas and Jansen*)?

The "virtual global village" has 270 hectares of arable land. Only 7.7 ha (3%) of this land are used for transgenic crops, including 4.5 ha of soybeans, 1.8 ha of corn, 0.9 ha of cotton and 0.5 ha of canola (James, 2000). The trend so far shows that agricultural biotechnology is mainly used in high-income countries. Even in industrialized countries, innovations in biotechnology are mostly used for medical purposes. But the opening-up of China and India to agricultural biotechnologies may change the pattern of the "global village" in this respect (see *Zehr*). In the long run, the utilization of biotechnologies may have a lot to offer to rural areas and to consumers. The balance between potential benefits and risks, including the security of genetic resources, is yet to be found. The emerging potentials of biotechnology for the majority of small farmers in developing countries contradict the critics in rich countries. Although biotechnology's potential for the resource-poor farmers is high, the poor performance of seed markets and inadequate information may hinder an adequate technology transfer (Tripp, 2000).

Rural growth and development cannot be built on agriculture alone. In the future, agriculture also in low-income countries will provide jobs for fewer people than today. It seems that diversification out of agriculture has become the norm even among African rural populations (Bryceson, 2000). Besides out-migration, new, non-agricultural income sources are needed in the rural areas. While diversification offers many opportunities, it also brings financial and personal risks that require consideration in innovative risk management.

Urban-Rural Linkages

Trends in globalization and urbanization enforce the tendency to divide urban and rural development into two independent paths. At the same time, they influence villages to take on "urban elements". The duality of urban-rural linkages determines the living conditions of people in rural areas and in urban centers. The villages of today may be integrated in the urbanization process and become the towns of tomorrow, different in size but similar in structure and functions (see *Töpfer*). This development depends, however, on policies and institutional arrangements. Therefore, a debate on "village versus city" is unproductive. The two have always been in a more or less complementary relationship. Rural areas supply cities with food, with natural resources, and nowadays increasingly with commercial products from small-scale industries as well. Cities provide jobs, services and income – which in turn flow into rural regions. The issue needs, however, a fresh debate for three reasons:

- In many parts of the world the complementary elements of the two have been undermined by bad policies detrimental to both the rural and urban populations. Improved governance, coordinated with the cities, must be achieved for villages and small towns (see *Yap and Savant Mohit*).
- There are new opportunities for more productive complementary interaction because of new technologies in communications and transportation systems.
- Natural resources upon which urban health and welfare depend are located in rural areas; thus, the city has to turn its attention to the village.

Infrastructure, i.e., transportation, communications, energy and services, is the backbone of improved rural-urban linkages. Its adequacy and efficiency is essential to providing beneficial linkages between urban centers and rural areas. The inadequate provision of infrastructure maintains the isolation of poor rural areas from national development (see *Shah*). Consequently, adequate investments in infrastructure are vital to raise rural productivity, improve the functioning of markets, enable environmental protection, and allow equal access to basic services, e.g., health care and education (see *Wanmali*). The roles of the public and private sectors in investment and maintenance of urban-rural linkages still need to be defined in many rural areas.

An Overview of the Volume

This publication focuses attention on the problems and the opportunities of rural areas worldwide. It is not a comprehensive review or even a "plan of action" for rural development. Rather it is a modest attempt to raise awareness and discuss a set of important issues about crops, jobs and livelihood in rural areas, such as:

- new and adapted technologies to utilize natural resources in a sustainable way, to increase the diversity of rural income sources and to improve rural life;
- creating more and new jobs, especially in the non-agricultural sector, through better education, infrastructure and improved access to new knowledge;
- addressing the problem of rural poverty and consequently improving the livelihood of rural populations, so that rural areas will remain beautiful not only for the short-term visitor but also for those who live there.

This volume consists of scientific contributions and policy statements as well as insights and examples from different points of view, such as that of a rural practitioner or of someone from the private sector. Each chapter of the book tries to follow this holistic approach, having one or several lead-contributions surrounded by insights and examples looking at the topics from different angles to round off the discussed issues. Due to the vast number of issues, the book does not claim to discuss all relevant questions exhaustively. The idea is to stimulate readers to get engaged in the debate on the future of villages.

This book follows the concept outlined above, arranging the contributions according to four interlinked principal themes:

- political frameworks,
- use of natural resources,
- new technologies,
- urban-rural linkages.

Each theme is subdivided by two or three chapters, is introduced by relevant experts and incorporates various contributions (see Table 1.1). In view of the outline discussed above and the introductions leading into each chapter below, the following overview briefly focuses on the fundamental questions about each issue and highlights some selective contributions from each chapter. Thus the readers can be tempted to start the book at different points, entering the volume from different angles.

The book begins by introducing three very different perspectives of the future of the rural areas. Reflecting the worldwide urbanization process, *Töpfer* highlights the crucial importance of urban-rural linkages and promotes regional development concepts for urban centers as well as for successful rural development. Based on the experiences of India, *Swaminathan* calls for the strengthening of the rural population's cultural, political and social abilities and for the support of their technological development and their efforts to maintain natural resources. *De Botero* develops a vision of the rural areas' role in a globalized world, stressing the importance of eradicating poverty, securing the global food supply, empowering all relevant stakeholders, regulating global markets, globalizing economic and social rights, and supplying the rural population with knowledge, access to information and technological opportunities.

Part I considers the political and institutional frameworks, which can support the development of rural areas for a brighter and more sustainable future.

One important question to be considered here is which policies should be pursued in order to make rural areas more attractive. The importance of this is shown in the contributions in the chapter, "Policies to Foster Rural Development". *Dresrüsse* stresses that not only the formulation but also the delivery of policies is crucial. He also highlights the fact that rural development policies have to adapt to the new challenges surfacing because of globalization. *Pandya-Lorch* discusses essential policies to stimulate agricultural and rural growth; these can be applied to most rural areas in low-income countries.

Policies can foster rural development, but for the improvement of rural living conditions, it is vital for the inhabitants of these areas to be involved in policy formulation and implementation. Therefore, the importance of rural organizations and their contributions to rural development are discussed in "Grassroots Movements". What should rural people actually do, and what do they need to improve their livelihood? How can their participation be increased? *Jatta* analyzes one advanced model for participatory development interventions that directly channels development funds to grassroots movements. The importance of (micro) financial resources for rural organizations as a means of overcoming development restrictions is described. Especially at the grassroots level, comprehensive approaches involving economic, social and cultural activities are essential. *D'Costa and Samuel* reports on a related example from the *Darewadi Watershed Development Project* in India.

Part II follows pathways to sustainable development. What can be done to ensure the sustainable use of energy, soil and water so that livelihoods are not endangered by resource degradation, and the resource base is maintained for future generations?

The chapter on "Sustainable Energy Systems" concentrates on the importance of the delivery of electrification to rural areas as one prerequisite of development. However, according to *Bartali*, the ongoing development of rural areas leads to the inadequacy of the commercial energy supply, which is often limited by inappropriate electrification technologies. *Mwakasonda* stresses the urgent need for most African countries to formulate their energy policies so that they can break away from their dependence on imported energy resources. They also need to minimize the use of biomass energy because of its ecological and social consequences.

Africa is also the focus of the next chapter, "Sustainable Soil Management." Here, *Mokwunye* emphasizes the importance of keeping – or, in the case of Africa, even improving – the fertility of tropical soils. Without achieving this goal, there will be a limit to the increases in productivity. Along with the fertility of soils, the coherent and transparent planning of land use for degraded landscapes is also a major issue. *Diallo* makes clear that the active participation of the local population in planning and implementing soil conservation projects is essential.

Table 1.1. Contributors and their Backgrounds*

	Science	Policy-Maker	Private Sector	Practitioners and NGOs
Vision & Strategy	Swaminathan	de Botero Töpfer		
Political and Institutional Framework				
• Policies to Foster Rural Development	*Pandya-Lorch* Ianbykh	Bohnet Das Scheele Schopen		*Dresrüsse* Sharma
• Grassroots Movements	Heidhues			*Jatta* A. Gupta *D'Costa & Samuel* Roy Geier Merckens & Shalaby Okoed
Human Actions and Natural Resources				
• Sustainable Energy Systems	Hennicke Bartali Mwakasonda		Deppe	C.L. Gupta
• Sustainable Soil Management	*Mokwunye* Vlek et al. Breman	*Diallo*	Krauss	Eger
• Enough Water for All	*Merrey* Zwarteveen	Dzikus		Polte Aeppli
Broadening the Technological Base				
• Information and Communication Technologies	Arunachalam Hudson Bayes	*Brun & Mangstl*	Ramani Heuermann Rwayitare	
• The Future of Agriculture	Pinstrup-Andersen Howard-Borjas & Jansen Lockeretz		Kern	
• Bio- and Crop Technologies for Agriculture	*Jung* Kalm Engels		Thomas & Dannigkeit Zehr	
Urban – Rural Linkages				
	Wanmali Yap & Savant Mohit Oucho Neu	Hundsalz		*Patara & Khosla* Shah

Note: * *Italic* names indicate lead contributions in the respective chapters.

Water issues, for instance international cooperation on transboundary water resources or international trade with water and water technologies, are much discussed in international fora. For the development of rural areas, issues of water conservation and management on the regional and local levels are essential as well. Hence, the contributions in the chapter, "Enough Water for All", concentrate on these problems (see *Merrey* and *Zwarteveen*).

Part III analyzes the potential of new technologies to broaden the rural economic base and thereby to influence rural development. Technologies not only play a vital role in agricultural production and in the input and delivery sectors but also in many other areas that affect the quality of life in rural areas.

Among the new technologies, one of the most outstanding areas is that of "Information and Communication Technologies". Their potential impact on rural life is discussed in the first chapter of Part III. In principal, ICTs could transform rural life even more than urban life, as the technologies drastically reduce the costs stemming from distance and remoteness. Market space is enlarged, and access to public goods is facilitated. However, in rural areas in developing countries, deficient access to ICTs and digital gaps must be overcome to receive these benefits. According to *Brun and Mangstl*, research and policy activities should react to the fact that so far a significant part of the global population, especially the rural population in developing countries, have not only been marginalized but fully bypassed by the adoption and utilization of these technologies. With a proper framework, ICTs can be of enormous benefit to the rural poor. *Hudson* argues that with modern technologies, especially ICTs, rural innovation may be stimulated, and thus, many problems in rural areas may be overcome. *Bayes* analyzes the example of the *Grameen Telecom* in Bangladesh, which leased cellular mobile telephones to women in the village-based, microfinance organizations. In this case, a positive effect on the empowerment of women and their households was evident.

Beside these potential effects of non-agricultural technologies on the social and economic lives of rural people, technological development and adoption in agricultural production is still the most significant issue for the present and near future. The key role of enhanced agricultural productivity in rural development, especially in the smallholder sector of developing countries, remains underestimated. In the chapter, "The Future of Agriculture," the potentials of new technologies and scenarios that may dominate agricultural production in the future are discussed. *Kern* investigates future agro-technologies and predicts that these new technologies will shape agriculture in the industrialized countries. Organic agriculture with low external inputs will not be able to solve the rural labor productivity problem. Because of its high labor demand per unit of output, it is not affordable to most small-scale farmers, as *Howard-Borjas and Jansen* concludes.

In "Bio- and Crop Technologies for Agriculture", the theoretical potentials of biotechnologies in crop and animal production are discussed (*Jung* and *Kalm*). The need of developing countries to raise agricultural productivity in order to meet the growing food demand also is discussed. According to *Zehr*, biotechnologies can contribute significantly to the needed increase in productivity in Indian agriculture. Besides the application of technology, the importance of genetic resources as inputs for modern breeding and for species conservation are considered as well.

In *Part IV*, changing "Urban-Rural Linkages" are addressed. Their restructuring for the benefit of the rural and urban populations is considered. Rural areas continue to be isolated from national and international markets due to the lack of adequate infrastructure. As *Wanmali* argues, until rural-urban linkages are improved, the rural economy and development cannot prosper. As a first step, improved rural infrastructure services are of immense importance. *Yap and Savant Mohit's* vision of the rural future includes the strong integration of rural areas in regional development. Globalization is confronting the rural areas, and only those areas with improved infrastructure and decentralized power will be able to meet its challenges. Experiences from India (*Patara and Khosla*) and from Africa (*Oucho*) call for development institutions and policies to be increasingly responsible not only for the improvement of hard infrastructure but also for providing basic soft infrastructure for financing, communication and transportation. The latter should be available especially to rural small enterprises to enable them to create off-farm based employment opportunities. These will be a key factor in improving the quality of life in rural areas.

General Conclusions and Implications

Four major themes are touched upon in this volume and can be seen as starting point for further debate:

- political and institutional frameworks to foster rural development;
- natural resources management and related actions;
- broadening the technological base of rural economies;
- improved linkages between urban and rural areas.

These four comprehensive themes are interlinked. Thus, more than one of these individual issues need to be addressed for the improvement of livelihood in a specific rural area.

Without favorable policy frameworks, all other efforts will not be able to develop their full potentials. Frameworks have to be established to facilitate participation in broader development, to signal opportunities for the long-term future, to encourage individual initiatives, and to stimulate investment. Especially young people have to be convinced that rural areas are places with prospects. Success is achieved in those regions where self-initiative and self-responsibility among the rural inhabitants is encouraged.

At the outset we asked a set of much-debated questions, which we repeat here and attempt to answer them in a nutshell on a preliminary basis with the understanding that each requires further study:

- *Will the villagers end up in (mega) cities?*
  No. So far only about 12% of the world population lives in megacities (more than 4 million), and the urbanization process seems to accelerate more in small towns than in big cities in the developing world. This trend may actually facilitate improved rural-urban linkages.
- *Will the rural areas be left behind in isolation, remoteness, and hopelessness by the fast development in the cities?*
  Not necessarily, because new infrastructure, especially new information technologies, are facilitating new bridges. It also depends, however, upon the capacity to build innovative institutions in rural areas.
- *Will the rich and immensely diverse cultures (including languages) of the world's rural areas vanish?*
  There seems a great risk that this may actually occur, and education policies and knowledge systems are challenged to address these issues. But possibly new rural cultures will emerge.
- *Will the emerging technological and institutional changes further increase the political insignificance of villagers?*
  This is not inevitable. Decentralization and improved local governance can amplify the "voice" of the rural people. In a context of improved access to rights and meaningful elections, rural people may increasingly take their own forceful initiatives, and political systems will have to respond.

As contributions from very different backgrounds in this volume show, the rural areas and their inhabitants successfully create and adopt innovations wherever possible. They are not backward, inflexible and averse to innovations, especially not to technological ones. Institutional innovations, however, are often hampered by political and legal frameworks. Furthermore, it seems evident that there is a great diversity of approaches to improve the quality of life in the rural areas. Keeping in mind that no universal solution exists, some general trends can be deduced from the contributions, especially for rural areas in low and middle-income countries:

1. Rural development has to be undertaken for and with the rural people and to be integrated into an overarching national and international development policy. This includes the responsibility that transparent, accountable, efficient local authorities have to interact with all groups of civil society and that national policies have to facilitate the supply of basic public goods in rural areas (e.g., health care, education, infrastructure). Without investment in these basic needs, no conducive environment in which rural development can flourish will be created.
2. Decentralization is one of the concepts necessary to foster rural development. It seems that Gandhi's idea that "... *the power has to move from the villages upwards...*" (see *Sharma*) remains relevant for many rural settings. However, decentralization involves costs and risks, and it requires capacity building in rural communities. Only by strengthening the traditional and newly established rural organizations will rural people be able to take the required initiatives.

3. Modern education and information systems have to complement, not contradict, the traditional knowledge systems prevailing in villages. Rural inhabitants will not be able to survive and cope with the challenges of a globalized world by cutting themselves loose from the urban knowledge and education systems. Access to modern education in combination with traditional knowledge is a key factor to ensure that rural inhabitants themselves are able to take part in modernization. Better access to technologies can help to bridge the information gaps, but the information flow should not be top-down from the global to the local levels or one-directional from the cities to the villages.
4. Natural resources for rural livelihood are increasingly scarce. Although rural inhabitants depend on these resources, they are undervalued, and degradation is increasing. Only by combining appropriate technologies as well as appropriate incentives and institutions can the utilization of natural resources be optimized with increased efficiency.
5. Modern technologies for agriculture and for other rural activities contribute to improving the economic and social situations of the rural population. Technological innovations fostering sustainable and productive agriculture have to be stimulated. This is and will continue to be the backbone of the rural economy for decades to come. Hence, a rejection of modern technologies may be counterproductive to improving the sustainability of rural areas. Instead of rejecting modern technologies, action has to concentrate on their safety issues, the compatibility of new and traditional technologies, and increasing the acceptability and adaptability of technologies to local conditions and capabilities. International agricultural research systems have an important role to play here.
6. Besides the importance of (small-scale) enterprises to sustainable rural livelihood, the potential of agriculture and agribusiness as engines for economic growth in rural areas should not be underestimated. In many rural areas, the production of food and other agricultural products as well as the supply of renewable energy will continue to be sources of sustainable incomes for a long time to come.
7. The urban centers and the rural areas are interconnected in many ways. Much attention is paid to migration, but the rural-urban linkages involved in water use, waste disposal, pollution, information and rights have not been properly addressed by policy, research and industry. The future developments of rural and urban areas are interdependent, and this relationship has to be considered in all future rural development concepts and activities.
8. "Voice" and capacity are needed to clearly articulate public demands and to stimulate action. Overcoming rural poverty involves not only issues of knowledge, technology and good politics but also of power, rights and participation.

Despite all the efforts to improve rural livelihood, poverty clearly will characterize most rural areas in the next decades. Still, there is opportunity and hope for villages in the future. The diversity of approaches to improve the quality of life in rural areas together with the inventiveness of the rural people provide great opportunities for the future development of these areas. The village is not the place where one has to stay if he or she has not yet "made it" in life. The village in the 21st century has the potential to be an exciting place for crops, jobs and livelihood.

References

Berdegue, J.A., T. Reardon, G. Escobar, R. Echeverria (2000): Policies to Promote Non-Farm Rural employment in Latin America. ODI.

Bryceson, Deborah (2000): Rural Africa at the Crossroads: Livelihood Practices and Policies. ODI Natural Resource Perspectives #52.

Cleaver, Frances (1999): Paradoxes of Participation: Questioning Participatory Approaches to Development. Journal of International Development, 11, pp: 597-612.

Dethier, Jean-Jacques (ed) (2000): Governance, Decentralization and Reform in China, India and Russia. Kluwer Academic Publishers, Dordrecht.

Fan, S., X. Zhang, L. Zhang (2000): How does public spending affect growth and poverty? The experience of China. Global Development Network.

IFAD (International Fund for Agricultural Development) (2001): Rural Poverty Report 2001 – The Challenge of Ending Rural Poverty. IFAD, Rome.

James, Clive (2000): Global Status of Commercialized Transgenic Crops: 2000. ISAAA Briefs, No. 21 – 2000 preview.

Moore, Mick, Jennifer Leavy, Peter Houtzager, Howard White (1999): Polity Qualities: How Governance Affects Poverty. IDS Working Paper #99 Institute of Development Studies, Brighton.

Postel, Sandra (1999): Pillar of Sand: Can the irrigation miracle last? Worldwatch Institute.

Tripp, Robert (2000): Can Biotechnology Reach the Poor? The Adequacy of Information and Seed Delivery. Paper presented at the 4th International Conference on: The Economics of Agricultural biotechnology. Ravello, Italy, August 24-28, 2000.

Wells, Jill (2000): Construction Wood Markets in Four African Towns. Research Report to UK Department for International Development (DFID).

2 The Crucial Importance of Urban-Rural Linkages

Klaus Töpfer

We are presently witnessing the gradual transformation of our societies from rural to urban. By next year, we estimate that half of the world's population will live in urban areas. In two decades, close to 60% of this population are expected to live in towns, urban conglomerations and megacities. Still, rural areas currently accommodate the majority of the population in many countries in the developing regions of Africa, Asia and Latin America. However, most countries in Africa and Asia are experiencing rapid urbanization rates. A growing number of people are leaving the isolation of village life in the hope of escaping poverty and finding employment in the cities.

Cities are – and will remain – the centers of global finance, industry and communications. They are home to a wealth of cultural diversity and political dynamism and are immensely productive, creative and innovative. However, they also have become breeding grounds for poverty, violence, pollution and congestion. Unsustainable patterns of consumption among dense city populations, concentration of industries, intense economic activities, increased motorization and inefficient waste management all suggest that the major environmental problems of the future will be urban. At least 600 million urban residents in developing countries – and the numbers are growing – already live in housing of such poor quality and with such inadequate provision of water, sanitation and drainage that their lives and health are under continuous threat. For many millions of people around the world, urban living has become a nightmare, far removed from the dreams of safety and prosperity held out by city visionaries. This is especially true for the young people, who will inherit the urban millennium.

Not only do we live in an urbanizing world, but we also are experiencing an unprecedented urbanization of poverty. In most cities of the developing world, up to one half of the urban population lives in "informal" slum and squatter settlements, which are neither legally recognized nor serviced by city authorities. The informal parts of the city do not enjoy many of the benefits of urban life, including access to basic services, health care and clean running water. Residents live in constant fear of eviction and most do not have access to formal finance and loan schemes which could enable them to improve their living conditions. However, this invisible majority is indispensable to the economy of the city.

Rural areas and the prospects of survival in the villages of most developing countries are not much better. Low levels of agricultural productivity, poor market conditions and limited access to basic infrastructure and services prevent most of the population in rural areas in Africa, Asia and Latin America from overcoming poverty. The situation is often made worse by the pricing systems for agricultural products which subsidize the urban consumer at the expense of the farmer. As a result, people continue to abandon their villages and migrate to the urban areas, often in the vain hope of finding other forms of livelihood. Rural development and the future of living conditions in rural areas need to be understood and addressed in the context of the continuing urbanization process. Experience from many countries, particularly the least developed countries, demonstrates that policies to stop the migration from rural to urban areas have had only limited success.

Since the industrial revolution and the corresponding growth of towns and cities, there have been two dominant views regarding the rural-urban divide: an anti-urban view and a pro-urban view. These stereotypes have persisted until today, and they have significantly influenced national policies, investments in infrastructure and services, and international development cooperation.

With the help of world conferences like the United Nations "*City Summit*" in Istanbul (1996) and "*Urban 21*" in Berlin (July 2000), perceptions about urban development are changing. At the *City Summit* four years ago, the potentials of well-managed and planned urbanization for economic growth, social stability and environmental protection were recognized. At the same time, the attention of governments and of civil society at large was drawn to the fact that, in many parts of the world, rural areas risk being further isolated from the mainstream of economic, social and cultural life.

The risk of further isolation and the future prospects of rural development in combating poverty and malnutrition in a globalizing world were major concerns of the Federal Government of Germany when organizing the "*International Conference on the Future and Development of Rural Areas*" ("*Rural21*") in Potsdam in June 2000. That conference adopted a final statement which underlined that rural areas and conurbations are partners in regional development.

At the World Conference "*Urban 21*", the Secretary General of the United Nations pointed out that global and local issues are more and more interlinked. Well-functioning cities are essential for successful rural development. Mr. Kofi Annan called for stronger local authorities and partnerships with civil society, as cities take on new responsibilities for managing markets, communication and transport with their hinterlands.

The time has come to put aside the urban-versus-rural controversy as it is unproductive and gives a wrong sense of alternative development options. If we want to outline strategies for the future of villages, we need regional development concepts which strengthen the links and complementaries between cities and rural areas. As a result of structural changes in infrastructure and employment, the villages of today will become the towns of tomorrow in many countries.

Better access to infrastructure and services along with communication and employment opportunities outside the agricultural sector is transforming the livelihoods in today's villages. The conventional view of rural development being equivalent to agricultural production no longer reflects the reality of rural areas in most parts of the world. As rural areas have diversified their economic development patterns away from agriculture-based employment opportunities, urban-rural linkages have become increasingly characterized by the economic demands of the service and industrial sectors. Increases in agricultural productivity through such advances as mechanization, agrarian reforms and genetic improvements have influenced the decline in demand for traditional forms of rural employment.

This continuing trend creates employment demands outside the agricultural sector, both within rural areas and in managing the flow of goods and services between settlements in urban and rural areas. Market interactions and their impact on employment patterns will become dominating factors in the development of rural areas and the management of urban-rural linkages. In a number of countries, explicit policies promote rural industries and commerce in order to create employment for surplus labor in rural areas. These policies are also expected to reduce rural-to-urban migration.

In the future, villages and cities are likely to lose their traditional distinctions. Villages will take on urban elements such as better access to infrastructure and services, communications technology and diverse employment opportunities. At the same time, expanded urban areas will become agglomerations of distinct communities with semi-rural characteristics. In many parts of the world in the future, settlements will vary in size but be similar in structure and functions. As this likely scenario will impact strongly on the future of villages, we need to reflect on some conventional planning policies which were formulated in the past to guide development in rural and urban areas:

1. We need to have a fresh look at policies which aim at reversing the urbanization trend. We can observe today that the past decades of investing in rural development dramatically increased food supply in many parts of the world, but it did little to stop the migration of people to urban areas in search of employment.
2. We need to combat persistent poverty in rural areas of many developing countries, as well as in countries with economies in transition. More effective regional development initiatives are needed which reduce isolation and establish better communication links between villages and towns.
3. We need new types of local government institutions which transcend traditional administrative boundaries between cities and rural areas. Production linkages between urban and rural areas require efficient regional development authorities.
4. The globalizing economy places new challenges to the ability of rural areas to compete in international markets. Without adequate regional development concepts, there is the danger that globalization will further marginalize poorly developed rural areas.
5. Better environmental accounting is needed in managing the linkages between a city and its rural surroundings. How can the costs of water supply, energy production, waste management or the creation of recreation areas be attributed in well-functioning, urban-rural relationships? Closer urban-rural linkages have multiple impacts on the ecology of rural areas as well as of cities.

A city's environmental impact on the ecosystem of its region is principally the result of its demand for natural resources and products from forests, rangelands, farmlands and aquatic systems from outside its boundaries, but also of urban waste being dumped in rural areas. Without a continuing supply of energy, fresh water and agricultural products, many cities would rapidly decline and would have reduced employment opportunities for their residents. In the past, the size and economic base of a city was largely determined by the size and quality of the natural resources in its surrounding region. If local ecosystems were degraded, the prosperity of the city suffered and its viability was in danger. In other words, a city's ecological footprint remained relatively local. Today, prosperous cities in the North use the entire planet as their ecological hinterland for food, energy and natural resources. If consumers in London or New York get their supplies from a wide range of countries in all regions of the world, how can a link be established between urban consumption patterns and their ecological impacts? The ecological footprints resulting from the consumption patterns of global cities can no longer be managed within the framework of urban-rural linkages.

Well-functioning cities are essential for promoting rural development. Well-functioning cities rely on forms of urban governance which are transparent, accountable and operate through public-private partnerships with the business sector and the different interest groups of civil society. In international development cooperation, we observe a new focusing of attention on the importance of working with responsive and accountable institutions of governance. Like other organizations of the United Nations system concerned with international development, HABITAT is concentrating its efforts on supporting development agendas at the local level which enhance transparency, promote people's participation and guide private sector investments to combating poverty in urban and rural areas.

The "*Global Campaign for Good Urban Governance*" is a new strategy by HABITAT to promote a common understanding of the importance of good governance in managing our cities in the context of rapid urbanization and declining living conditions. Urban-rural linkages are an essential element for improving the performance of cities in the 21st century. The prospects for villages and for rural development are fundamentally linked to the performance of cities within defined regions and even beyond. The second "*Global Campaign*" in the work of HABITAT focuses on the promotion of secure tenure. Equitable and secure access to land tenure, in urban as well as in rural areas, is a precondition for investment, promotes good governance and social justice, and has a direct impact on combating urban and rural poverty.

3 The Independence of the Village

M.S. Swaminathan

On August 15, 1947, my country, India, became independent from colonial rule. On that day, the man whom we call "*the Father of Independence*", Mahatma Gandhi, did not come to Delhi. He was not there to join in the celebrations. He was in a place called "*Noakhali*", now in Bangladesh, preaching communal harmony and peace. Gandhi sent a message to the new rulers in Delhi, saying that India lives in its villages and that our independence will be complete only if we achieve what he called "*Gram Swaraj,*" which means the independence of the village, i.e. the local self governance. This was the only message he sent.

Today it is still important to encourage the rural population, strengthening their cultural, political and social abilities as well as their technological development and the stewardship of their natural resources. Dialogue is therefore needed - not a dialogue of the heads of states and the heads of governments, but of people who care about others and are concerned about the future of humankind.

Four major groups of issues must be considered in relation to the future of the villages: (i) broadening the technical base, (ii) human actions and natural resources, (iii) political institutional frameworks, as well as (iv) urban and rural linkages. The crucial question is, "*How do we bring these issues together in an interacting mode of organization?*" The topic calls for a new social contract between us and those who are underprivileged, for we all know the extent of the inequity in the world today.

In the center where I work in Madras, now called Chennai, we started a voyage nine years ago. We started an adventure by bringing the relevant issues of rural development together. We call it a "*bio-village,*" and it is a model for village development. The term implies human-centered development, where human beings are the purpose and means of development. The concept has three major elements, all flowing together and each with its own horizontal dimensions. The following short description of the main elements of our "*bio-village*" serves as an introductory example for the contributions in the following chapters. These attempt to bring together the main issues of the future of the village:

- The first dimension is to work for a *demographic transition*: to lower birth and death rates. This is done with an awareness of the population-supporting capacity of the ecosystem. It depends on such criteria as the extent of land and water resources, the number of animals in the area, and how many people a particular village can support. Ultimately, only an awareness of the population-supporting capacity of the ecosystem and the conviction that children must be born for happiness and not just for existence will lead to a sustainable demographic transition.
- The second dimension is to overcome the increasing *divide*: There is an increasing divide between those maintaining genetic resources and those having access to technologies that benefit from the fruits of the technical revolution. A

modern genetic revolution is taking place in many fields, from medicine, to agriculture and veterinary science. Rural people urgently need to gain access to genetic technologies and to benefit from the maintained natural resources. Until now, however, their access to modern technologies has been denied. Besides the genetic divide, a digital divide also exists. However, it is better understood and more talked about, largely because it is more transparent in terms of information, understanding and communication. We have tried to overcome the divide through what we call *"rural knowledge centers."* These are computer-supplied, internet-connected, locally operated and managed centers in rural areas. These centers are run by mostly illiterate women, and they are performing enormously well. In the past, we thought illiterate meant ignorant. We now know, however, that wisdom cannot be equated with literacy. These illiterate people can take to the new technologies like fish to water. We have seen that these *"illiterate persons"*, especially women, can master the modern communication technologies very well. Therefore, there is no excuse for the digital or for the genetic divide.

- The final dimension is to promote *job-led economic growth*: India, China, Bangladesh and other countries are intensively populated. They are population rich but land hungry. These countries cannot afford to have the same kind of jobless economic growth as in the West. Job-led economic growth is essential. It must be based on small-scale enterprises and microenterprises and include services such as microcredits and other inputs available in very small units. In today's world, unfortunately, global macroeconomic policies are designed to make microenterprises non-viable, because they are based upon Ricardo's famous principle of comparative advantages. There is a conflict between small-scale microenterprises and microcredit on the one hand and, on the other hand, mass production technologies where efficiency is measured by downsizing in terms of personnel. The challenging question remains, *"How do we make global macro- and local micropolicies that mutually reinforce their countries?"*

4 A New Vision of the Rural World's Role in a Globalized Environment

Margarita Marino de Botero

As we all know, the world is changing rapidly, and global development is bringing forth new trends that have a crucial impact on the rural world. Globalization has redefined development in such a way that issues such as justice, ethics and responsibility seem unfashionable. Fundamental aspirations of societies, except for those dealing with the economy, have been completely marginalized.

Present national policies that are implemented under pressure from multilateral development banks or international financial institutions and require structural adjustments have provoked massive cuts in essential governmentally funded services. Services that are vital for social well-being, especially of people living close to subsistence, taking away what few social advantages developing countries have gained over the last few decades. When this happens and when national governments are released from their social responsibilities, social exclusion, uncertainty and vulnerability increase, primarily affecting the rural world.

Most governments in developing countries are unable to uphold a coherent national development policy due to fears of negative effects on their fragile connection to the world's market. The dramatic cut in internal social investments in most countries experiencing multiple transitions, from demographic to political ones, has triggered an economic crisis that is aggravating poverty, increasing unemployment, reducing incomes and thus creating wider differences between the urban riche and the rural poor. Today, the primary concern of these governments is to not be left behind by globalization. The governments' loss of autonomy and the failure of their policies to remedy the growing gap of inequality within their countries' makes introducing social concerns into development policies difficult. Thus, the process by which people are marginalized is accelerated. What kind of development do we really have?

Today, according to the World Bank and UN statistics, more than eighty countries have a lower per capita income than they had a decade ago. On the other hand, there are more resources and wealth in the world today than at any other time in history. The increasing concentration of information technology and wealth has created huge differences among and within countries, and essential issues of distribution and equity have been neglected.

Though much of the world has urbanized, the majority of the world's poor continue to live in rural areas. By 2010, the existence of some 47% of the population in the developing countries will still depend mainly on agriculture. 70% of Asians live in rural areas and the percentage is even greater in some African countries. So it is vital that these people have access to land, credit, services and markets. They should also be able to participate not only in the local market, but also to take advantage of the larger regional and global markets, to meet their needs and to sell their products. So far, developing countries have not been able to fairly insert rural areas into urban markets nor their national markets into global ones. The later is made evident by the fact that the OECD countries, with only 19% of the global population, are responsible for 71% of global trade in goods and services, 58% of foreign direct investment and represent 91% of all internet users.

The failure of the old institutions, traditional social policies and economic models has not deterred the same people who promoted them from designing new ones, this time shaped by structural reforms. Leaving aside the fact that most of our social architecture stands as a barrier limiting the distribution of equal opportunities. The growing gap between the modern urban industrial sector and the rural one, along with the weakening of the linkages that used to exist intensify the problem. The situation of mayor rural areas in the third world is made worse by some unresolved problems:

1. The lack of consideration in national development policies given to the interrelations between poverty, adequate employment opportunities and income in rural areas. It is seldom recognized that eradication of extreme poverty is crucial for any development process and for creation of wealth in a nation as a whole. Increased agricultural productivity could solve the problems of the rural poor, especially if linked to the ties that exist within the country's economy and the macro economical structures.
2. The lack of agrarian reform, credit and technological assistance strengthens the inequality of millions of landless peasants.
3. The late appearance of environmental concerns in national development policies that have allowed the widespread degradation of natural resources and further reduced land productivity and increased poverty.

Today, the world recognizes that poverty is growing and spreading, both in the north and the south. However, in developing countries, the poorest of the poor still live in the rural areas where job opportunities are few, and the breakdown of ecological systems, social exclusion and violence are common. The result is a large-scale movement of people, of economic and ecological refugees to the slums of the cities. Conventional growth has failed to build up a strong rural sector and to significantly impact the social situation. Nowadays, there is a new trend: the promotion of biotechnology by international institutions such as food, seed and drug multinationals. The question of the uncontrolled use of biodiversity from the south and that of traditional property rights of indigenous people arise here. Both the academic and the political world is divided on the question of the opportunities and risks of modern technologies and their impacts on small-scale farmers and the rural world in general.

The incompetence of governments and international institutions in tackling these issues has served to make the NGO community more visible. NGOs are now expected to work not only as activist for structural change, but increasingly as service providers, even as substitutes for the state itself. It is acknowledged that NGOs have been able to awaken the greatest awareness of the issues of cultural and ecological diversity and the value of social participation. They have caused a fundamental transformation in mainstream developmental thinking. Their different ways of working with communities have certainly put a stop to the unwanted trend towards "development monoculture" and have highlighted the importance of new issues such as inter-generational equity, corporate responsibility and environmental citizenship. Sometimes as harsh critics of existing conditions and other times engaged in and building social movements and solidarity groups that are increasingly getting people out on the streets to protest.

The existence of a strong civil society certainly should give us considerable hope for the future. However, even if NGOs mobilize millions of people, and it is said that they touch 20% of the world's poor, the result of their work is uneven, and their contribution to social change is less substantive and durable than desired. A global social movement connecting different networks and organizations needs to be based on more than wishful thinking. More strategic and systematic work on common agendas and understanding of their political impact is vital. If civil societies want to indeed lead the world to real and lasting social and economic transformations and sustainable development then they must accomplish more than a patchwork of local, isolated projects. Whether they can influence the world strategies or not, there is an urgent need to help them restore the vitality of regions, to foster institutions able to solve local problems, and to concentrate on creating educational systems that can help translate local experiences into policies, so that both the knowledge and benefits of the new advanced science and information technology can be shared.

Actually, the focus of rural policies has shifted to a new vision of the role of the rural world in global politics. While forging an international consensus on the minimum standard of equity conditions in future development schemes, some major international institutions have convinced themselves that the problem with universal entitlement to livelihood and participation in common issues is that they are not available to the rural poor.

Our vision regarding the new role of the rural world can be summarized as follows:

- There is a need to introduce agreements to define public policies in sectors, which are strategic to national development, such as the rural one. This requires strengthening national systems and insuring its efficiency in achieving more participatory democracy.
- There is a need to improve the performance of rural communities and agricultural producers to eradicate poverty and to achieve food security.
- To prove that the wealth of human knowledge can serve the needs of the poor, we have to equip the rural world with scientific and technical knowledge, with access to information and to learning opportunities, and to construct alliances and coalitions.
- It is urgent to highlight the new visibility of traditional knowledge in the search for new products and to recognize the rights of indigenous people when sharing the benefits of countries with high biodiversity.
- Furthermore, to share knowledge, local institutions and supporting centers of learning and dialog have to be built and gender aspects as well as community facilities have to be reinforced.

- Means have to be explored for the successful integration of the complementary roles of villages and cities, of the links between the agrarian and the industrial sectors.
- Whether international and national consumers are willing to pay higher prices for products that are organically and ethically produced remains to be seen. The issue of green markets, social labor and fair-trading should be promoted. Some examples of green labels have contributed to better working conditions for workers.
- The much talked about restructuring of the international financial architecture must create a permanent dialog between different stakeholders to empower them to function equitably. This will happen only if we accept the need for differential treatment of disadvantaged societies in a number of key areas.
- We have to recognize the fact that economic, ecological and societal systems are interdependent, and that the reduction of disadvantages and inequalities is the best way of protecting the world's security and peace. In other words, we need to regulate equity in global markets and globalize economic and social rights improving solidarity.

As many stakeholders as possible should be given a voice when it comes to changing the direction of development. Innovative and applicable public policy can create a world in which all people are socially responsible.

I. POLITICAL AND INSTITUTIONAL FRAMEWORK

Agenda 21 Principle 5, Earth Summit, Rio 1992

Principle 5
All States and all people
shall cooperate
in the essential task
of eradicating poverty...

Policies to Foster Rural Development

Introductory Statement

Michael Bohnet

Although the urban issues of migration and of population increase are dominating overall political discussion, the rural areas are dominating production for world food demand and for the supply of natural resources. Furthermore, with more than 50% of the poor still living in rural areas, rural poverty endangers political and economic stability. Continued worldwide development is only sustainable if the challenges and obstacles of rural life are taken into consideration more intensively. What political framework safeguards cultural and institutional values, making rural areas more attractive? Does decentralization, one of the leading catchwords at present, lead to "good governance" per se? And which framework will support technological innovations for improved rural livelihood? Is it possible to guarantee basic needs (public goods) such as health, education, and infrastructure to the inhabitants of rural areas simply by improving the effectiveness of national governments? Hence, this session is focusing on policies which foster rural development, and all contributions are pursuing these questions.

As *Dresrüsse's* contribution highlights, even more important than good policy formulation is its delivery. But foremost, the rural areas and their local agenda for policy priorities have to be seen in light of the global issues of ecological, economic and social sustainability. Future rural development has to adapt to these new challenges.

The discussion has shown that we have to depart from linear thinking and move to systemic thinking. Based on this approach, some crucial and generally agreed upon policies to stimulate agricultural and rural growth are formulated by *Pandya-Lorch*.

In addition, policies should promote effective participation of all citizens, include the respect of all human rights, and strengthen democracy on the basis of the village. In this context Indian experiences are discussed (*Das*), and the example of the history of India is reflected (*Sharma*).

Ianbykh's contribution adds rural employment as further essential for rural development. According to *Ianbykh*, Russia's ongoing reforms in the agro-food sector need to focus on rural employment. During the current transitional period, the rural population of Russia is facing severe unemployment and living conditions below the subsistence level. Rural poverty there can only be fought with the development of self-employment in private household plots, market integration, access to inputs, services, and financial funds. Furthermore, the introduction of transparent agrarian policy is necessary.

One generally accepted policy is to create and develop efficient and effective markets, especially for agricultural products, and to replace inefficient state market companies and inappropriate government regulation. In this context *Scheele*, referring to the political discussion in the European Union, develops general insights regarding the management of local public goods.

Aside from the international issues, *Schopen's* contribution shows that a strong conceptual discussion about agricultural and rural development also exists in German and European politics, including the issue of the multifunctionality of agriculture in the rural area. Rural areas and agriculture should not only create agricultural products but also promote environmental conservation and protect the so-called cultural and social heritages.

Further isolation of rural areas and the opportunities of rural development in improving livelihood were major concerns of the *Potsdam Declaration* of the *International Conference on the Future and Development of Rural Areas* ("*rural21*"). This declaration indicates opportunities, methods and strategies for the sustainable development of rural areas and possible ways to implement them, highlighting the importance of the development of endogenous potentials and the exchange of experience with other regions. Extracts from the declaration can be read as a final point.

Summarizing the different aspects of this session, the main function of policies to foster rural development is to strive for the village to be both a place of work and a place of home. Furthermore, the village provides an opportunity to experience nature. The citizens are shaping the village, and the village can also shape the future. Finally, the village is also a home to remember, a place to keep our intellectual and cultural heritage alive, and a place to preserve this cultural heritage. The saying holds true: *People will not look forward to prosperity who never looked backward to their ancestors.*

5 The Need to Reconcile Village and Globalization

Günter Dresrüsse

We who were born in country places
Far from the city, and shifting faces;
We have a birthright no man can sell
and a secret joy no man can tell.[1]

Where Do We Stand?

Deryke Belshaw, one of the great rural development experts and theoreticians, observed back in 1993 (Belshaw, 1993) that the debate on the importance of integrated approaches in rural areas made him feel more as if he had come full circle than made any real progress: from the community development approach of the sixties; through the basic needs, food security programs and integrated rural development of the seventies and eighties; from resource management and sector investment programs back to the future; to the communal funds and self-reliant community investment only too familiar to us from the era of village development funds and, well, community development.

After a number of detours down overly economic and technocratic avenues[2], everyone nowadays fortunately regards humankind as being at the center of development efforts in the rural areas. What we rejected at the end of the eighties as too complex and unmanageable in rural development is currently being revived in the form of complex program initiatives. These aim both at horizontal integration (connecting different sectors) as well as at vertical integration (connecting the micro or village level with the regional meso level).[3] On the macro level, the initiatives seek to establish a sociopolitical basis for these actions and a framework for development.[4] Rural development is thus rehabilitated.[5] In any case, the GTZ largely resisted the pressures and deliberately continued to

---

[1] Poem: Farmers humming in the Indian countryside. Obaidullah Khan.
[2] "*The World Bank was originally the International Bank for Reconstruction and Development, dominated by the values and methods of engineers concerned with infrastructure (things) and economists concerned with numbers (to which people are reduced)*" (Chambers, 1993).
[3] "*...as development economists concentrated primarily on the macroeconomic and microeconomic features of developing countries and largely ignored the meso-economic, the intermediate, institutional features which mediate between the macro and the micro.*" (Adelmann, 2000)
[4] A very good example of this is program planning for the rural sector in Peru (GTZ, 2000).
[5] See also the article by Eger, Müller, Pfaumann in the most recent edition of GTZ's *Akzente*, GTZ 3/2000 (Eger, 2000).

develop and put forward its regional rural development concepts.[6] Nevertheless, even though the World Bank, the Inter-American Development Bank (IDB)[7] and a number of bilateral donors stress the importance of rural development, their words have not been followed by deeds. As so often the case in development cooperation, financial statistics are not the appropriate yardstick to measure success.[8] The issue is rather qualitative: *"Are we now in a position to implement in rural areas development concepts that carry conviction with the people concerned (politicians and inhabitants alike), because they can see the immediate benefits?"* One indication of this is whether or not the concept can be used as a model and replicated without the financier being part of the demand equation, i.e. the concept is not only attractive to the recipient in association with its funding.

What can we put forward in terms of main lessons learned as we refute with greater confidence the arguments of the cynics and take upon ourselves the laborious task of trying again, this time more successfully, or expect others to do so?

Lesson One: More than good policy formulation we need good policy *delivery*.

Lesson Two: Cooperation and coordination are factors of strategic importance for dynamism in rural areas. Ownership is not therefore a "new and fancy" instrument of development planning. It is rather the result of intensive debate among stakeholders and the reconciliation of their divergent interests in important decisions which not only reflect the will of the majority of rural people but also protect the rights of minorities.

Lesson Three: The village in the twenty-first century reconciles the global issues of ecological, economic and social sustainability with its local agenda for policy priorities.

What do we learn from Lesson One? Good rural policies do not necessarily mean better rural welfare! Or, good policies require good governance as well. This first lesson is one we really should have learned by now: We have continued to reformulate agricultural and rural policies; we have helped to answer the questions about *"WHAT has to be done in rural areas?"* in order to improve the welfare of the rural population. And in doing so we, as GTZ experts, were often struck by the fact that the macroeconomic debate was conducted without reference to the microeconomic conditions in which businesses were making or losing money in

---

[6] The first (conceptual) volume on regional rural development placed people very firmly at the center of development efforts. Not even the positive evaluations of this concept were enough to ensure that RRD retained the position which, with the benefit of hindsight and on the basis of the current debate, it clearly deserved within development cooperation and technical cooperation (GTZ: LRE kurzgefasst, 1988, p.13 ff.).

[7] Rural development was one of the key topics at the Annual Meeting of the IDB's Board of Governors in New Orleans in March 2000.

[8] "*... the World Bank was founded to provide foreign savings to provide additional finance when domestic savings are insufficient to finance the necessary investment push. However, the tangible inputs are the handmaidens of development, not the ultimate source of development and certainly not the appropriate criteria for quantifying development achievements.*" (Adelman, 2000)

the rural areas.⁹ Where things have gone wrong, feedback from the bottom up has clearly been no better than mediation from the top down. Policy impact became visible only when the failure was evident. We at the GTZ were unable to contribute our wealth of experience to the debate at the political level. Moreover, this topic was subject to a taboo that the German Federal Ministry for Economic Cooperation and Development (BMZ) maintained for years. It was said to be the preserve of the policy dialogue conducted by the BMZ (but often delegated to the World Bank as the most competent partner in the policy dialogue). Inevitably, the people conducting the dialogue were not "*close enough to the ground*" to be able to take account of the implications of a policy and its implementation problems.

Policies continued to focus exclusively on the WHAT questions. And we read little about the HOW: the processes, mediation, participation and the appropriate feedback policy. Little work has been done on these subjects.[10] We in the GTZ, together with a large number of partners, have often felt little support in our efforts to give equal weight to work on participation, partnership and process orientation, i.e. the so-called soft topics, alongside the important WHAT topics.[11]

In addition, many of the policies brought with them institutional cutbacks. Already weakened, those working on policy delivery in many cases now had no partners left at all. The new "actors" (the private sector and civil society) were seldom surrounded by the legal and social frameworks or the human resources they needed to bring about a smooth transition. Apart from the grassroots experts and numerous, alas often overly cynical national partners, this was noticed by hardly anyone in the early days. The numerous and largely unsuccessful structural adjustment exercises have a great deal to teach us in this respect.

"*The withdrawal of the state from the provision of services to agriculture is justified if the market in agricultural products also offers to suppliers whose position in the market is*

---

⁹ "*Business is conducted within the economy. In the case of Brazil, the observer can easily lose sight of this simple truth. ... many important posts in the executive are filled by economists, people who are familiar with macro-economics but may never have seen a company from the inside in their entire lives. ... dealing with the economic policy debate in Brazil is an assembly job — the macroeconomic discussion is frequently divorced from the discussion on the real economy and a link has to be established between the two.*" (Meyer-Stamer, 2000; translated quote)

[10] There is no other possible explanation for the recent "spectacular discovery" of "partnerships," "ownerships," "participation" and "inclusion" in the World Bank concepts (CDF/PRSP) "*We have learned the critical importance of the capacity of developing countries to help themselves to own, frame and implement development strategies...*" (Wolfensohn, 2000, p. 3); yet again, there is little enough explanation of "WHAT that is," but there is definitely too little debate on "HOW it works." But far more important in this context is the Bank's Knowledge Management Initiative. It gives grounds for hope that it will be possible in future to mobilize and combine far more systematically the know-how of individual actors and knowledge bearers (resource persons), as the new Chief Economist of the World Bank, Nick Stern, put it, "*This calls for a different role for the Bank: one that puts still more emphasis on learning and knowledge*" (Collier, Dollar, Stern, 2000, p. 3).

[11] Comment of a banker colleague: "*You GTZ people are always holding your cosy workshops!*" Our response is to point to our positive experience with participative planning in our projects (Müller (1999): Planificando el Uso de la Tierra).

weak, an assured income from which the services of private suppliers can be financed. As long as the competitive disadvantages of small farmer suppliers — low productivity / high production costs, poor product quality, unattractive products, lack of access of market and price information, low market volume — cannot be eliminated, the state's withdrawal leaves behind a vacuum which holds little attraction for private suppliers and drives small farmers still further towards the margins of the market. There can be no doubt that for a transitional phase at least, the state for all its newly defined mandate will not be able to abdicate its social and political responsibilities entirely." (GTZ-Peru, 2000; translated quote).[12]

It has too often been forgotten that policy delivery entails institutional change, along with the rethinking of roles and mandates, which is vital to implementation.[13] This is now recognized, including its importance for the sector's innovative potential. Here too, major changes must be taken into account through a new institutional set-up.

In summary we can say that the conclusion to be drawn from Lesson One is that policy must be organized as a learning process for everyone involved, institutionally and societally. While policy design is the intellectual challenge, it must "deign" to consider the day-to-day experience of policy implementation and to regard it as the positive and necessary learning process leading to innovation and progress in the rural areas.

How Do People Organize Village Life in the 21st Century?

If our partners want to organize these learning processes then the second lesson must be put into practice: rural development has too long been limited to the agricultural sector, concentrating in many cases on the promotion of agricultural production and processing. That is not surprising since many a ministry of agriculture was transformed into a ministry of rural development then, in the eighties and nineties, reverted back into a ministry of agriculture. In addition came ministries for environmental protection and resource management (too often at the instigation and with the support of a large number of donors). The competencies were lacking to find the counterparts to look after the well-being of people in rural areas in all facets of their lives, not just in specific aspects such as the economic (cattle farming, cropping, forestry, credit, etc.), social (health, education, drinking water, culture, pensions and youth support etc.) or cultural. This led to widespread

---

[12] *"The current view is that a balanced mix between state and market are required for development. What matters more than the government-private sector mix is the distribution of economic and political power to which policies and institutional behaviour respond."* (Adelmann, 2000).

[13] *"Markets depend upon a complex array of public institutions ... Services with large externalities or market failures such as transport, telecommunication, or power infrastructure, require regulation. This role for government as the builder and provider of institutions for the market economy will be the theme of the World Development Report 2001. ... The main lessons we draw from the experiences of the 1990s are that public action in areas highlighted above is both more important and more difficult than was appreciated at the start of the decade."* (Collier, Dollar, Stern, 2000, p.2).

isolation in the development of the rural areas. We found it increasingly difficult to make people see the importance of an integrated approach for the development of a country, and marginalization was the inevitable result. [14]

Meanwhile, many solutions were implemented in rural areas through different kinds of access to people, namely through their daily-life environments, through support to their local government organizations, and through management of the services necessary in this context. In this way people's needs were again dealt with holistically and using a systems approach. It became immediately obvious that a major part of the task consisted of the ability to reconcile the divergent interests of various actors at the local government level and to link these needs "upwards" and enforce them.

This brings us to the center of the decentralization debate: it is precisely the upward linkages that are problematical in many cases in Latin America. There, good relations with higher levels make the local actors "bridgeheads" of the national administration in the rural areas. They know how to procure resources and how to go about the necessary coordination, etc. Or else they are oriented downwards, know the problems and potential of their populations and current initiatives at that level, but often only at the expense of a lack of acceptance "higher up." For example in Colombia, a number of new citizens action groups have emerged, but the elected mayors are regarded by many people "higher up" as merely the political arm of the guerrillas.

So as we see, it is not only at the national political level that we are up against divergent interests. No, the local government and village level has its own complex political context requiring skillful interest management.

As well as shaking off the limitation to the agricultural sector, it is imperative that actors in the rural areas reach agreement on new cooperation and coordination mechanisms within the transformation process necessary to their development. These are learning processes, which can be improved with techniques and instruments, and we must address these as a matter of priority. To that end, the GTZ has in the last twenty-five years undergone a wide-ranging learning process of its own, together with partners and would like to share this information with other organizations, many of which are still in the early stages.[15]

Just like the inhabitants of villages in the twenty-first century, so too the experts of the GTZ have learned that communication, the exchange of knowledge and coordination are basic prerequisites for developing these complex systems and for getting through the different stages of organization and decision making in the

---

[14] "*In essence, the agriculture sector lost its prominent role in the development strategies of the region's countries because it failed to put forward a strategic vision to attract government support for a sector with a crucial role to play in growth with equity, and due to the pressing need for resources for macroeconomic reform programs during the past decade. Ultimately, the shrinkage of the agriculture portfolio is a reflection of inadequate technical leadership, both by international agencies specialized in agriculture and the Bank itself*" (IBD, 2000, p.15).

[15] Bosque Secco in the Dominican Republic is a good example. Here the GTZ succeeded in organizing forest farmers, enabling them to fight for bargaining power. As a result, the farmers' incomes have risen and a contribution has been made to sustainable forest management (cf. Heblik, 2000).

rural areas. It is a chance to turn a variety of experiences into the correct policy for the regional or local context. [16]

It sounds banal, but the reconciliation of divergent interests, power structures, protection of minorities, sustainability and distribution is a highly sensitive process that still has to be introduced and practiced in many societies. As part of that process, the state must learn its substitutive role and consequently give up many functions, accepting new roles. Local government actors as well as other local participants must not only learn to take on responsibility — usually they are only too keen to do so — but they must also be able to exercise this responsibility. That is to say, they need the skills to handle and operate funds, accounts, plans, budgets and mediation processes. Investments must be channeled into this learning process as a matter of priority. As always, the key is education and training oriented to the concrete needs of the population. It is difficult to understand why in Latin America of all places this aspect of developing the rural areas has been neglected.

From 21st Century Village to Global Governance. Are They Reconcilable?

Why is the need to reconcile village and globalization brought home to us so often by the media, from Jolo in the Philippines to famines in Africa and outrages against the environment in Latin America or Eastern Europe? Is this of any significance for the inhabitants of the twenty-first century village? We believe it is. For one thing, modern communication brings them into contact with the rest of the world. Secondly, their own actions have an ever greater impact on such overarching issues as climate protection, biodiversity, water use or combating desertification. The local horizon is not broad enough to take proper account of the consequences. We encounter the same dilemma if we come from the other direction: how can world market movements in raw materials or changes in consumption patterns be communicated and explained so that small rural agricultural production and processing enterprises in the twenty-first century village can react flexibly, responsibly and with sound business strategy in order to stabilize and raise their income? Just as we are discussing and working on ways of improving interaction between the macro, meso and micro levels, we come under pressure from the phenomenon of globalization and other supra-regional challenges. This involves the twenty-first century village not only with global trade and environment issues but also with the protection and sustainability of land extending beyond national boundaries (protection of mangrove forests in coastal zones, resource degradation in Chaco Americano, etc.).

---

[16] "*In the discussion to date [on devising rural development programs in Peru] three principal aspects have been mutually obstructive: ....only a radical change of perspective can offer a way out of the dilemma. The heterogeneity of the ongoing projects is no longer perceived as an impediment but rather as an opportunity and potential for working out a program; ...*" (GTZ-Peru, 2000; translated quote).

Do we need regional and global institutions for this purpose, about which there is a great deal of discussion at the present time (global governance)? Or is *"intellectual reflection on global governance no substitute for good governance"*?[17] Governance here includes not only national but also local levels; it involves not only framing policy but shaping, implementing, changing and adapting it. In this process, the twenty-first century village will also require the political will to bring about a reconciliation between city and countryside. *"In order to survive, the urban centers of Latin America need clean drinking water, cleaner air, preservation of biodiversity, educational services, food supplies and, above all, the social buffer function at times of crisis. All services that are being taken on by rural areas."* (Eger, Müller, Pfaumann, 2000; translated quote). Are the mayors and people of the twenty-first century village aware of this important general macroeconomic function they are taking on? And are they demanding the necessary funds to pay for these services?

Increasingly, communication and mobility are blurring the traditional distinction between city and countryside. While the city-dweller in many developing countries in the past decades has financed the development effort through numerous indirect transfers of money from the rural areas, the transfers should now start to flow in the other direction. As a result of the reorientation I have described above, the dynamism of the rural areas will then become an important engine of national economic development.[18]

References

Adelman, Irma (2000): Fifty years of economic development: What have we learned? Paper presented at the Annual Bank Conference on Development Economics-Europe, Paris, 26-28 June 2000.

Belshaw, Deryke (1993): FAO's new Approach to Rural Development: Micro-Activity. In: A Strategy Vacuum? In: entwicklung + ländlicher raum, 2/93 Frankfurt.

Chambers, Robert (1993): Rethinking Rural Development: Whose Reality Counts. Zentrum für Regionale Entwicklungsforschung, Schriften 51, Münster, Hamburg.

Collier, Paul, David Dollar, Nicholas Stern (2000): Fifty Years of Development. Paper presented at the Annual Bank Conference on Development Economics-Europe, Paris, 26-28 June 2000.

---

[17] Further, *"Proponents of global governance back up their demands with a number of erroneous diagnoses. The "disempowerment of the nation state" by globalisation is usually at the heart of their argument. ... Yet economic integration does not just steamroller over rulers. ... Governments determine the scale of cross-border trade, foreign investment, immigration and emigration and thus the degree of integration into the global economy.... As at the peak of globalisation (1870 to 1913), governments still retain economic autonomy: protection of property, contractual guarantee, provision of public goods, environmental protection or social state."* (Sally, 2000; translated quote).

[18] *"Economic development, as distinct from mere economic growth, combines: (1) self-sustaining growth; (2) structural change in patterns of production; (3) technological upgrading; (4) social, political and institutional modernisation; and (5) widespread improvement in the human condition."* (Adelmann, 2000).

Eger, Helmut, Ulrich Müller, Peter Pfaumann (2000): Die Wiederentdeckung des ländlichen Raums. In: Akzente 3/2000, GTZ, Eschborn.

GTZ (1988): Ländliche Regionalenetwicklung. LRE kurzgefaßt, Schriftenreihe der GTZ, Nr. 207, Eschborn.

GTZ-Peru (2000): Strategiepapier für den Sektor Ländliche Entwicklung der Deutschen Technischen Zusammenarbeit in Peru (unveröffentlichtes Arbeitspapier).

Heblik, Daniela (2000): Bäume statt Kakteen. In: Akzente 3/2000, GTZ, Eschborn.

Meyer-Stamer, Jörg (2000): Über den Verlust von Dekaden und den Verlauf von Lernkurven. Wirtschaftlicher Strukturwandel und die Irrungen und Wirrungen der wirtschaftspolitischen Diskussion in Brasilien. In: Ibero-Analysen, Heft 3, Berlin.

Müller, Ulrich (ed) (1999): Planificando es uso de la tierra. Catalogo de herramientas y experiencias. Herausgegeben für Deutsche Gesellschaft für Technische Zusammenarbeit (GTZ), Eschborn.

Sally, Razeen (2000): Staaten sind gefordert. Der Welthandel braucht keine neuen Institutionen. In: Die Zeit, Nr. 30, 20.7.2000.

Wolfensohn, James D (2000): Rethinking Development - Challenges and Opportunities. Tenth Ministerial Meeting of UNCTAD, Bangkok, February 16, 2000. http://www.worldbank.org/html/extme/jdwsp021600.htm.

6 Policies to Stimulate Agricultural and Rural Growth

Rajul Pandya-Lorch

The rapid urbanization of the developing world is considered by many to be the key demographic feature of the late twentieth and early twenty-first centuries. Projections suggest that the developing world's urban population, after doubling from 1980 to 2000 from 1 billion to 2 billion, will increase by another 70% to reach 3.4 billion in 2020 (UN, 1998). While the magnitude of the urbanization underway is overwhelming, it should not be forgotten that 2.9 billion people currently live in the rural areas of the developing world. Their numbers have doubled since 1950 and are expected to increase during the next two decades, albeit by much smaller increments, to 3.1 billion (Table 6.1). With the exception of Latin America, the developing world is still very much a rural world.

Rural areas continue to lag behind urban areas with regard to virtually all social indicators. The proportion of the population that has access to basic necessities of life such as safe water and sanitation, health services, and education is much smaller in rural areas than in urban areas. Poverty remains a major problem in rural areas; available data suggest that about 1 billion rural people, over one-third of the rural population, live in poverty (Jazairy, Alamgir, and Panuccio, 1992). About 90% of these rural poor live in Asia, primarily South Asia, and sub-Saharan Africa.

Will rural areas be left behind as the world rapidly urbanizes, modernizes, and globalizes? Will the gap between rural and urban people widen even more? What will it take to spur sustained broad-based economic growth and development in rural areas? This article will focus on how agriculture can contribute to poverty alleviation and economic growth in rural areas. Most rural people in the developing world depend on agriculture for their livelihoods. If not engaged in their own agricultural activities, then they rely on non-farm employment and income dependent in one way or another on agriculture. About 54% of the male labor force and 62% of the female labor force in low and middle-income countries are engaged in agricultural activities (World Bank, 1999). In many low-income countries, agriculture contributes the major portion to rural income with a share ranging from 30 to 90% (von Braun and Pandya-Lorch, 1991). Agricultural expansion is a catalyst for broad-based economic growth and development in most low-income countries. Being interconnected with the non-farm economy, agriculture generates considerable employment, income, and growth in the rest of the economy, especially the rural (Pinstrup-Andersen and Pandya-Lorch, 1995).

Table 6.1. Rural Population – 1950, 2000, and 2020

	Number of Rural People [in millions]			Proportion of Total Population in Rural Areas [%]		
	1950	2000	2020	1950	2000	2020
Africa	191	510	670	85.4	62.2	50.9
Asia	1,158	2,302	2,316	82.6	62.4	50.4
Latin America and the Caribbean	97	127	125	58.6	24.6	19.0
Developing Countries	1,407	2,918	3,098	82.2	59.5	48.0
Developed Countries	367	284	227	45.1	23.9	18.7
World	1,774	3,201	3,325	70.3	52.6	43.3

Source: United Nations, 1998

Key policies to stimulate broad-based agricultural and rural growth have to be discussed. While the specific policies and strategies required will vary from country to country, and indeed from locality to locality within a country, action is needed in several key areas to facilitate and sustain broad-based agricultural and rural growth in developing countries.

Invest in Poor People

Developing countries must invest in poor people to improve their well-being, increase their productivity, and enhance their access to remunerative employment and productive assets. Access to education, particularly for female children, is vital; recent research by Smith and Haddad (2000) finds that improvement in women's education accounts for over 40% of the reductions in child malnutrition that occurred between 1970 and 1995 in the developing world. Governments, local communities, and non-governmental organizations (NGOs) should assure access to primary education for all children, particularly female and rural children; to primary health care, including reproductive health services, for all people; to clean water and sanitation services; and to training for basic literacy and skill development in adults. They should work together to strengthen and enforce legislation for gender equality. Improved access for the rural poor, especially women, to productive resources can be facilitated through land reform and sound property rights legislation, strengthened credit and savings institutions, more effective labor and land markets, and infrastructure. Direct transfer programs, including programs for poverty relief, food security, and nutrition intervention, remain necessary in many countries, particularly in the short term. Support must be maintained for effective famine early warning systems and other disaster preparedness and management systems. Widespread local, national, and regional instability and armed conflict contribute to the persistence of poverty and misery in rural areas. Until priority is given by all parties to addressing the causes and consequences of conflict and to preventing their outbreak in the future, peace in rural areas, often the theater for armed conflicts, will remain elusive and with it broad-based growth and development.

Strengthen Agricultural Research and Extension Systems

Agricultural research and technological improvements are crucial to increase agricultural productivity and returns to farmers and farm labor. This in turn reduces poverty and meets food needs at reasonable prices. Accelerated investments are needed for yield-increasing crop varieties, including hybrids that are more drought-tolerant and pest-resistant, and for improved livestock. Yield-increasing production technology such as small-scale irrigation and irrigation management systems and techniques such as integrated pest management also demand attention. Also, the needs of small farmers, particularly women, must be addressed.

Modern science must be effectively put to work on the problems and constraints of poor small-scale farmers in developing countries. While few of these countries can afford to develop the tools of biotechnology, even fewer can afford not to use these tools as they become available. Biotechnology, if appropriately focused on solving small-scale farmers' problems, can help to increase productivity and reduce production risks. The use of biotechnology is even more effective when combined with traditional research methods, better agronomic practices, and better markets and policies. In developing countries, delivering the potential benefits of agricultural research in general and biotechnology in particular to small farmers and poor consumers will require a combination of expanded public investment and innovative public-private partnerships. It will also require governments in both industrial and developing countries to develop effective safety regulations, to create and enforce appropriate intellectual property rights legislation, and to enforce antitrust legislation to counter excessive concentration in the life science and seed industries.

Many of the poor countries most dependent on productivity increases in agriculture significantly underinvest in agricultural research. Available information suggests that per capita agricultural research expenditures in low-income countries are one-tenth the amount of those in high-income countries (Pinstrup-Andersen and Pandya-Lorch, 1995).

National and international investments in agriculture in developing countries have been low and, in recent years, even declining. Agriculture has, by and large, been neglected, and in many countries it has been taxed indirectly and/or directly. A stagnant or deteriorating agricultural sector has in many instances contributed to stalled employment creation and income generation in rural areas. Low-income developing countries must sharply expand their investments in agricultural research. Priority should be given to redressing the balance between scientific personnel and other expenses. In many low-income countries, particularly those in sub-Saharan Africa, available funds per agricultural researcher are insufficient to assure efficient use of the available human resources.

Effective interactions between farmers and research institutions are essential for disseminating research results and technology and for ensuring that research priorities reflect the needs of farmers. In some countries, the private sector and farmer cooperatives effectively perform these extension functions. However, because of its public-goods nature, agricultural extension for small-scale farmers producing staple foods must continue to be provided by the public sector. Public-sector extension in developing countries has a mixed performance record, and innovative strategies and technique will be required to assure effectiveness in the future.

Promote Sustainable Intensification of Less-Favored Lands

A significant proportion of the rural poor live in "less-favored" areas, which are characterized by difficult climatic conditions, poor infrastructure, and inadequate services (Hazell, Jagger, and Knox, 2000). These lands abound in the developing world; according to the CGIAR (1999), "favored" lands account for only 10% of the developing world's agricultural area, implying that less-favored lands account for the remaining 90%. About two-thirds of the total rural population live in the less-favored areas. Some of these have good agricultural potential, especially if roads and irrigation are made available.

Conventional wisdom calls for governments and donors not to invest in less-favored areas, the argument being that investment in high-potential areas generates more agricultural output and higher economic growth at lower cost. However, growing numbers of people are challenging this conventional wisdom. They argue that governments and donors should invest more to increase agricultural production in less-favored areas (Hazell and Garrett, 1996). While migrating out of the village may be a solution for some of these areas in the long run, few countries are in a position to effectively absorb large numbers of poor people fleeing the less-favored lands. For the most part, the problems of less-favored lands will have to be solved within those areas.

According to Hazell and Garrett (1996), the key elements of an agricultural intensification strategy for less-favored areas include the following: (i) improve technologies and farming systems, with better integration of annual crops and perennial crops, with farm trees, and with livestock at both the farm and landscape levels; (ii) ensure secure property rights and effective institutions so that farmers have incentives to pursue sustainable farming practices and invest in improving and conserving resources; (iii) manage risk effectively through agricultural research, safety net programs, and credit and insurance markets; and (iv) provide the appropriate policy environment in order to insure that the right production signals are sent to farmers.

In addition, incentives must be provided to farmers and communities to restore degraded lands and protect natural resources. Local control over resources must be strengthened, and local capacity for organization and management improved. Farmers and communities must be provided with incentives to implement integrated soil fertility programs in areas with low soil fertility. Also, comprehensive water policy reforms must be undertaken to make better use of existing water supplies by providing incentives to water users, improving procedures for allocation, developing improved technology for water supply and delivery, providing secure water rights, and reforming distorted price incentives (IFPRI, 1995).

Develop Efficient and Effective Markets

Many developing countries are privatizing their agricultural input and output markets, replacing inefficient, poorly functioning state marketing companies and excessive, inappropriate government regulations with private sector marketing agents. It is essential that this process results in efficient, effective, and competitive markets. Agricultural systems will be competitive only if all components, for example, input markets, production, and output markets, are efficient and effective. The role of the state is to create an environment conducive to competition among private agents while assuring access to productive resources by the poor, enabling them to compete on equal terms.

Market infrastructure that serves the public good, such as market information, roads and other rural transportation facilities, electricity, and communication facilities, should be developed and maintained. Other tasks for government include removing institutional barriers to the creation and expansion of small-scale credit and savings institutions and making them available to small traders, transporters, and processing enterprises. Such institutions are also effective in helping the rural poor manage risks and generate income. Without access to good roads, telecommunications, and strong credit and savings systems, farmers cannot be sure they will be able to sell their output or get the inputs that they need.

Improve Capacity of Developing Country Governments

More effective national and local governments are essential. The economic, social, and political conditions they create or support determine the success of the efforts of other key actors such as farmers, the private sector, and NGOs in agricultural development. Each country must identify the appropriate functions of its government vis-à-vis other parts of society. The government's capacity to perform these functions must be strengthened while it relinquishes those functions better performed by others. Predictability and transparency in policy making and enforcement are critical. Improved security and personal safety in rural areas are prerequisites for sustained growth and development. Where armed conflicts are occurring or are imminent, priority must be given to conflict resolution. States play an important role in assuring that conditions necessary for competition in private-sector markets are present. Governments must also invest in or facilitate private-sector investment in education, health care, agricultural research, infrastructure, and other public goods to improve well-being and livelihoods.

Conclusion

If governments in the developing world are concerned that their rural areas will be left behind as the world rapidly urbanizes, modernizes, and globalizes, and if they are concerned that the economic gap between rural and urban people will widen even more, then they must make a concerted effort to invest in stimulating and sustaining broad-based economic growth and development in rural areas. They could focus their investments on non-agricultural activities, but in most rural areas the generation of significant non-farm income and employment depends critically on having a dynamic agricultural sector. There is considerable evidence that if agriculture stagnates, the local economy has no engine to drive non-agricultural activities. Therefore, governments throughout the developing world must intensify their efforts to develop, implement, and enforce appropriate policies to support agricultural and rural growth. Within the context of rapid urbanization in the developing world, policymakers must recognize the importance of rural-urban interdependence to the livelihoods of both (Tacoli, 2000) and develop a set of strategies for nurturing and benefiting from this interdependence. It is time to redress the balance of priorities and investments between rural and urban areas and peoples. Neglecting or not giving due attention to rural areas and rural people in the urbanization rush will condemn many millions of people to misery and, over time, shift the locus from rural to urban areas. Problems of rural poverty, hunger, and misery must be tackled and solved in the area of origin.

References

von Braun, J. and R. Pandya-Lorch (1991): Income Sources of Malnourished People in Rural Areas: Microlevel Information and Policy Implications. Working Paper on Commercialization of Agriculture and Nutrition No. 5. IFPRI, Washington, D.C.

CGIAR - TAC Secretariat (1999): CGIAR Study on Marginal Lands: Report on the Study on CGIAR Research Priorities for Marginal Lands. Marginal Lands Study Paper No. 1. FAO, Rome.

Hazell, P. and J.L. Garrett. (1996): Reducing Poverty and Protecting the Environment: The Overlooked Potential of Less-Favored Lands. 2020 Vision for Food, Agriculture, and the Environment Brief 39. IFPRI, Washington, D.C.

Hazell, P., P. Jagger and A. Knox. (2000): Technology, Natural Resources Management and the Poor. Draft prepared for the International Fund for Agriculture Development.

IFPRI (International Food Policy Research Institute) (1995): A 2020 Vision for Food, Agriculture, and the Environment: The Vision, Challenge, and Recommended Action. IFPRI, Washington, D.C.

Jazairy, I., M. Alamgir and T. Panuccio (1992) The State of World Rural Poverty. IFAD, Rome.

Pinstrup-Andersen, P. and R. Pandya-Lorch (1995): Agricultural Growth is the Key to Poverty Alleviation in Low-Income Developing Countries. 2020 Vision for Food, Agriculture, and the Environment Brief 15. IFPRI, Washington, D.C.

Smith, L.C. and L. Haddad (2000): Overcoming Child Malnutrition in Developing Countries: Past Achievements and Future Choices. 2020 Vision for Food, Agriculture, and the Environment Paper 30. IFPRI, Washington, D.C.

Tacoli, C. (1998): Rural-Urban Interdependence. In: Garrett, James L. and Marie Ruel (eds): Achieving Urban Food and Nutrition Security in Developing Countries. 2020 Focus 3. IFPRI, Washington, D.C.

United Nations - Department of Economic and Social Affairs, Population Division (1998): World Urbanization Prospects: The 1996 Revision. United Nations, New York.

World Bank (2000): Entering the 21st Century: World Development Report 1999/2000. World Bank (in conjunction with the Oxford University Press), Washington, D.C.

Agenda 21 Principle 21, Earth Summit, Rio 1992

Principle 21 The creativity, ideals and courage of the youth of the world should be mobilized... to ensure a better future for all.

creativity, ideals, courage = **better future**

7 Decentralization and Active Community Participation in India

Anita Das

In India, the last decade has seen efforts made for increased political participation of people both in the urban and rural areas, especially for the economically and socially weaker sections such as members of the scheduled castes, scheduled tribes, and in particular women. Catalyzation for this work came from the parliament, civil society and women's groups. The 73rd and 74th Constitutional Amendment Acts passed in 1993 were major expressions of progress on the issue, providing for peoples' participation in rural and urban local bodies. These amendments to the constitution are landmarks both because they prescribe the representation of the scheduled castes and the scheduled tribes in proportion to population and even more so because they reserve one-third of the total number of seats available for women from all social strata.

The consequent amendment to the State Acts by the different state governments, the holding of elections and subsequent devolution of powers to these local bodies, and the subsequent performance of these elected bodies have given varying results in the different Indian states. A comparative analysis of the performance of local bodies in different states would no doubt yield interesting results because of a variety of factors such as prevailing rates of literacy, awareness, education, social structures, status of women, the commitment of the political bureaucracy, capacity and willingness to accept and allow social political transformation.

Even without embarking on a comparative analysis of the performance and results in the different states, the impact of the 73rd Constitutional Amendment Act of 1993 on people's political participation in the rural local bodies, e.g. Panchayats in the state of Madhya Pradesh, has been clearly significant.

According to the 1991 census, the state of Madhya Pradesh, lying at the heart of central India, has an area of 443 thousand square kilometers and a population of 66 million people. The scheduled tribes (ST) make up 23% of the population and the scheduled castes (SC) 14%; both are disadvantaged segments of society. 77% of the state's population lives in its 71,000 villages.

Consequent to the 73rd Constitutional Amendment Act in 1993, Madhya Pradesh was the first state to amend its Panchayat Act and to hold elections. By August 1994 elections according to the new provisions of Constitution had been held, with nearly 500,000 representatives of the people in the three tier Panchayat Raj. After the completion of a full tenure of 5 years, the second round of elections to the PRIs was held during January 2000.

The emergence of a new platform for widespread political participation has brought about phenomenal change in the rural scene of Madhya Pradesh. In place of the earlier form of governance comprising of only people's representatives chosen for the national parliament (i.e. Members of Parliament whose total number in Madhya Pradesh is 56) and representatives elected to the State Legislative Assembly (whose total number on the basis of Legislative Assembly Constituencies is 320), a vast number of the elected representatives are now in position in the 3rd tier of government in the Panchayat Raj. The three tier Panchayat Raj consists of the Gram Panchayats with 474,351 representatives, the Janpad or intermediate level Panchayats with 9,556, and the Zilla Panchayat at the district level with 991. In addition, the earlier system of government was made up of the central bureaucracy at the level of the national government, the state bureaucracies at the level of each state government and within that the divisional, district and block level bureaucracies, which enjoyed vast executive powers. These bodies have undergone considerable changes in powers and functions as a result of the introduction of the third tier of governance, i.e. the rural local bodies. Although, the change in power structure was not so perceptible at the central and state levels, the situation at the district, block and village levels has been dramatic with the devolution of powers and functions in many government departments to the Panchayats.

The 1994 Panchayat elections proved to be a turning point in the way the state had been administered and managed for the last 50 years. Powers, rights and duties, which had traditionally belonged to the state bureaucracy, were now transferred to the district level and to the elected representatives. One-third of these representatives were required to be women living in rural areas. This is an unusual phenomenon considering that rural women have almost always remained outside the process of decision making even in their own families. Women's lives are determined more by social custom and precepts than by self-determination. Except for women from the poorer households who are forced to seek employment on other people's land, on construction sites or by collecting forest produce, the women in rural areas have always been prevented from contact with the outside world. Even visiting government officials would seldom talk to the rural women; the officials limited their contact with the village community to the men. Thus there have always been strong barriers not only for the rural women to interface with the outside world but also for outsiders and officials to communicate with these women because of social custom and tradition.

In the early part of the 5-year tenure for the elected Panchayat representatives, the bureaucracy was grappling with the trauma of transference of powers to these representatives, which was seen as threatening in itself. Consequently the issue of dealing with the newly elected women took a back seat to other problems in spite of the fact that involvement and preparation of women PRI members had been emphatically prescribed. There were attempts to hold orientation sessions for women PRI members by governmental and non-governmental agencies, but the efforts were inadequate to cover the large number of women who were elected. Centuries of deprivation of opportunities cannot be overcome so quickly. While there are shining examples of success, the majority of women were usually guided or rather controlled by the husband or the son, or some male relative in performing the role for which they were elected.

It is indeed interesting to examine what kinds of performances emerged in the second tenure of the elected representatives in terms of women's participation and actualization of their roles with PRIs. It should be mentioned in this context that the strong institution of family in India permeates the process of all decision making. It is therefore no surprise that the opportunities the constitution has bestowed on Indian women are largely realized in accordance with decisions made by the family. When asked how they decided to run for office in the first round of elections, many women responded that they were told to put a thumb impression on or to sign a form given to them by their male representatives. On winning the election, in several cases it was the husband who was carried out in a victory procession and was supported by the community while the woman sat at home. Surprisingly, this was the position even when the woman was a graduate and an elected president of a district Panchayat in the state of Haryana. In the second round of Panchayat elections, family consultation would also have played a large role in determining whether or not the woman should run for office.

Shortly before the second round of elections in Madhya Pradesh, a very successful woman Sarpanch of the Shajapur district who had even won a prize for being the best performing Panchayat in the district was asked why she was not running for election again. Her reply was that in this round the seat was reserved for a man, and therefore her husband would run. When it was pointed out to her that there was no reservation for men and that this time it was an open seat for which she also could run, she expressed that her preference was to stand down in favor of her husband. Making decisions on the basis of mutual consent within the family is a characteristic feature of women in the rural areas. There have been instances of acute power struggles within families on account of the reservation for women in the PRIs, but such cases have been few. The overwhelming force in women's participation is men, mostly male relatives acting as proxy for women in the Panchayats. This sort of willing suspension of rights and power in favor of a male representative is widespread, even more so among women from wealthier families than those coming from the SC, ST or economically weaker social strata. Well-off rural woman are more subjected to social barriers on account of gender. While no special enthusiasm has been displayed by women coming from the normal strata of society, as evidenced in such women being elected only to the reserved seats, women belonging to the deprived groups such as the scheduled castes, scheduled tribes and other backward classes have aspired to and succeeded in winning more than the reserved seats. It is these signals, which give vibrancy and real meaning to the constitutional mandate on political participation to all segments of the population.

It is however amply clear that in order for people to participate in political and public life, education, awareness and experience are important. These things determine the quality of decision making and follow-up action that is required by the PRIs.

Gram Swaraj

After taking the first major step of decentralizing the government at the district and sub-district level through the PRIs in 1994, in 2000 the state government is actively explored the concept of *"Gram Swaraj"* or village democracy. This means, in other words, direct governance by the people instead of by the people's representatives. By empowering the Gram Sabha (which is comprised of all voters in a village) instead of having the powers vested in the Gram Panchayats, people in these villages are able to make their own decisions about the environment, sanitation, forest management, bodies of water, habitation, schools, primary and preventive health care, common property resources as well as common facilities pertaining to their own livelihoods and support systems. This can be done through committees of villagers, formed by themselves according to their requirements. They would be free to have their own procedures for work and accountability, which the villages had been doing over centuries at their own initiative and according to local wisdom. This system has been disturbed by the present-day model of development, where things move from the top down and disregard local wisdom and traditional ways of problem solving. The new system of Gram Swaraj envisions people making their own choices and decisions. Every family can be involved in some aspect of community decision making. In other words, the Gram Sabha uses structures, which involve every family in the village in making and implementing decisions. Thus, the process becomes one of self-government instead of the many being governed by a few. The role of government and field level bureaucracy would be to act as facilitators, giving information and technical support. Active involvement of the people in their own development should increase its pace, be more people friendly, sustainable and result oriented. The idea presented here is basically that if there is a welfare project for rural women, let them manage it and run it. Bureaucratic structures doing it for them have less chance for success as the people involved are not direct stakeholders, have poor access or confront social barriers.

Experiences of Village Committees

The following experiences illustrate that the village committees are able to steer decision making, implement the decisions and benefit from them. The programs have at the same time led to better results not only in their primary target area but also in other development parameters. Thus it is thought that Gram Swaraj would help the process to be more "demand driven" than "supply driven." However, this is still at the conceptualization stage and is mentioned here as an indication of change in the state government's approach to governance and development of rural areas.

Joint Forest Management

An area where policy and institutional shift in favor of the people has brought about a perceptible impact on their lives is the Joint Forestry Management (JFM) project of Madhya Pradesh. This state has the highest forest cover in India, and a large part of rural life revolves around the forests there. The richest forest reserves are found in the tribal homelands. The access to and use of these forest resources has been the center of conflict between the tribal people, the private contractors and the state forest department officials. These forests are rich in timber and non-timber forest produce (NTFP). Timber being a nationalized produce, trade is highly restricted. However, the NTFP has traditionally been collected and sold by tribal people who are economically exploited by the traders in the process. The tribal people in for quick monetary gains tend to over-harvest these resources without concern for regeneration, or protection of the resource base.

The NFTP, collected mainly by tribal women and children, provides considerable supplementary income to impoverished tribal families. Head-loading of firewood to supply to nearby towns is another source of income to poor households. In the forest areas, the government is committed to meeting the requirements of the local people in terms of the forest produce through the system of "*nistaar*," wherein the fuel and fodder requirements of villagers are met at lower rates.

It is thus evident that the tribal people who have traditionally depended on the forests for their habitat and livelihoods are in a precarious way. They are trapped in a situation of diminishing forest wealth and excessive state control, directly affecting their livelihoods.

In 1988, the government of India came out with the National Forest Policy. It recognized and placed emphasis on people's traditional and symbiotic relationship with the forest and the need to incorporate their protection and the safeguard of traditional rights in the conservation of forests and wild life. In 1990, the government of India also issued directives for a Joint Forest Management (JFM) approach. In this there was a shift from a purely departmental management system to one, which involved the people living on the fringes of the forest in a community, based, participatory management of the forests. In 1995, the Government of Madhya Pradesh formulated a resolution and introduced JFM in select areas after training their officials. They in turn formed Village Forest Committees (VFC) and Forest Protection Committees (FPC) at the village level.

Through extensive interactions, and using participatory rural appraisal techniques, there has been a very substantial mobilization at the village level in areas around the forests with varying degrees of success. The result of an impact study by the forest department in a village of the Hoshangabad district of Madhya Pradesh highlights the increase to the total village income (mainly from agriculture and forestry) and the reduction of seasonal migration and of the rate of interest on loans.

Watershed Management through People's Participation

Madhya Pradesh, in spite of an abundance of natural resources, remains a state of developmental paradoxes and economic backwardness. Of a total geographical area of 44.3 million hectares, not more than 20 million hectares are cultivated. 80% of the total crop area relies on rainfall, with the corresponding possibility of wide, unpredictable variations in agricultural production from year to year. The relentless exploitation of natural resources and erosion of soil, vegetative cover and ground water due to increased biotic pressure has resulted in a fragile ecosystem. Underdevelopment, rural poverty and extensive unemployment among the farm and non-farm workers is a common feature of these villages. Thus livelihood of rural people in rain-dependent areas is precarious, leading to heavy migration out of certain areas. These rain-dependent areas, which are susceptible to drought, have been the subject of government attention for more than two decades. Central government programs sought to reduce the frequency of drought and to improve water retention, soil conservation and environmental balance. The programs also tried to increase the income of the economically weaker sections by giving them wage employment. In 1994 the state government decided to change the entire approach to watershed management by creating a new institutional framework, namely the Rajiv Gandhi Mission for Watershed Management. This would formulate a fresh approach to watershed development enshrined in the mission's goal as *"sustainable use of material resources by an empowered and aware community."*

While the mission objectives remained largely the same as in the earlier schemes, they did add the maximizing of people's participation in concept planning, implementation and maintenance of soil and water conservation, and of focusing on the disadvantaged groups within the committees through equitable distribution of resources and sharing of benefits. The mission strategies as spelled out were notable:

- *Participation* is the most appropriate tool for empowering the community. People's participation is facilitated right from the planning stage through the stages of implementation and evaluation.
- Acting as *facilitators* for the activities to be carried out at the village level, government agencies or NGOs can ensure the involvement of people at every level. In this way, the villagers develop a sense of ownership for the activities carried out at the their village levels.
- *Participatory evaluation*, in which the community itself, involving all stakeholders, evaluates its own activities. This leads to a SWOT analysis (strength, weakness, opportunity, threat) of the entire range of activities carried out in the area, which in turn leads to rectification and replication.
- The activities which are planned or for which inputs are given are designed to ensure that the area becomes *drought resistant.*
- *Community regulations* ensure the regulation and control of common property resources. They provide for the proper handling of equity issues as well as for the formulation of conflict resolution mechanisms.

- *Women's participation* is ensured by self-help groups, by thrift and credit groups for women, and by at least 30% participation from women in village watershed committees.

It is interesting to observe that with more or less the same level of funding as under other plans already operated by the government of India, the establishment of this new institutional framework brought much better results. The success derived from focusing on collaborating and working directly with people and from specifically involving women and economically marginalized groups such as landless laborers. Even more important was the empowerment of the local community through decision making about such common concerns as equitable distribution of resources and the sharing of benefits.

The convergence of various schemes pertaining to watershed development under the umbrella of the Rajiv Gandhi Mission facilitated a functional process with coordination at every level. Working from one platform enabled an effective interlink with other development agencies in the field of agriculture, forestry, horticulture, fisheries, sericulture etc. The Watershed Mission has succeeded in districts like Jhabua in making a substantial difference to the fractured ecology of the micro watersheds. At the same time, it has made a point about the capacity building of the community and micro level institutions.

Partnering with Rural Communities for Primary Education

Although free and compulsory primary education to all citizens was a constitutional goal since Independence, the state was lagging woefully behind in its fulfillment because of meager resources and the consequent poor outreach. Low enrollment, especially for girls, and very high drop out rates were a recurrent feature.

As part of an overarching policy framework of decentralization and active community participation, the government of Madhya Pradesh made a radical break and promoted the Education Guarantee Scheme (EGS) in 1997. Its goal was to provide primary school facilities for every child in the state in a timely manner. Under the EGS, the government guarantees the provision of a primary schooling facility for the children in any area where there is no such facility within a kilometer. This should take place within a period of ninety days of receiving a request for such a facility by the local community. The program operates on a decentralized basis through the collaboration of the state government, the local body/panchayat and the community.

The EGS created a primary schooling facility in every area of the state through a partnership between the state government, local government (panchayat) and the community. A new paradigm of community-centered primary education in a rights-based framework has come up. The fact that forty primary schools opened every day of 1997 showed the extent of the demand.

Rural Women's Empowerment through Self-Help Groups

Another initiative to help rural women help themselves is the promotion of self-help groups (SHGs). The situation of women in rural Madhya Pradesh is characterized by a high degree of poverty, near illiteracy, being an adversely low proportion of the population, high MMR (maternal mortality rate), low social status typical of a paternalistic society, and having little access or exposure to the outside world. These rural women, already marginalized, are further burdened by the mechanization of agriculture and by their isolation from information and technological advancements. The aim of government development programs for rural women in the last two decades has been to address the problems of maternal and child health. All programs that were supply driven did not get off the ground. In the early 90's, the policy makers admitted this, and thus policy shifted from the government formulated program to the encouragement of rural women to form self-help groups. In these groups, women came together for developing saving plans and meeting the microcredit needs of their members through a process of self-determination and mutually agreed upon rules of transaction. Such groups have been particularly successful. They have given women a sense of purpose, opportunities for mobilization and networking, and most importantly, the confidence to articulate their own aspirations and visions for development for their children and communities. With reduced dependence on external forces, particularly on the extortionate money lenders, the rural women in the SHGs have greater economic security, especially in times of hardship. There are innumerable success stories of rural women from deprived households coming into their own and even moving into microenterprises, a sure sign of greater economic confidence.

Conclusion

It can thus be seen that the initiatives mentioned in this article flow both from the Constitution of India, in terms of the 73rd Constitutional Amendment Act, and from the political leadership's recognition that in Madhya Pradesh, a state, predominantly rural and with quality of life indicators below the Indian average, decentralization was the key to activating rural communities and ensuring their participation. The development experiences cited here have shown that in ground level implementation of the programs, the involvement of the concerned people in all stages from planning to implementation vastly improves the quality and quantity of results. Above all, this type of engagement of the villagers in the process of development results in phenomenal capacity building, in the development of rural leadership and in the removal or reduction of inequality of opportunities inherent in poverty-ridden areas. The most important contribution of this approach probably would be the harnessing of the villagers' rich potential of human resources and indigenous knowledge systems together with an accelerated access to modern-day knowledge, information and technology and letting them work together in the process of governance.

8 Gandhi and Village Governance

Santosh Kumar Sharma

Most current global problems emanate from the exploitation of the majority of the people, the environment and the economic system by a few. Democracy is now commonly recognized as the only vehicle for realizing a sustainable society. Democracy is however often improperly structured in both developed and developing nations, which leads to exploitation of (1) the poor nations by the rich nations, and (2) the poor by the rich in all nations. Exploitation in governments other than democracies is much higher. So long as such exploitation continues in any nation, social conflict and environmental degradation will continue to plague the world. For preventing exploitation and promoting global sustainability, we need to understand the true meaning of democracy. *Even more important, we need to evolve a process for realizing it.*

Basic Structure of Universal Democracy

Political science has been mostly studying democratic experiences in different time periods and in various nations, *but it has not offered any normative definition of democracy.* It has also completely overlooked India's 4,000 years of democratic ethos and practice, documented in various ancient scriptures containing considerable wisdom about the institutions and ethics of democratic governance. The basic structure of universal democracy depends on the following defined instruments:

Democracy: Democracy can best be defined as "how the majority of the people would like a nation to be governed." No one can challenge this simple definition.

Power Structure: Given the choice, the common people will first retain resources for the local governments for handling all local matters such as administration of justice, police, education, health care, land, water systems and forests. They would then devolve the remaining resources to the state and national governments for providing (1) higher level infrastructure, (2) support to local jurisdictions with inadequate resources, and (3) coordination but not interfere in local matters.

Sovereign Rights of the People: To prevent abuse of authority by their elected and appointed servants, the people would institute effective transparency mechanisms covering their sovereign rights to information, consultation, participation and referendum. The people will thus make every government accountable to them through these transparency mechanisms and not to the higher level government.

Separation of Executive, Legislative and Judiciary Branches: The people will also want the executive branch to be directly accountable to them and not via the elected body as in the Westminster system. The executive, legislative and judiciary branches shall thus be distinct and separate with exclusive local, state and national jurisdictions. This will prevent the national chief executive from assuming dictatorial powers as has happened in some third world nations.

Multi-Stakeholder Upper Houses: In the past, upper houses were instituted to protect feudal and other vested interests. One major recommendation of the Rio Conference on Environment and Development held in 1992 is that multi-stakeholder councils should be set up in every nation to moderate decision making for sustainability. Such councils can be effective only if they are part of mainstream governance. The people will, in view of their growing concern for sustainability, institute such multi-stakeholder upper houses at the local, state and national levels, consisting of representatives of various interest groups such as disadvantaged communities, farmers, labor, industry, women, religions, NGOs, academics and professionals.

Bureaucracy Accountable to People: The people shall also want each government to have its own slim bureaucracy accountable to the people. One method of ensuring such accountability is that the departmental heads in every government should be on contract, appointed and impeachable by the concerned elected body. Though they will function under the elected chief executive, they will be directly accountable to the people and protect their subordinate staff from political abuse.

Along with certain rights regarded fundamental to democracy, this can be said to be the basic structure of universal democracy. The global society needs to endorse this definition of democracy and launch campaigns to get it instituted in all nations.

Democratic Experience of India

India has a 4,000-year old, highly sensitive democratic ethos once practiced by Ram, the monarch of the epic Ramayan, whose rule called *"Ram Raj"* symbolized good governance. This ethos, documented in various Hindu and Buddhist scriptures, basically postulates that local entities, such as villages and *janpad* (districts or municipalities), control local resources and decision making. The state provides infrastructure, ensures complete transparency, respects public opinion, coordinates and assists, but does not exploit.

Grassroots governance plays a pivotal role. The village parliament consisting of all adult men and women is the supreme authority. Thus *women have been franchised in India for 4,000 years, whereas they were enfranchised in the West only in the twentieth century.* The village parliament elects the village head and councilors *through secret ballot*, usually for one year terms, and can remove them at any time for misconduct. It assigns land and other environmental resources *on village lease*, ensures that no one abuses them, and sees that the needs of all are met.

Most monarchs respected such local empowerment. However, over India's long civilization, feudal interests and self-seeking priesthood connived in making holes in the social fabric. Somehow, caste based on profession (the West too has its "Smiths" and "Carpenters") became linked with birth, and the village lease was converted into feudal tenancy. Still, village democracy existed and was vibrant in most parts of India as recently as the nineteenth century. In his famous lines recorded in 1830, Sir Charles Metcalfe, then Governor General of India, observed that these tiny self-sustaining village republics largely contributed to the happiness and prosperity in India and that he dreaded everything that had a tendency to break them up. The colonial rule deliberately destroyed them, transferred land, forests and other resources under the control of state governments managed through district collectors and made the people tenants of the state or intermediaries.

Ideological Options

The ideological options available to humanity are (1) Western capitalist democracy, (2) Soviet socialism, (3) Third world socialist pseudo-democracy, (4) Chinese fascism, and (5) Gandhian democracy. Western democracies are good, except for unbridled rights permitted to individuals for resources, business and wealth. This leads to unsustainable consumption patterns and exploitation of weak nations. Soviet socialism has collapsed. Third world pseudo-democracies, based on colonial institutions and a confused economic system, have faltered. Facing bankruptcy, they are now drifting towards a foreign investment based neo-colonialism. Many praise the Chinese model. With a single party rule, rightist economic model, and a powerful war machine, it truly represents neo-fascism fostering widespread abuse and human rights violations.

Gandhian democracy has great similarity with the best functioning democracies of the world, such as the Swiss. Gandhi added some powerful features for containing consumption and promoting social justice and equity. It therefore is post-modern. In Gandhian democracy, power flows upward from the people. Under the surveillance of empowered local communities, it nurtures ethics, responsible business practices, and sustainable consumption patterns. It promotes self-reliance and truly symbolizes the ideology for the third millennium.

Gandhian Ideology

Gandhi has been voted a man of the millennium in Internet polls conducted by Time magazine of the USA and by the Times of India. He is perhaps the only social philosopher of the twentieth century who understood what ailed the world and who offered a holistic package in the first half of the twentieth century for global sustainability at a time when there was little realization or concern about it. Many scholars have observed that the world did not understand Gandhi since he was born 100 years ahead of his time. In his book "*The Third Wave*" published in the 1980s, Allwin Toffler has a chapter titled "*Gandhi and Satellite*" in which he sees the emerging information technology as truly Gandhian, leapfrogging the third world into information villages.

Praising Gandhi for being voted a man of the millennium, Dilip Patgaonkar, editor of the Times of India, observed: *"What then accounts for the Mahatma's commanding popularity as revealed in this poll? The answer surely lies in the Mahatma's appeal as a quintessential moral force. India may have rejected the path of ahimsa, satyagraha and swadeshi.... But all along India, somewhere in its inner recesses, is also alive to the fact that what Mahatma Gandhi preached and practised was deeply rooted in values that have stood the test of time in all cultures and civilisations."*

Gandhi wanted to re-connect India with its vibrant democratic ethos. He firmly believed that restoration of the authority of the village parliament would rejuvenate the village economy and gradually rid the villages of all social ills. Most social leaders of India at that time, such as Swami Vivekanand, Rabindra Nath Tagore, Subash Chandra Bose and Aurobindo Ghosh, voiced similar sentiments.[1]

Fig. 8.1. Gandhian Democracy

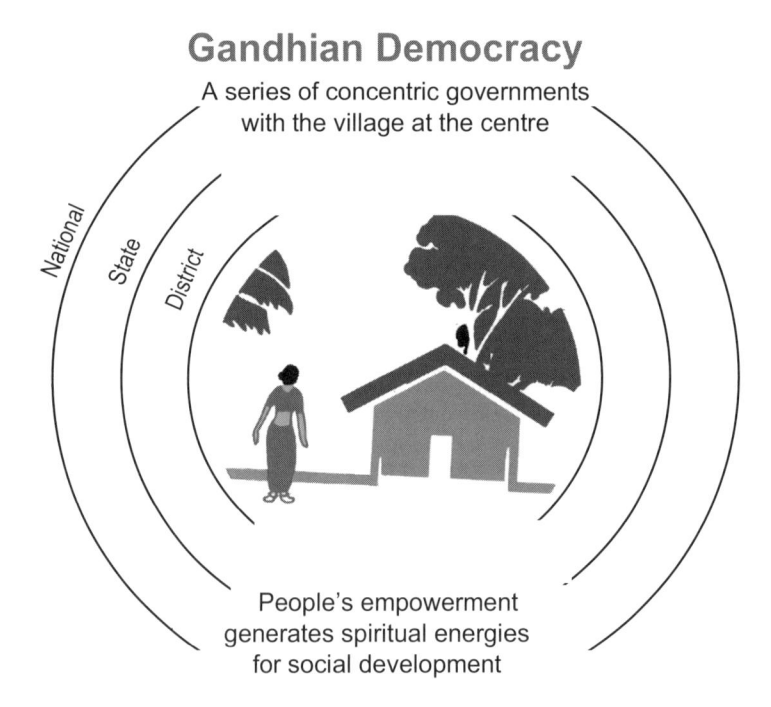

---

[1] Gandhi's political philosophy has been ably documented in the book *"Gandhian Constitution for Free India"* by Prof. Shriman Narayan with acceptance foreword by Gandhi published in 1946. Edited, annotated and republished by People First, it can be accessed on our website www. peoplefirstindia.org.

Exploitative World Constitutions

Britain is often called the mother of democracy. Without intending to denigrate its people, it needs to be understood that Britain is truly the mother of feudal and colonial exploitation. Britain's powerful feudal lords first compelled the monarch to institute a parliament. They then made it titular and the parliament, not the people, supreme. Basic principles of management dictate that the executive, legislative and judiciary branches should be distinct and separate. In the Westminster system, the executive and legislative powers are mixed up. In a small country such as Britain, politics was aligned in two parties. The system is ostensibly working satisfactorily. In most countries, since every legislator is a potential minister, the system fosters abuse of authority, jockeying for power, horse-trading, instability, jumbo cabinets, and bribing of legislators. Legislators who are supposed to be watchdogs over the executive become the mad dogs of democracy!

In pre-war Germany, instability fostered by the Westminster system led to Nazi fascism. In France and later in Sri Lanka, recurring instability compelled the leadership to switch to a mixed-up system — Westminster with a directly elected President.

When India attained independence, the Indian leadership chose to retain the faulty Westminster system and exploitative colonial institutions, centralized, non-transparent and bureaucratized. To this, it added the Soviet practices of centralized planning and a controlled economy, thus creating a disordered economy in a confused polity. Such an exploitative, mixed-up system has, in the 50 years since independence, led to a general social, environmental, economic and political degradation. Most third world nations suffer from this evil.

The institutions of governance in most developed nations are largely democratic, which accounts for their high level of social development. However, their consumer-driven economic systems are wasteful and exploitative, driving them to over-consumption of resources and exploitation of poor nations. Governance in both developed and developing nations needs to be corrected for sustainability.

Sovereign Rights of the People

Any sovereign when appointing a representative to act on his or her behalf, would insist on (1) being kept informed, (2) being consulted, (3) participating in decision making, and (4) making her or his own decisions. In a democracy, these become the sovereign rights of the people to (1) all information except that restricted by the society in the public interest, (2) consultation through public hearings, (3) taking part through participatory councils, and (4) decision making through referendums.

The government should be required to furnish a list of categories of information that it wants to restrict in the public interest. The courts can scrutinize on behalf of the society whether the restriction of any category is indeed in the public interest or not. Public hearings should be mandatory for all urban and regional projects exceeding a specified size, for example 1,000 square meters in land area or 10 meters in height. These meetings should be held at all sites affected by the project. Hearing commissioners identified through an independent process should conduct the public hearings. Their reports should be made public and taken into consideration when deciding the matter. Participatory councils can be constituted for issues of concern to the community such as education and environment. To curb financial improprieties, citizens identified through an independent process should be invited as members on sensitive committees, especially those for defense contracts. Decision making through referendums is the supreme sovereign right of the people, intrinsic to democracy, and exists even if not provided for in a constitution. It is inherent to the people by virtue of their being sovereign, and, as such, is a legally enforceable right. However, judges may be reluctant to issue a writ recognizing it. Referendum is a right of the people to overrule their elected servants, not a right of the representatives. Dictators have often abused referendums, for example in Pakistan and the Philippines, to legitimize their rule for life. Such abuse should not be permitted. The usual procedure for referendums poses, however, problems in countries with high illiteracy such as India. But even referendums held by elected representatives to empower the people, for example in France in 1980s and recently in the United Kingdom, are technically irregular.

Contemporary democracies need an institutional mechanism for processing referendums and overseeing that the other sovereign rights of the people are properly instituted and accessible to them. Such an institution would promote sustainability by correcting faulty institutions, policies and practices, and by preventing the abuse of authority. It is proposed that the following section be added to all international covenants such as the Charter of Human Rights and the Earth Charter.

Conscience Keeper of the State

Notwithstanding the initiatives taken at the Rio Conference on Environment and Development in 1992, there has been hardly any decrease in the rapid depletion of the earth's resources and the resultant damage to its life support systems. The primary reasons are over-consumption of resources and greenhouse emissions by the rich nations, increasing poverty and population, and depleting forests and water systems in poor nations. Globalization has given a boost to growth in world economies. It has also increased the removal, movement and consumption of the earth's resources. Increasing conflict and human rights violations within and between nations has added to the woes. Unless some key strategic initiatives are taken soon, the situation may become irreversible early in the next century. A key strategic issue is the formation and implementation of an institutional mechanism for ensuring that the representatives of the people do not make decisions based on power, business politics and kickbacks. This mechanism should also enable local communities to control resources and decision making for containing consumption and promoting social justice and equity.

Human nature normally has a tendency to accumulate wealth and indulge in wasteful consumption of resources, often at the expense of others. Misuse of authority and corruption among those in power compounds the problem. Ethics and spirituality motivate people to curb such tendencies. They, however, tend to be personalized and fail to impact the society as a whole. Religious bigotry fosters social conflicts and often tends to make religions counterproductive. Contemporary global society needs to design institutional mechanisms that optimize desired attitudes towards social and environmental issues based on universal values. Such mechanisms should make personal benefit subservient to societal good, without scuttling individual creative and entrepreneurial initiatives.

In the past, socialist and third world nations have attempted such efforts. Their methods however undermined the initiatives of individuals and became counterproductive. Some socialist countries have now opened entrepreneurial initiatives, but the societal initiatives are still controlled by the state. They have fascist characteristics fostering exploitation and disharmony. The key to promoting sound attitudes lies in true empowerment of local communities so that all participate and no one is excluded. This will facilitate development of values of the local community *as a whole* to bear on social and entrepreneurial initiatives. This needs to be institutionalized in the mechanisms of democratic governance.

Most industrialized nations have given unbridled freedom to ownership over land and other environmental resources, entrepreneurship, accumulation of wealth, and consumption. Most third world nations suffer pseudo-democracies based on anti-people colonial institutions, such as centralized resource management, non-transparency, overbearing bureaucracy, and weak local governments. *After electing their representatives, the people in developing nations have little say in governance.*

To prevent abuse of the society and of environmental resources, contemporary nation-states need a new institution, an independent Sovereign Rights Commission, with authority to direct referendums *except on issues fundamental to democracy or the integrity of the nation.* There can, for example, be no referendum on making the state theocratic or a region seceding.

Better than the royal priest of bygone days, such commissions will function, like Gandhi, as the conscience keepers of the state, considering the value of the society as a whole. They should correct faulty institutions, policies and practices and oversee that the sovereign rights of the people to information, consultation, participation and referendum are properly instituted, and are accessible to the people, and ensure that exploitative or wasteful decisions are not taken. Such an institution will provide a legitimate, non-violent process for properly democratizing the institutions of governance and correcting faulty policies, practices and decisions of, and abuse of authority by, the elected and appointed servants of the people.

To prevent abuse of authority by such a commission under political or business pressure, the law should also provide that *if ten per cent of village and/or urban neighbourhood assemblies in any local jurisdiction resolve for referendum on any issue, the referendum will become mandatory.*

The global society needs to exert pressure on all nations, developed and developing, to institute Sovereign Rights Commissions, to facilitate an egalitarian global society based on ethics and perceptions of local communities as a whole, and for global sustainability. The empowerment of the people through such an institutional mechanism will truly generate their latent spiritual energies for self and social development and preservation of the earth.

Legal Right to True Democracy

In the pseudo-democracies in most third world nations, the exploitative constitutions are considered sacrosanct and are regarded as weightier than the people. A privileged few have ascended above criminal justice while the vast majority, poor and illiterate, have no social justice! An anti-people constitution imposed on the people and authenticated in their name in violation of their trust lacks legitimacy. It is similar to an attorney obtaining a thumb impression to transfer his client's property to himself. The client can get the transaction declared fraudulent. A constitution authenticated in violation of trust is thus, according to the law, fraudulent. It can now be legitimized only by the sovereign people through referendums. The people have the legal right to demand a process for referendums.

The people, sovereign in democracy, have the legal right to a writ for an institutional mechanism for directing referendums or for a referendum, an intrinsic right, on instituting such a mechanism. If it is ruled that they do not have such a right, then it would be as if a Frankenstein had been created that was destroying its creator, the people, who were impotent to do anything about it. A constitution that makes the sovereign impotent is bad law. It is accepted law that for every wrong there should be a remedy. The people thus clearly have the right to institute democratic reforms through referendums.

9 Rural Incomes and Employment in Russia's Transition Phase

Renata Ianbykh

Russia's ongoing reforms in the agriculture and food sector place the issues of the employment and incomes of the rural population first and foremost. In 1998, the rural population in Russia was 39.6 million people (or 27.0% of the total population of 146.7 million). It is now in a transition period and faces many difficulties, with unemployment and living below the subsistence level being the most acute problems. According to GOSKOMSTAT (State Committee on Statistics of Russian Federation), 47% of the rural population had incomes below the subsistence level in 1998. The official rural unemployment rate is 4.7%, but the actual one exceeds this figure in some regions by 4 - 5 times (Working Materials of SRLPP Project, 1999).

In the time of the Soviet Union, large collective and state farms (*kolkhozi* and *sovkhozi*) provided employment, social services and incomes to most of the rural population. Today, about 80% of the large farms are insolvent and must restructure their business and management. With ineffective farms, agricultural production is steadily declining, and the demand for labor is shrinking. Millions of agricultural workers have been or will have to be laid off (see Table 9.1).

From 1990 - 1998, large farms reduced the level of production 2.6 times and the number of employees by 4.4 million people. In 1998, the average monthly salary in agricultural enterprises was 469 rubles (US $ 16.8). This was only 44,5% of the average salary in Russia. It was obvious that agricultural enterprises would try to get rid of the burden of surplus labor. Thus, the collapse of the system of planned agriculture, which led to the deterioration of large farms, is the first reason for the rise in unemployment. The second reason is the low participation of the rural population in the development of individual private farms which was introduced in 1990 by the federal law, "*On Private Farming in Russia*". It was assumed that rural people would jump at this opportunity to start individual farming at their own risk. This did not prove to be the case. Today, individual private farms use only 6.9% of the total agricultural land, and they produce 12.7% of the country's sunflowers, 7.1% of the grain, and 5,4% of the sugar beets. The number of individual private farms in 1999 was 261.1 thousand, and the average land plot was 55 hectares (Goskomstat, 1999:351).

Table 9.1. Amount of Population Employed in the Agricultural Sector

	1985	1990	1991	1992	1993	1995	1996	1997
National economy [million people]	74.9	75.3	73.8	72.1	70.9	66.4	66.0	64.6
Agricultural sector [million people]	11.0	10.3	9.9	10.1	10.1	9.7	9.3	8.6
Employed in large farms [mln. people]	10.3	9.3	9.7	8.9	8.9	5.8	5.3	4.9
Employed in agriculture, share in national economy [%]	14.6	13.6	13.2	14.0	14.3	14.7	14.0	13.3

Source: Goskomstat: Agriculture in Russia, 1995:28; 1998:16. Ministry of Agriculture and Food of Russian Federation: Agro-industrial complex of Russia in 1998, 1999:49.

Private Household Plots

In the absence of traditional job sources, rural people have started to look for alternative opportunities and have begun to develop their private household plots (*lichnye podsobnye khozyaystva*). Private household plots (PHPs) emerged in the early 70s when rural citizens, the workers of the *kolkhozes* and *sovkhozes*, were allowed to work in their kitchen gardens to help provide their families with food. Food shortages had become a dominant feature of the Soviet economy. Of course, the sizes of the PHPs were strictly limited as was the amount of time that agricultural workers were allowed to work on their "private" land.[1]

During the current transition period, the importance of PHPs has increased significantly. PHPs have become not only sources of subsistence but also for many rural families provide their only reliable sources of income. Today, 16 million PHPs (of which 14 million are in rural areas) all over Russia provide more than 50% of the total agricultural production (see Table 9.2).

PHPs contributions to the total agricultural production in Russia vary according to the different products. They account for 92% of potato production, 77.1% of vegetables, 60% of poultry and meat, 48.8% of dairy production, 29.2% of eggs and 56% of wool. The average size of a PHP is 0,4 ha (Goskomstat, 1999). However, some specialists find that when all the land plots that are used by PHPs (including lands of municipalities and collective enterprises) are taken into account, then the average size is 1.7 ha (Uzun, 1999).

Despite the important new role of PHPs in the rural economy, the government policy towards them has not changed. PHPs are not considered to be subject to governmental regulation, and they therefore are excluded from all state programs for support. For example, all funds from the State Subsidized Credit Fund for

---

[1] The PHP land was private only in the sense that nobody could cultivate it and use the results of production except the family members. Officially it was state land given to the rural families for temporary use.

Agriculture go to large farms, regardless of their effectiveness. PHPs do not have access to state leasing, investments and guarantees. PHP owners have no status. They cannot be registered as self-employed and thus cannot pay into pension, medical and other social funds. This means that PHP owners do not have the social guarantees that other employees enjoy. They cannot apply for pensions or medical insurance. Some of the PHP owners assume fictitious jobs so that they can maintain records of how many years they have worked and can pay taxes, but their major source of income comes from working in their PHPs.

Some Russian academics and agrarian policy makers believe that the absence of clear governmental policy for PHP development is a major constraint to improving the income-generating potential of many rural households (Petrikov, 2000; Uzun, 1999; Shmelyov, 2000). Providing for the development of PHPs, the state could solve many of the social security problems of both the rural and urban populations. *First*, people self-employed in PHPs do not need unemployment money. *Second*, the migration from rural to urban areas could be prevented, and the state could economize on the costs of new buildings in the cities.

From what has been said above, one can conclude that the introduction of a transparent agrarian policy for PHPs is essential. It could be accomplished by passing a special federal law concerning the development of private household plots where the status and rights of the owners are clearly defined. Some regions (Tyumen, Omsk Oblasts) have already passed such regional laws and approved special programs for the development of PHPs. These programs include measures on PHP integration into the local and regional markets. They allow access to input supply, service provision, and joint usage of machinery and marketing. The programs also help with the organization of agricultural cooperatives, technological modernization of the small farming sector, the creation of financial funds to support PHPs, the development of local self-government, and access to NGOs.

Table 9.2. The Share of Main Types of Agricultural Producers in Land Use and Agricultural Output, %

	1990 [%]	1995 [%]	1998 [%]
	Share in land use*		
Large farms	98.1	81.7	83.7
Individual private farms	0.1	5.0	6.6
Private Household Plots	1.8	4.7	5.4
	Share in agricultural production		
Large farms	26.3	50.2	40.6
Individual private farms	0	1.9	2.1
Private Household Plots	73.7	47.9	57.3

Source: Goskomstat, 1999:16, 18.
Note: * The sum of the columns is not equal to 100%, because other land-users also exist (for example, northern tribes using polar pastures, etc.).

To stimulate the economic activity of the rural people, the plots of land in PHPs must be enlarged. Land is not a scarce resource in Russia, and large farms do not have the means to use it all. Moreover, after the land was privatized, rural citizens each received an average of 5 - 7 hectares of land which they could use as they wanted.[2] In March 1996, President Yeltsin issued a decree allowing the PHPs to expand up to the land share size. But the legal and practical procedures associated with this are still in their infancy. Some regions (e.g., Leningrad) have limitations on the size of PHPs. PHP owners are increasingly expressing their desire to expand their production, and this means the expansion of the all types of land: arable land for crop production and pastures and hayfields for increasing the number of livestock. Thus, the 14 million PHPs in the rural areas have great potential for self-employment and for generating income. They also help to preserve rural lifestyles and cultural traditions and give the people some control over the vast Russian rural territories. The importance of rural family's private household plots in the transition from a planned to a market economy is immense. Recognizing this, the government of the Russian Federation is supporting the Sustainable Rural Livelihoods Pilot Project, which was started in September 2000 under the umbrella of the Russian-British Development Program.[3]

Sustainable Rural Livelihoods Pilot Project

In the Russian Federation, many projects dealing with land privatization and farm reorganization problems have been started, and many have failed. Only some of them have touched the difficult issues of rural poverty, unemployment and the economic survival of depressed rural areas. The ultimate goal of the Sustainable Rural Livelihoods Pilot Project (SRLPP) is to secure sustainable livelihoods through the creation of economic coping strategies and income-generating activities, hoping to provide employment and increase the incomes of the rural population. To do this, the following 8 tasks must be accomplished:

- the development of economically-viable, market-oriented collective enterprises;
- the enlargement of plot sizes for PHPs;
- the stimulation of commodity production in family households;
- the equipping of individual private farms and PHPs with modern small machinery and appliances;
- the development of alternative off-farm businesses;
- the microfinancing of family farms via locally created municipal funds and credit cooperatives;

---

[2] Every land share owner could sell his or her land share, lease it, give it as a present, make it a stock in the charter capital of a large farm, withdraw the land plot to start individual private farming, or expand the PHP. Land could be used only for agricultural purposes.

[3] The project was started in 2 counties (raions)– Lodeinoye Pole Raion of Leningrad oblast and Novosilsky Raion of Oryol oblast. In this very paper we will use mainly Lodeinoye Pole Raion data.

- the support of the rural communal and social infrastructures; and
- the development of rural self-government.

The work should be started with an information campaign, because rural people are still not aware of their constitutional rights concerning land and farm property. Legal procedures should be developed and tested to facilitate the transfer of land shares from large-scale collective agricultural enterprises to private plot holders. The process of farm reorganization should be continued. The introduction of third party arbitration courts could help to explain procedures and clarify local legal issues on land and farm assets to the rural citizens. The activity of PHPs could be stimulated by improving the provision of services (i.e., input supply, access to farm machinery, transport and marketing of produce). These also include:

1. more effective cooperation between private plot holders and the functioning remains of the large parent farms,
2. the creation of cooperatives by PHP owners, and
3. the establishment of reliable, transparent and fair trading arrangements with private market integrators which meet the needs of private plot holders for the marketing of their products.

Diversified rural income-generating activities for small-scale rural producers include the development of new activities for both off and on the farm. Rural tourism and the development of national handicrafts are possibilities. To support such activities, the *Fund for the Support of Rural Development* was created to provide microcredit. Now small-scale rural producers have access to credit and to advice about legal matters, taxation and starting businesses. As a result every 10th rural family living in Lodeinoye Pole Raion (county) has benefited from the project.

The *Fund for the Support of Rural Development* (FSRD) is a very important component of the SRLPP. The FSRD is the first microcredit fund created for the needs of PHPs and small rural businesses in Russia. It has tested new approaches. First, the FSRD relies mainly on local sources of funding. Lodeinoye Pole Raion has a considerable amount of timber resources, and the municipal administration can direct some of the "forest" money to the FSRD. Second, the FSRD uses different interest rates according to the size and social importance of the business project.[4] This makes loans affordable for rural families. Third, the local budgets take part in the FSRD, subsidizing the difference between the market and subsidized interest rate. Fourth, there is no minimum limit for loans, so a borrower can apply even for modest loans (for example, US $ 20) for the household. Fifth, the credit terms are favorable since the length of agricultural production in every case is taken into account. Sixth, the FSRD staff helps the borrower to prepare a competent business plan. And last, the responsibility of borrowers on loans taken from the FSRD is collective. In a 9 month period, the FSRD extended credit to 95 rural citizens in an amount of 1 million rubles (US $ 32,000) (see Figure 9.1).

---

[4] The loans are given at ¼, ½ and ¾ of the Central Bank interest rate.

Fig. 9.1. Structure of Loans from the Fund for the Support of Rural Development

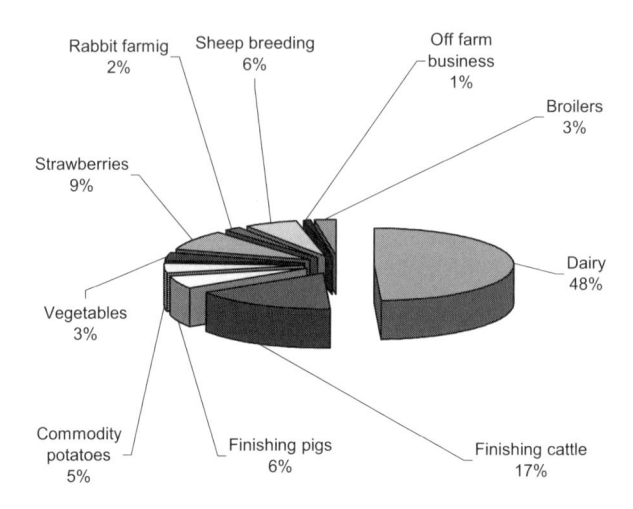

- Rabbit farmig 2%
- Sheep breeding 6%
- Off farm business 1%
- Broilers 3%
- Strawberries 9%
- Dairy 48%
- Vegetables 3%
- Commodity potatoes 5%
- Finishing pigs 6%
- Finishing cattle 17%

The maintenance of the communal infrastructure is also a key problem for rural areas. In the Soviet Union, social infrastructure was kept up by the *kolkhozes* and *sovkhozes*. Now they financially cannot take this responsibility and have transferred it and the social assets to the local municipalities. The problem arises when the local municipalities cannot support the rural infrastructure because of their own lack of resources. SRLPP tries to elaborate new approaches. Improved local support for the maintenance of rural infrastructures and for social assets could be achieved through better coordination between farm businesses, private plot holders (or their cooperative structures) and local government structures (village councils, volosts, raion administrations, and other raion level service providers).

Case Study: Employment and Incomes of Rural Population in Lodeinoye Pole Raion of Leningrad Oblast[5]

Lodeinoye Pole Raion is one of the 17 raions of Leningrad Oblast (province) in the northwestern part of the Russian Federation. It is situated 250 km from the

---

[5] This case study is based on the implementation of methodical approaches elaborated for SRLPP realization (2000) by the research group. The greatest contribution was made by *Uzun, Petrikov, Shagaida, Ianbykh*, et al. The calculations were made by *Rodionova and Ovchintseva* (2000).

capital of Leningrad Oblast, St. Petersburg. The population fluctuates at around 40,000, of which the rural population accounts for 10.5 thousand people or 26%.

The raion belongs to the poorest in the region. The rural poverty level (the percentage of the population with an income below the subsistence level) in the raion is 92%. Rural families have very small incomes. The estimation of the aggregate income is based on data from household records, official statistics and estimates by the local residents of costs of output produced on private household plots. The aggregate income includes the issues listed in Table 9.3.

We obtain the gross average monthly income per capita by dividing the sum of the annual incomes from the mentioned sources by the number of local residents. It is RUR 669 a month, which is 69% of the subsistence level.[6]

The *average income* per household in the rural areas of the Lodeinoye Pole Raion is equal to RUR 1,736 a month and is calculated by dividing the aggregate income from all mentioned sources (on the basis of available data) by the number of households.

The food security status in the Lodeinoye Pole Raion is estimated as the ratio of the gross average monthly income per capita (RUR 669) to the portion of the subsistence level value relating to foodstuffs - in other words: standard basket of goods – (RUR 542).

RUR 669 : RUR 542 x 100% = 123%. This means that the aggregate income of the population is 23% higher than the prices for food products comprising a subsistence level diet. Only 47% of the population live in families whose income is not less than the part of their subsistence level income spent on food.

Table 9.3. Rural Residents Income in the Lodeinoye Pole Raion, 1999

Sources of Income	Per person [RUR */ month]	%
Wages	343	51
Pensions	145	22
Income from private household plot farming and livestock keeping	81	12
Collecting and sale of found wildings	83	12
Social benefits (child care allowances, etc.)	9	1
Unemployment payments	6	1
TOTAL	667	100

Note. * In 1999 the ruble (RUR) / dollar ratio was 27.8

---

[6] The subsistence level value for the Lodeinoye Pole Raion is calculated based on the consumer basket as fixed in the Leningrad oblast. The content of the consumer basket is multiplied by corresponding average prices in Lodeinoye Pole raion. The monitoring of the prices is conducted by statistics of the Lodeinoye Pole raion. The average subsistence level value for the rural population of the Lodeinoye Pole Raion in 1999 amounts to RUR 969.

The *level of employment* of the rural population is calculated as the ratio of the number of residents involved in economic activity to the total population of the working age.

The total number of the population of working age is 5,311 persons. The number of residents involved in economic activity is 623 (the number of workers in agricultural enterprises) + 1,397 (the number of rural residents employed in other enterprises) + (registered self-employment: 65 private farmers + 100 individual entrepreneurs registered in the rural area) + not registered self-employment: 446 (estimated number of people employed on private household plots) + 243 (estimated number of people employed in collecting of found wildings) = 2,874.

The level of employment of the rural population is 2,874 : 5,311 x 100% = 54% (including estimated self-employment). The level of employment excluding estimated self-employment is 41%.

Level of self-employment: The calculation of not registered self-employment on private household plots is carried out by comparing the annual labor costs of output produced on private household plots to the labor norms (2,004 hours per year with the norm of a 40 hour work week). As a result, we have the estimated average annual number of workers on private household plots. Actually, not registered self-employment pertains not only to the working age population but also to pensioners and children. For many people with official jobs, this is auxiliary employment; for those without such jobs, it is primary.

The *value of self-employment* in collecting found wildings is calculated as a ratio of the time spent by rural residents on collecting wildings to the labor norms (2,004 hours).

Registered self-employment (65 private farmers + 100 individual entrepreneurs = 165) + not registered self-employment (446 estimated number of people employed on private household plots + 243 estimated number of people employed in collecting of found wildings = 689) = 854 : 5,311 x 100% = 16%.

Unemployment assessment: In the information provided by the Raion Employment Center, the unemployment rate as of early 1999 was over 9%. Information from household records proves that in early 1999, the number of people without jobs was three times as high as the number of people registered with the employment center. The total number of formally unemployed was 1,031 + 492 = 1,523 people (29%).

Thus the rural unemployment level in Lodeinoye Pole Raion is enormous. Taking into account that neither household records nor official statistics have information about the employment of 25% of the rural population (who in all probability also are unemployed), one can assume that the total unemployment level is more then 50% of the people of working age (see Figure 9.2).

The data from Lodeinoye Pole Raion prove that rural unemployment and poverty are two dominant problems facing rural development in Russia's depressed regions. The state has to launch special programs aimed at solving these problems. One effective measure is to stimulate the development of self-employment in private household plots. This should be supported by the creation of microcredit funds, rural consultant centers, involvement of NGOs, and third party arbitration courts which could provide the rural population with financial resources and the necessary economic, legal, social and technological information.

Fig. 9.2. Rural Employment in Lodeinoye Pole Raion

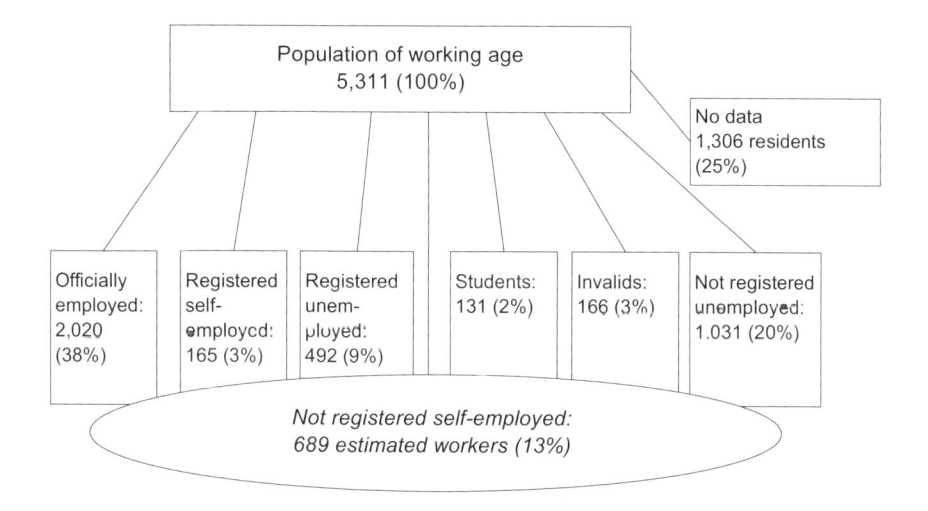

```
                        Population of working age
                             5,311 (100%)
                                                          ┌─────────────────┐
                                                          │ No data         │
                                                          │ 1,306 residents │
                                                          │ (25%)           │
                                                          └─────────────────┘

Officially   Registered   Registered   Students:   Invalids:   Not registered
employed:    self-        unem-        131 (2%)    166 (3%)    unemployed:
2,020        employed:    ployed:                              1.031 (20%)
(38%)        165 (3%)     492 (9%)

              Not registered self-employed:
                689 estimated workers (13%)
```

References

Decree of the President (1996): N 337, April.
Goskomstat State Statistical Committee of Russia (1996): Agriculture in Russia in 1995.
Goskomstat (1999): Russian Statistical Yearbook. Moscow.
Goskomstat State Statistical Committee of Russia (1999a): Agriculture in Russia in 1998. Moscow.
Goskomstat State Statistical Committee of Russia (1999b): Agricultural Activity of Households in Russia. Moscow.
Ministry of Agriculture and Food of Russian Federation (1999): Russian Agro-Industrial Complex in 1998. Moscow.
Petrikov, A. (1999): Big problems of small households (in Russian). In: Agrarnaya reforma: economika i pravo (2), pp. 2-5.
Petrikov, Uzun, Rodionova, Ovchintseva, Ianbykh et al. (2000): Providing Employment and Increasing Incomes of the Rural Population (in Russian). Materials of Manual for Working Groups, forthcoming.
Rodionova and Ovchintseva (2000): Estimation of the Poverty Level of the Rural Population in Lodeinoye Pole Raion of Leningrad Oblast (in Russian). In: Working Materials of Sustainable Rural Livelihoods Project (not published).
Shmelyov, G. (2000): Role of Households in Russian Agrarian Sector and Their Cooperation. (in Russian). In: Agrarnaya reforma: economika i pravo (3), pp. 5-6.

Uzun, V.Ya. (1999): Land Privatization and Farm Reorganization: Ideas, Mechanisms, Results and Problems (in Russian). In: Farm Reorganization: Their Effectiveness and Sustainable Growth. Golitsyno workshop materials. Moscow, pp. 36-54

10 Policies to Manage Local Public Goods in an EU Context

Martin Scheele

The term local public goods can have different meanings in different circumstances. It can refer to the availability of extension and information services, to education, and to medical care; it can stand for the preservation of the environment and rural amenity values. The term can refer to food security or even to intangible concepts such as ensuring equitable income distribution or equal employment opportunities. In economic theory these goods and services are known as "public goods"; in applied policies they are called "rural development measures" or, in the context of WTO, "non-trade concerns".

Public goods have in common that they are not satisfactorily supplied by the established markets. One of the key characteristics of public goods is that, up to a certain degree, they can be consumed jointly without any negative spillover among users. However, this feature may well be vulnerable to congestion effects. Another feature of public goods, which is closely linked to that of joint consumption, is the technical difficulty in excluding "free riders" from consuming them. Finally, on the supply side, it is the fact that such goods and services are not divisible that sets limits to an individual engagement to ensure their supply.

What makes these public goods "local" is the limited number of consumers and suppliers who act in a regionally defined context. Especially when it comes to environmental issues, the public good at stake is often not transferable and shows site-specific characteristics. Decisions about the supply of local public goods should ideally be taken at the local level. However, particularly in cases where national or supra-national funding is needed, these levels also have a role to play in the decision making process.

Since no markets exist, the "public" nature of local public goods suggests the need to pay attention to the adequate "management" of those goods. One way or another, management means "social interaction". In the case of public goods, "management" stands for social interaction beyond market transactions. Management of public goods means that people at all levels of social life, the community, the village, the region, the nation, or the world, have to act together in order to "correct" the allocation of resources towards an enhanced supply of public goods.

Demand and supply of local public goods are often subject to transactions between urban centers and rural areas. The nature of these transactions may vary by region, by economic and natural conditions, as well as by established institutions and traditions. These transactions might involve a flow of financial resources from urban to rural areas, which in return provide services such as preserving the quality of nature and landscapes, providing recreation sites for city residents, or ensuring a social basis for people working in the cities. There are also cases where the rural areas finance "local public goods" on their own or are even net contributors of benefits to urban centers. In most cases links between urban centers and rural areas are considered to be mutually supportive rather than to be antagonistic.

The management of local public goods requires answers to a series of key questions such as: What are the public goods relevant in a certain local context? Which resources can be mobilized? What is the appropriate quantity of a certain public good and how can a decision about it be made? What are the appropriate supply mechanisms, and who will bear the costs?

As the natural, economic, structural, and institutional conditions vary among countries, regions, villages, and communities, there is no single answer to each of those questions. At best, some common insights can be developed regarding general principles. Apart from this, the global dialogue about experiences gained under particular circumstances might give food for thought to find some answers.

The Issue of Public Goods in the Social and Economic Fabric of Rural Areas

The issue of public goods is reflected in Agenda 21 and its concept of sustainable development. Agenda 21 suggests a widening of the perspective from economic objectives to social and environmental issues that markets do not cover. The public goods aspect is not restricted to social and environmental issues. It includes economically relevant issues such as favorable institutional conditions, education, credit availability, marketing channels, information, and communication networks. All of these factors are important to private success in the market place.

There is a certain interdependence between the different elements of sustainability. Strengthening the economic viability of rural areas is the very basis for providing the means to preserve the social and environmental functions of rural areas. Social implications result from the successful management of changes in rural employment patterns, the diversification of economic activities, and the promotion of local products, services, craft activities, and tourism.

The preservation of the environment is also a precondition for developing long-term economic potential in rural areas. The ecological integrity and the scenic value of landscapes are essential for making rural areas attractive for the establishment of enterprises, for places to live, and for the tourist and recreation businesses.

Rural areas provide living space not only for the people who work there but also for many commuters working in the cities. They help to ease centers of dense development, particularly through the preservation of the ecology, as buffer zones, and as recreation areas. As these different functions show, rural areas are very complex places. However, many of their valuable features are not necessarily inherent and therefore must be emphasized in rural development policies.

Urban-rural interaction plays a key role in this context, although obviously topics and issues vary among countries. Whereas in developing countries agriculture plays a leading role in the social and economic fabric of rural areas, its relative significance in most rural areas of industrialized countries continuously declines.

In the political discussion in the European Union, concerns about local public goods in "rural areas" refer to a multiplicity of landscapes as well as of functions according to the pattern of land use. The main issues are the institutions and the infrastructure forming the basis for economic and social structures and shaping the roles of farming and forestry, handicrafts, and small, middle and large companies.

Safeguarding the Rural Environment and Landscapes

Modern landscapes encompass a vast range of visual features and amenities and represent complex, site-specific eco-systems. They bear the form and composition of a farming heritage. This legacy is evident in many diverse features, such as the pattern and size of fields, the extent and type of grasslands, the existence of landscape features, and the use of terracing, cropping rotations, and settlement patterns.

The ecological stability of rural areas also is shaped by the farming past, which has influenced the evolution of diverse species of flora and fauna. Indeed, the significance of farming is a *positive* force for the development of biological diversity, including a mosaic of woodlands, wetlands, and extensive tracts of open country in which agriculture naturally developed.

However, the agricultural processes by which the farmed landscape and biological diversity evolved should not be seen as a deliberate attempt to create cultural values for later societies. The impact on the environment has been a largely unintended consequence of efforts by farmers to overcome technological challenges limiting their productive capacity, in particular in relation to use of water and nutrients.

Technological efforts to overcome these challenges initially gave rise to the cultural landscapes later to be valued by society. However, further advances in technology, especially in the last century, have led to detrimental impacts such as wind or water erosion, water pollution, loss of biodiversity and the destruction of rich landscapes which had once been created.

A particularly acute and universal problem has been the marginalization of certain farming systems in less fertile areas resulting from the increasing pressures of competition. Even within the economically developed countries, the continuation of farming systems which are integral to the conservation of the landscape and its biological diversity may be threatened.

In some instances, because of the anthropocentric origin of high nature value landscapes, maintaining their ecological value requires the continuation of those agricultural activities which respond positively to the site-specific environmental requirements. Here, the economic and social challenges must be confronted together with the environmental ones. In the absence of policy instruments to mitigate impulses from the market, development will be driven solely by farmers forced to focus on narrow economic concerns when deciding whether or not to adopt the most efficient techniques which are not necessarily environmentally benign. For all but a few philanthropic commercial farmers, the provision of public goods will hardly enter the equation.

Establishing Functioning Allocation Mechanisms

An efficient provision of local public goods requires that scarce resources be combined in such a way that welfare (in terms of an aggregate utility function) is maximized. Resources have to be allocated according to a wide spectrum of competing and changing demands including those relating to public goods.

Moving towards an improved allocation of resources requires, first of all, to establish the "rules of the game," i.e. to specify and distribute user rights and duties. Only on the basis of such an institutional arrangement can the reallocation of resources be achieved through individual or social contracts and transactions.

However, even then, an under-supply of public goods can result. This will be the case in particular where, due to the absence of markets, their value is not visible. Producers lack the incentives to reallocate resources and factors of production under their control to contribute to the provision of public goods. In the case of the absence of markets, welfare enhancing reallocations of resources can be achieved only through social interaction as the primary allocation mechanism for the provision of public goods.

Social interaction in the provision of public goods would be established as a bottom-up approach where the number of partners involved is small, the issues at stake are of local importance, and the local constituency has the necessary resources to bear the costs of local public goods. However, with goods of a highly public nature (i.e. jointly consumed and not able to exclude "free riders") and the need to mobilize external resources, the relevant allocation mechanism would be the broader context of democratic decision making at the regional, sub-national, or national level. It could also become an issue for agreements at the supra-national level.

Of course, even a more centralized process of decision making must not lose sight of the local context. An appropriate interaction between general framework and local implementation plays a key role in ensuring the accountability of the supply mechanism.

The need to ensure coherence between local activities and the wider context of a certain economic and social system is not limited to the question of how *policies* can best meet local needs. Sometimes, market related solutions to perceived public goods problems might appear if the perspective is broadened beyond the local context. The flow of resources between sectors and economies through trade can equalize differences arising from regional constraints in the use of resources. This also applies for trading inputs and know-how. Trade can help by providing partial substitutes for natural resources in high stress areas (for example in the form of human capital, i.e. know-how) or finished products.

Who Will Bear the Costs of Providing Public Goods and why?

As with any transaction, the costs of providing public goods and services fall on those who "buy" a certain reallocation of available resources. Those who can claim the property rights to the resources or factors of production used have the advantage of receiving the related factor income. Decisions about who has to bear the costs of providing public goods have to respect existing property rights or, where these do not yet exist, define and assign them.

Managing public goods often requires, as a first step, a decision on the initial distribution of user rights to the resources or factors of production needed for providing public goods. Such a decision first stipulates the distribution of income generated by the use of these resources. Secondly, this decision results unavoidably in a certain initial allocation of resources.

In spite of the undeniable allocation effects of decisions concerning the distribution of user rights, no economic argument can be employed to define what a desirable initial distribution of user rights would be. Every initial allocation can be changed afterwards through transactions with the income distribution effects of the initial decision remaining in place. Therefore, contrary to the frequent claims of economists, distributing the costs of public goods is not an economic question in the first place, but one that depends on equity considerations, institutional traditions, and the given balance of power.

The specification and distribution of newly defined property rights is particularly necessary in cases of currently developing scarcities in natural resources. Here problems result from an incomplete definition of user rights. Following the above reasoning, no straightforward answer is at hand about who should acquire those rights and who should bear the costs of adjusting a certain allocation of resources to higher levels of environmental quality:

- In cases where the "environmentalists" hold the property rights, pollution creates direct costs for the polluters if such property rights are appropriately protected against third party misuse, for instance through fines.
- In cases where the "polluters" own the property rights, they will have an interest in responding positively to financial incentives offered to them under the condition that they improve environmental quality. Not complying with such a request would cause a loss of income.

Both solutions mean an "internalization" of environmental costs; both lead to efficient solutions, albeit resulting in different outcomes regarding the distribution of costs and benefits.

A workable approach to addressing the problem of incomplete property rights for the use of natural resources and the preservation of the environment is the concept of "reference levels". The "reference levels" result from institutional settings (definition and assignment of user rights) corresponding to a certain state of user-environment interactions. This concept is referred to in the EU agro-environmental measures as "good farming practices".

It is assumed that, up to the reference level of "good farming practice," individual user rights do not cover affecting the state of the environment. Therefore, the costs of causing (not legally authorized) environmental effects fall on the individual. However, environmental targets would be seen as going beyond the "reference level" if their achievement involves privately owned factors of production or interferes with already established user rights which then require an appropriate remuneration of owners in order to cover the costs incurred or income lost.

The Issue of Extra-Regional Transfers

The above argument about the neutrality of cost-shifting assumes that either preferences for public goods can be perceived clearly or that costs and benefits of supplying public goods remain in the same spatially defined context of political decision making.

If those who make decisions about the desirable level of public goods are not confronted with the costs of providing them, then a systematic incentive to overstate preferences for certain public goods would lead to serious efficiency losses. This reasoning is used in the "*principle of fiscal equivalence.*" This principle suggests that an undistorted demand for public goods can be achieved only if those who enjoy the benefits from a certain public good actually have to bear the costs of providing it.

The economic rationale of this principle seems straightforward. However, political reality includes the broader perspective of social cohesion. Equity and other redistribution considerations have meaning in social action and political decision making and result, through enhancing acceptability, consensus building, and political stability, in positive welfare effects. Where extra-regional transfers are involved, decision making should go beyond the institutional context of the users of local public goods. Decisions should involve representatives of those providing funds and should reserve the right to check on how the money is spent.

Interregional transfers are needed where local actors, communities, villages or regions do not have the financial resources to provide the necessary level of public goods. Strengthening cohesion through interregional transfers is also in the interest of donors as economic improvement brings increased opportunities for all partners. Of course, cohesion should be more than helping the weak to meet their consumption needs. In a dynamic perspective, the provision of infrastructures, services, and human capital as well as the preservation of natural resources, landscape amenities, and culture values should create the basis and stimulus for an economic take-off leading to self-sustaining development.

EU Policy Framework for Rural Development

In the European Union, the issues of local public goods and non-trade concerns are addressed by policies established under the Structural Funds as well as by the Rural Development Policy, part of the Common Agricultural Policy (CAP). The latter were introduced by the Agenda 2000 and aim at achieving more competitive, multifunctional agriculture and sustainable development of rural areas.

Agenda 2000 included a significant reform of Common Market Organizations with a further shift from price support to direct payments. With the intent to support the adjustment of farmers to a changing economic environment and to contribute to rural development, Agenda 2000 included a major overhaul of rural development policies which was further developed to become the second pillar of the CAP.

This new integrated rural development policy will operate across the Union and accompany the changes brought about by agricultural reform. It is based on a common Framework Regulation that has to be implemented by member states through co-financed Rural Development Programs at the appropriate geographical level.

The objectives of the Rural Development Policy are to enable local and regional actors in rural areas to (i) respond to new demands for services in the environmental, tourist, cultural and amenity sectors, (ii) develop the knowledge and infrastructure for a more diversified and value added agricultural and food production, and (iii) ensure balanced territorial development which includes an equitable distribution of employment opportunities and economic activities.

Managing the provision of local public goods relates in concrete terms to the following types of measures:

- Measures to help strengthen the competitiveness and viability of rural areas, to develop a higher value added agricultural and food production, and to contribute to more diversified economic activity.
- Measures to improve the quality of life and opportunities in rural areas and to ensure a balanced territorial development with a particular focus on changes in agricultural structures and the distribution of employment opportunities and economic activities.
- Measures to promote good environmental practices as well as the provision of services linked to the maintenance of habitats, biodiversity, and landscape in response to new demands for services in the environmental, tourist, cultural and amenities sectors.

The particular features and requirements of a region call for policy responses that take into account the specific regional conditions. Therefore, allowing for a sufficient degree of discretion in the application of Rural Programs is the core of the EU approach in this field.

Conclusions from Practical Policy Implementation

Corresponding with the philosophy of the EU's approach to rural development is an awareness of the inherent limitations of generalizing conclusions. Nevertheless, an attempt is made below to identify some lessons that, at least from the perspective of EU policies, seem to have general meaning for the problem of managing local public goods.

The first point is the complexity of rural areas, their structures and their functions and their diverse economic performances. In providing public goods, this complexity carries a heavy responsibility to develop programs creating region-specific strategies. These strategies should reflect local opportunities and threats and should ensure that local potential is fully utilized.

Second, we need to take a long hard look at our different policy objectives, competitiveness, rural development, and environmental integration, and to ensure that the instruments available to carry out these objectives are being used effectively.

Third, we need to ensure that an appropriate combination of measures is used which reflects the broad objectives mentioned above. The implication of this is a need for clear targeting of resources to those policy areas where they will achieve the greatest impact.

Fourth, it is not only "what" is done that matters; the "how" and "who" elements are also important. In this respect good administration, transparency, monitoring and partnership are essential to successful programs. At the end of the day, unlocking human potential can be the key to transforming rural areas. Any region's greatest assets are its inhabitants, its organizations, its companies and the institutional capacity that they can marshal towards collective objectives.

Finally, the lesson of recent years is that exchange of experience has proved a key force in making rural development policies across the European Union ever more effective. The exchange of experience provides an important opportunity to better understand the needs and potential of rural development, including the provision of local public goods.

11 German Policy for an Integrated Rural Development

Wilhelm Schopen

The progressive globalization and internationalization of the economy cause an intensified competition among regions with industry being concentrated in the most favorable locations. In this competition, rural areas are faced with diverse ecological, economic and social challenges.

The Federal Government of Germany therefore set the goal of strengthening rural areas. Germany accepts the challenges ensuing from Agenda 21, which declared sustainable development as the guiding principle of all social and political action. As an integrated element of a rural development policy, agricultural policy is also called upon to document its contributions to the development of rural areas.

What Does Sustainable Rural Development Mean for Us?

The Federal Government of Germany wants:

- a differentiated shaping of rural areas as economic, residential and living areas as well as habitats,
- an improvement in economic efficiency, especially by creating jobs,
- a mobilization of endogenous development potential,
- an improvement in the competitiveness of agriculture and forestry,
- a responsible management of natural resources, thus an improvement in the ecological compatibility of agriculture and forestry.

Even if in macroeconomic terms agriculture and its up and downstream sectors lost in importance, they still play a central role in rural regions. Agricultural policies in Europe and in Germany have responded to this change in the role of agriculture for the development of rural areas. European agricultural policy increasingly sees itself as a policy for rural areas. The key target is the conservation of viable rural communities, encompassing competitive agricultural and forestry sectors. The decisions on Agenda 2000, also including Council Regulation 1257/99 on support for rural development from the European Agricultural Guidance and Guarantee Fund (EAGGF), created the prerequisites for an area-wide integrated support for rural areas.

As the second pillar of agricultural policy alongside traditional market policy, more use will be made of agricultural expenditure for spatial development and nature conservation. This new policy expressly recognizes that the field of agriculture undertakes various tasks for our society, including the preservation of the historical and cultural heritage.

The multifunctionality of agriculture is a key criterion of the European model of agriculture, which secures the preservation of cultural landscapes and their recreational functions as a service to society. It also ensures environmentally sound land management that is as close to nature as possible. Support for rural development comprises the promotion of the diversification of agricultural and semi-agricultural activities as well as the promotion of tourism and crafts, i.e. of additional income in rural areas. This highlights the interaction of many sectors in its approach.

Three fields of action characterize the range of measures. The promotion of investments is designed to safeguard or create competitive jobs in the agricultural sector, while fostering its adjustment to the requirements of environmental and nature conservation and animal welfare. Secondly, the attractiveness of rural areas as places to live and work is to be improved. The purpose of this is to enhance the attractiveness of rural areas compared to conurbations. In view of globalization, the benefits of the agglomeration of conurbations are increasingly important. Rural areas run the risk of trailing behind development. Village renewal is one of the diverse measures countering these trends. Thirdly, the interests of environmental and nature conservation as well as the protection of rural heritages are to be taken especially into account. Agri-environmental schemes constitute a key element in realizing this aim. They allow the rewarding of the introduction or maintenance of particularly environmentally friendly agricultural management practices, which go beyond the so-called good professional practice (e.g. extensive area management, organic methods of cultivation, long-term set-aside and contract nature conservation etc.).

The common thread running through the entire reform of rural development is the principle of sustainability. However, sustainability must not be confined to the environmental dimension alone. The challenge of sustainable development lies in maintaining economic performance and social balance, while at the same time conserving and strengthening the cultural heritage.

The implementation of measures to promote rural areas is tied to the development plans, which are elaborated by the *Laender (Geman states)* and coordinated with the authorities concerned and with management and labor. They therefore represent an important and essential step towards regionalization, participation and subsidiarity.

Rural21 – Conference on the Future and Development of Rural Areas

Germany sees the development of rural areas not only as a national and European task, but regards this issue also as a task of international dimensions. In many parts of the world, the rural population is still suffering from poverty, a low level of education and insufficient supplies. Rural exodus and unsettled land ownership issues create tension and illustrate that justice, democracy and peace are frequently inextricably linked with the future of rural areas and their inhabitants. This is why a specific and committed policy for rural areas is also indispensable in the international context.

For this purpose, the Federal Ministry of Food, Agriculture and Forestry held an International Conference "*rural21*" in June with the title "*Future and Development of Rural Areas.*" Participants from over 40 states of the world discussed strategies and measures for the development of rural areas in Potsdam.

Rural21 aimed at stressing the importance of rural areas for sustainable development as well as pointing out prospects and proposals for action for the future. *Rural21* brought policy makers, inhabitants of rural areas and non-governmental organizations together, demonstrating that the sustainable development of rural areas is a joint task for the future. At the conference, the topic of sustainable development was examined from all sides by various events and was supported by concrete examples, e.g. village development in the context of excursions. In spite of the varying initial conditions in the individual states, there was a broad consensus on sustainability necessarily being the model of any rural development.

A key element of *rural21* was the participation of non-governmental organizations in the speeches and discussions. This, too, is a kind of participation increasingly gaining ground in public conferences. A joint final statement, submitting key proposals for action, was presented as an important outcome of the conference (see Potsdam Declaration, Chapter 2). The conference participants agreed to contribute towards translating these political declarations of intention into practical action.

Village 2000 - Examples of Sustainable Rural Development

So far the theoretical claims and moral concepts of sustainable rural development have been discussed. In the following, concrete examples of how rural development can be implemented by the citizens concerned in selected villages in Germany will conclude this contribution.

Under the heading "*Humankind – Nature – Technology,*" EXPO 2000 showed new approaches to life, living and working in the century which has just begun. Essential was a vision of creating a viable world for future generations by an enhanced awareness of the management of natural resources. At the same time, EXPO was an international place for meetings and communication. The "*Global Dialogue*" fits into this philosophy. The program of "*Projects around the World*" also complements the EXPO 2000 events. The basic idea of this project can be characterized as follows: Even in our high-tech society, we cannot by any means give suitable answers to all questions. Concrete and vivid projects which could be experienced firsthand – e.g. Village 2000 – were to describe approaches to solving problems, conveying and recommending ideas. The World Exhibition provided an opportunity to prove that urgent problems are being tackled, that this work leads to concrete results and that sound solutions are also ready.

The EXPO project, *"Village 2000 - Examples of Sustainable Rural Development"*, a joint venture by the German Federal Government and the German *Laender,* is one of these 280 Projects around the World and the only project comprising twelve *Laender* at the same time.

At the initiative of the Federal Agriculture Ministry, the project was successfully realized in cooperation with the *Laender* Agriculture Ministries and primarily with the mayors of the twelve villages involved. The sustainability model was the decisive criterion for the selection of Village 2000. Practical examples are to give ideas about how the population can shape its living environment in a future-oriented way through help for self-help. Each of the twelve subprojects involved focused on a specific aspect of integrated sustainable rural development, thus making a crucial contribution to the entire project. Moreover, *"Village 2000"* clearly illustrates that the diversity of rural areas and their villages requires a differentiated approach to solving their problems.

The following topics exemplify the priorities in terms of content:

- Coastal and nature conservation with an autonomous energy supply:
  The North Sea island of Pellworm demonstrates how an environmentally friendly and autonomous energy concept is implemented in remote areas.
- The important role of women in a village:
  The village of Glaisin, Mecklenburg, shows the initiatives developed by women to create jobs and reactivate community life in the village, thus illustrating how women especially fill participation with life.
- Conversion of old rural dwellings:
  Netzeband in Brandenburg shows how farms can be put to new uses through the restoration and alternative use of rural dwellings no longer required for agriculture.
- Organic farming and direct marketing:
  Senst in Saxony-Anhalt is to demonstrate how the sustainability principle determines the development of a village through organic farming and direct marketing as well as ecological wastewater treatment.
- Ecological village development is also the theme of the village Steinheim-Ottenhausen in North-Rhine/Westphalia.
- The creation of new jobs through the start-up of new businesses is demonstrated in the Thuringian village of Körner-Völkenroda.
- The Saxon village of Dreiskau-Muckern shows how a village, which was already intended for demolition due to opencast mining of brown coal, was brought back to life through the initiative of its citizens.
- And Sternenfels in Württemberg shows how the creation of jobs in communications or the safeguarding of affordable homes creates new work opportunities.

These are just some examples of village development which could be complemented at will.

In summary, four thematic areas can be indicated to which the individual projects can be assigned:

- the village as a place of work,
- the village as home,
- the village as an opportunity to experience nature,
- citizens are shaping their village.

All of the rural communities involved in the EXPO project "*Village 2000*" have one thing in common: Committed citizens took the shaping of their environment and future in their own hands and helped make "*Village 2000*" a success in various ways. The citizens of the EXPO villages acted in the spirit of Agenda 21 as they responded to current social problems in an exemplary manner. "*Village 2000*" also proved that competence, diverse experience, profound knowledge and know-how is available in rural communities – and that everywhere there are local answers to the global challenge of sustainable development.

12 Potsdam Declaration Rural21

Participants from all over the world came to meet in Potsdam at the *"International Conference on the Future and Development of Rural Areas Rural21"* from June 5 - 8, 2000. They adopted the *"Potsdam Declaration Rural21"* of which the following is an excerpt[1]:

... Many international conferences and meetings at high and highest levels focused the attention of the global public, governments, international organisations and the civil society on these problems, reminded them of the urgent need for action and adopted the corresponding recommendations, targets and measures. *Rural21* – held on the occasion of the World Exhibition EXPO 2000 – is part of this process and aims at indicating chances, ways and strategies for the sustainable development of rural areas and possibilities to implement them.

Rural21: Towards a Sustainable Development of Rural Areas – What Do We Want?

Appropriate framework conditions must be created and secured so that rural areas can develop as multifaceted areas for living as well as for economic and cultural activities. In order to secure the sustainable development of living conditions, answers to the manifold social, economic and ecological challenges must be found.

The key to a viable and sustainable development of rural areas lies in the development of specific prospects, the development of endogenous potentials and the exchange of experience with other regions. Development strategies must adequately reflect the diversity of starting conditions as well as the opportunities and bottlenecks for development. It must be possible for regional and local actors to respond to their problems with as much flexibility as possible. *Rural21* identified the following fields of action for the sustainable development of rural areas:

- combating poverty, securing food supplies, overcoming inequality;
- investing in people – creating more and better jobs;
- guaranteeing access to productive resources, settling conflicts over land use;
- conserving natural life support systems, integrating environmental aspects into all policy areas;
- establishing a balanced partnership between urban and rural areas;
- creating an efficient infrastructure and ensuring access to it;
- securing an efficient, multifunctional agriculture and forestry;
- ensuring good governance and participation.

---

[1] The full text can be read at: http://www.verbraucherministerium.de/landwirtschaft/laendl-raum/rural21.htm

Conclusions – What Do We Have to Do?

With the help of successful and practical examples, results and experiences *rural21* has highlighted how the problems of rural areas can be tackled and how their development chances be realized. Much will remain to be done in the future to ensure a self sustained and sustainable development of rural areas.

The diversity of natural and socio-economic starting conditions, the complexity and interdependency of problems and their origins require an integrated, comprehensive approach joining all forces available. All population groups concerned must be included and actively involved in the search for ways to cope with the problems to be solved.

As conference participants who prepare and implement decisions in our countries at the local, regional and national levels in the administration and civil society as well as in academic institutions, we are called upon and committed to contribute to political will being transformed into practical action. In doing so, we should focus our efforts on the following priorities:

- Combating poverty and food insecurity by education and training, especially of women and young people, which is an indispensable prerequisite for income and employment.
- Promoting employment in rural areas in agriculture and forestry, including their upstream and downstream sectors, inter alia, by diversifying and using non-agricultural sources of income and employment opportunities, and also in other economic sectors.
- Ensuring a legally safe access to and availability of productive resources for agricultural and rural activities, including loans, technology and market access.
- Strengthening efforts to maintain, develop and sustainably use natural life support systems in rural areas, including the conservation of biological diversity.
- Balancing economic, social and ecological development between rural and urban areas, including their rural and urban settlements, by a reconciliation of interests, cooperation and partnership.
- Establishing and maintaining efficient, environmentally sound infrastructures for the production, processing and marketing of products and improving social services in rural areas.
- Taking into account the multifunctional role of agriculture and forestry and their contribution to balanced development prospects for rural areas, including their natural and cultural landscapes.
- Initiating, guiding and coordinating local and regional development processes...
- *Rural21* participants agree that the international discussion and exchange of experience about the opportunities and ways of the sustainable development of rural areas should be continued and intensified.

In order to maintain and develop rural areas, they require a specific policy to the benefit of their inhabitants and of the settlements and landscapes shaped by them with its irreplaceable potential of natural resources!

Grassroots Movements

Introductory Statement

Anil Gupta

The grassroots movements can be seen as a means of improving participation in decision making by the rural population, in particular by the poor. Only in improving the ownership of rural development activities can their sustainability be ensured. The key impacts of grassroots movements have been (a) the increased economic emancipation and social integration of specific sections of the population (women and youth) into the society as well as the acceptance of the rural population in general, (b) pressure on the state for becoming more accountable towards the urges of the people at grassroots, and modify policies accordingly (c) make research and development systems more sensitive to the problems of the people at grassroots, (d) help rural poor become actors in their own development and (e) learn from the innovations and solutions developed by grassroots innovators on their own without outside help. As has recently been seen, grassroots movements also play a significant part in fostering good governance at the national and international levels.

The following contributions highlight specific aspects of the over-all agenda regarding grassroots movements. Especially the importance of NGOs in the non-formal sector and their growing social capital for innovations is discussed. Each contribution offers a unique perspective, including those of the environmental and educational movements in different countries.

The Sahelian Areas Development Fund Program, which is the most advanced model of participatory development interventions tested by IFAD for directly channeling development funds to grassroots movements, is introduced. The program is built on mutual trust between all stakeholders, encouraging groups to initiate their own investment ideas, to provide equity capital according to their capabilities, and to seek complementary funding from other sources (*Jatta*).

The further process of Grassroots Movements has to take into account the consideration that all policies should adopt a gender perspective to ensure that development and dividends are shared equally, ensuring secure livelihoods (*Okoed*).

Grassroots movements as means to overcome rural financial market failures is analyzed. The informal livelihood support groups of the local communities are not as well appreciated today as they were some years ago, or at least two decades ago. For some strange reason these rotating savings and credit associations have not become a building block in the entire global microfinance movement, and yet they remain a very important institution at the local level providing means of livelihood to the poorest people and contributing to the reduction of the income gap (*Heidhues*).

Grassroots' innovation is a further investment in a better rural future. The emphasis is on what people already know in rural areas, on their knowledge systems, and on their ways of doing things. It looks at their already existing capacities for innovation, for producing new ways of thinking and of acting, new technologies, new ways of marketing, etc. The idea is to make the rural societies dynamic not through the process of receiving knowledge either through modern communication technologies or through formal systems of education but through the creation of new systems of knowledge based on local perspectives, on already existing systems of knowledge and on local capacities for innovation (*Roy*).

The organic agricultural movement is presented as a grassroots and bottom up movement, not just to bring about technological change but also a cultural value change where nature is respected (*Geier*). The paradigm and reality shift which is emerging in European consumption patterns nowadays involves increased access to healthy and organic foods as well as planning not only to produce healthy crops but also to provide more jobs and better livelihoods in rural areas. This gives organic farming understood as a grassroots movement a prospective outlook for the 21st century and beyond.

Two concrete examples from Egypt and India demonstrate the capacity of grassroots movements with comprehensive approaches, covering economic, social and cultural activities. All these fields of activity and project parts are interrelated, forming a network and cross-fertilizing each project part. So many different innovative aspects within the project parts can be realized. (*Merckens and Shalaby* and *D'Costa and Samuel*).

The contributions presented in this session give me faith in the spirit of enterprise, in the spirit of creativity, in the spirit of innovation at the grassroots level. This spirit combines what are called the "5 E's": equity, excellence, efficiency, environment, and education. There will be no sustainable development without resource mobilization and judicious utilization of the existing local resources by the community through locally evolved practices, institutions and management structures. It is high time that the formal, national and international development organizations encourage fuller participation of the rural population, especially the rural poor in the rural development process worldwide. It is, however, equally important to make clear to the local active society that grassroots movements on their own are not able to sustain their achievements. Only in cooperation with national and international organizations can experiences be shared and further innovations be stimulated.

Honey Bee Network may serve as an example. Honey Bee Network has helped provide a sort of informal platform to converge creative, but uncoordinated individuals across not only Indian states having varying cultural, language and social ethos but also in 75 other countries around the world. What it is trying to do in a rather quiet manner may transform the way the resources in which poor people are rich are used in future. These resources are their knowledge, innovations and sustainable practices. The Honey Bee Network evolved twelve years ago in response to a personal crisis. While I had grown in my career, received awards[1], recognition and remuneration for writing about knowledge of innovators and other knowledge experts at grassroots, very little of this gain had actually been shared with the providers of knowledge in concrete terms. Much of my work was in English language till that time. I had tried to share the findings of my research with people; but it had not been institutionalized in local languages. Likewise, I had tried to acknowledge the knowledge providers; they still had remained broadly speaking, anonymous. It was obvious that my conduct was not very different from the conduct of other exploiters in society. They exploited in land, labor or capital markets. I exploited the poor in knowledge market. It is at this stage a realization dawned that something had to be done to overcome this ethical dilemma. The Honey Bee as a metaphor came to rescue one day. Honey Bee does what we, intellectuals, don't do. It pollinates the flowers and takes away the nectar of flowers without impoverishing them. The challenge was, to define the terms of discourse with the people in which they will not complain when we document their knowledge, they will have the opportunity to learn from each other through local language translations, they will not be anonymous and they will get a share in any wealth that we may accumulate through value addition or otherwise. Honey Bee Network has brought lots of volunteers together who share this philosophy partly or completely and who want to link up with an immense source of energy and inspiration available with the grassroots innovators[2].

---

[1] The Honey Bee Network (http://www.sristi.org) has also received many awards and recognition. Apart from Pew Conservation Scholar award to Prof. Gupta in 1993, the Far Eastern Economic Review chose SRISTI and Honey Bee Network for Asian Innovation Gold Award in 2000 9 Oct 26, 2000.

[2] The Honey Bee Network was founded with the help of Prof Vijay Sherry Chand, Jyoti Capoor, and many other friends. Later Kirit Patel joined and made an immense contribution. Kapil Shah, Rakesh Basant, Amrut Bhai, Riya Sinha, Srinivas, Dilip Koradia, Murali Krishna, Alka Raval, Chiman Parmar, Praveen, Mahesh Parmar, Hema Patel, Shailesh Shukla, T.N. Prakash, P. Vivekanandan, Sudhirender Sharma, and many others have contributed to the growth of Honey Bee Network.

Agenda 21 Principle 22, Earth Summit, Rio 1992

Principle 22
Indigenous people
and their communities
and other local
communities
have a vital role in
environmental
management
and development...
States should
recognize
and duly support
their identity,
culture and
interests...

13 The Active Rural Society in West and Central Africa

Sana F.K. Jatta

Due to the nature of the mandate of the *International Fund for Agricultural Development* (IFAD), which is to combat hunger and rural poverty in low income, food-deficit regions and to improve livelihoods of the rural poor, the need to build strong partnerships with grassroots movements throughout its intervention zones was quickly realized. Both the mandate of IFAD and the need to build partnerships to attain it are immediate concerns today for the villages of West and Central Africa because of the desperate situations in which they find themselves.

The Sahelian Areas Development Fund (SADeF) Program is the most advanced model of participatory development interventions tested by IFAD for directly channeling development funds to grassroots associations in a flexible manner. It is built on mutual trust between villagers, the government of Mali and IFAD. The program encourages groups to initiate their own investment ideas, provide equity capital according to their capabilities, and seek complementary funding from SADeF Program resources or any other sources they may deem appropriate within an agreed framework.

IFAD was established in 1977 as a specialized agency of the United Nations with a single mandate, unique among the International Financing Institutions (IFIs). Its raison d'être is solely "*to combat hunger and rural poverty in the low income food-deficit regions of the world and to improve the livelihood of the rural poor on a sustainable basis.*" In other words, it *shall* seek out the poorest of the poor in their *villages, hammocks*, and *nomadic camps*, however remote these may be. Once they are located, it *will* endeavor to organize them into groups, professional or otherwise, with a critical mass that will facilitate helping them, in collaboration with other stakeholders, to develop the capacity to achieve sustainable improvements in their income, access to food, and improve their general livelihood.[1] That mandate is still as relevant today at the start of the new millennium as it was when IFAD was created over 20 years ago. It is of special concern for the rural poor of the villages of the sparsely populated Sahelian, desert (in some cases), and invested forest zones of West and Central Africa.

Consequently, the companionship between IFAD and grassroots movements in West and Central Africa is as old as the Fund itself. It is mutual and self-fulfilling. The partnership was given a new lease of life in the mid-1980s when, after reviewing implementation experiences of the first generation of its projects, IFAD quickly concluded that more should be done to build up a sense of ownership by the rural stakeholders of project investments.

After a decade of trial and error with various forms of associations and groups up until the mid-1990s, there was still a feeling that more needed to be done to

---

[1] See IFAD Update No. 7, dated February 2000: Editorial.

improve their overall performances as vehicles of development. With the exception of a few cases, a number of IFAD projects failed to establish sustainable groups capable of pursuing their own development agendas beyond the projects. A consensus soon emerged that there may have been a flaw in the adopted strategy, at least in certain instances where some of these groups turned out to be artificial at best.

Yet, it was obvious then as now that for sustainable development in remote areas there is no alternative to working with serious grassroots organizations. Consequently, in the mid-1990s, IFAD carried out another review and evaluation of the results of the groups' development activities under its projects in the Sahelian and forest zones of West and Central Africa.[2] Some of the conclusions have since led to a major reorganization of the way IFAD does business with grassroots movements in the region. One exciting innovative operation being implemented in Mali for a year now is the Flexible Implementation Agreements and Schedules for village infrastructure development and individual productive investments under the Sahelian Areas Development Fund (SADeF) Program. This program was designed according to the capacity of villagers to provide equity capital and their real investment needs for which they must be initiators and sponsors.[3]

Some Relevant Characteristics of West and Central Africa

For obvious reasons, the constraints that hamper the proper functioning of homogenous groups or associations are accentuated in the Sahelian and forest zones of West and Central Africa. These regions are extremely remote due to the virtual absence of access roads. Also their natural environments are generally harsh and fragile, with a dearth of development potentials in some instances and the prevalence of diseases in others. Since the drought of the early 1970s, which decimated the livestock herd, the Sahel zones have been exposed to recurring droughts. In the forest zones, agricultural practices and timber logging are leading to massive deforestation, environmental degradation, and soil and water erosion. Consequently, rural village societies in these regions tend to need, more than most, grassroots associations to overcome the harsh living conditions and to improve livelihoods. Yet, they are mostly still isolated and individualistic societies with pockets of heterogeneous groups and sub-groups living side-by-side without any easily discernible forms of alliances. Furthermore, they would need compensatory development assistance in order to help protect the environment in which they live and thereby generate global benefits.

Under these circumstances groups take on a special significance in rural development in the light of the obvious challenges and potentials. In some

---

[2] The full report of the study is available on demand.
[3] See IFAD Update No- 7: *The Cornerstone of Agricultural Development in Mali – People's Empowerment and Ownership.*

instances they provide interesting traditional forms of associations. These could be closely studied in order to identify those major characteristics that make them succeed and prosper for possible replication elsewhere. But in other cases they provide examples of seemingly insurmountable constraints that consistently hamper the proper functioning of groups of any kind, which generally tend to require a common purpose and some sort of homogeneity. However, in all instances most of the actions necessary to ensure sustainable livelihoods in these harsh conditions cannot be carried out without grassroots movements and institutions.

Empowering Grassroots Movements

Working Hypothesis on Grassroots Movements

The working hypothesis all stakeholders embrace now is that grassroots organizations are the most appropriate, effective, and efficient means of reaching the rural poor and ensuring their full participation in development activities. Nevertheless, it is important to bear in mind that when a group is not organized for a specific development purpose it can turn rather quickly into a wildly political machine, open to abuse by some unscrupulous politicians. Therefore, right from the start we should ensure the future autonomy and integrity of group members and their leaders. Only in this way can we guarantee the effectiveness of the groups in carrying out development activities and the sustainability of the investments once external funding ceases.

Participatory project implementation strategies are in fact merely a means of enabling beneficiaries to have a stake in project achievements. IFAD has now incorporated participation concepts in practically all its projects. A survey revealed that around 37% of its projects had a participatory element in the period between 1974 and 1984; that figure now stands at over 90%.[4] Under these new trends rural sponsors will initiate, plan, implement, monitor, and evaluate their own village and individual productive investments, which will be co-financed by outside donors. The necessary technical advisory services will be provided by private or public technical assistance and non-governmental organizations on a competitive basis.[5]

Thus, the group development approach favored by IFAD is a carefully thought-out and evolutionary strategy aimed at finding an appropriate way to reach the different segments of its target groups. The ultimate goal is, of course, to achieve sustainable increases in on-farm as well as off-farm production and incomes in rural areas, as well as to raise general living standards there. Therefore, the aim of the groups and associations established under IFAD-supported projects have

---

[4] See the IFAD Brochure on "*Participation: - People Behind the Projects*". IFAD, September 1997.
[5] See the SADeF Program in Mali briefly described in the IFAD Update No. 7, dated February 2000.

always been twofold: to represent, and thereby empower, the beneficiaries in a responsible manner; and to assist with funding and implementation of project activities while ensuring the long-term sustainability of the investments.

Therefore, the underpinning working hypothesis of IFAD's operations is that both economic well-being and non-tangible social well-being, in the simplest and purest forms of empowerment of the group, are required in unison to alleviate poverty of the soul, mind, and body, and to improve the chances for a sustainable livelihood.[6]

Achievements in Promoting Grassroots Movements

A recent assessment of the Groups Development Components of up to 23 ongoing IFAD supported projects in West and Central Africa noted that virtually all of them made provisions for "*all project activities to be targeted on beneficiary groups and associations that will be established.*" A close analysis of the results obtained in promoting groups development in IFAD's projects and programs in West and Central Africa revealed the following encouraging conclusions:

- Using traditional labor sharing, solidarity and age groups as a strong basis and providing material, financial and human-power support, it is possible to establish well-structured and strong community-based associations and groups that can be representative and effective in empowering their members.
- Certain grassroots groups specialized in adult literacy training and created their own independent "consulting" service in the domain. Other Specialized Users' Groups took over operating the infrastructure of water management schemes or organizing produce marketing and inputs supply arrangements following the withdrawal of public institutions in the aftermath of Structural Adjustments Programs.
- Grassroots organizations have been shown to be: (i) an appropriate framework for promoting deprived groups economically and socially; (ii) encouraging the active involvement of their members in decision making and in training sessions; (iii) a means of negotiating and obtaining local investments; (iv) an effective way to manage rural resources and community facilities; (v) an appropriate support for rural self-promotion and local initiatives; (vi) a satisfactory means of backing independent financing and local development endeavors; and (vii) an opportunity to promote individual efforts and to encourage small-scale income-generating activities.
- The groups development approach reinforces social cohesion in village communities by encouraging the integration of underprivileged segments, notably women and youth, into the day-to-day affairs of the community.
- The microfinance grassroots associations in charge of Village Development Funds and/or Banks have introduced the notion of savings in the form of cash

---

[6] See Klemens van de Sand "*IFAD's Decentralised Approach to Governance*" in the UN Chronicle, Vol. XXXVII, Nov. 1/2000, p. 90-94; and "*The Central Article*" in the IFAD Update No. 6, dated September 1999: *Performance and Governance – Confronting the Issues*.

as opposed to other traditional, riskier savings mobilization practices such as livestock, which could be easily decimated by the harsh climatic conditions prevalent in the region.
- Groups also enable their members to have access to credit through mutual group guarantees, which provide reassurances to the credit operator who can rely on peer pressure when appropriate.

These positive results partly explain the reason that donors, IFIs and bilaterals as well as NGOs have adopted the approach and make it an integral part of their procedures. Unfortunately, however, the approach has other shortcomings which warrant careful consideration whenever it is used in development projects and programs. A major flaw in the approach is that the sustainability and economic viability of the quasi-totality of the groups formed under external funding is entirely linked to the continuation of such support. In most cases once external funding ceases the groups are also disbanded and disappear. More efforts are, therefore, needed to overcome these design and implementation weaknesses.

Some Impacts of Grassroots Movements on Projects

The introduction of the group approach in projects has induced important changes in the conduct of development actions in rural areas and in the social organization of the affected village communities. In the *first place* all key players and stakeholders in rural development (governments, donors, project teams, villagers, civil servants and consultants, etc.) now accept the rural poor as full time actors in their own development. They recognize their associations and groups as serious interlocutors. This has eventually led to the abandoning of the heavy-handed, top-down approach used by several governments and their representatives in working with rural people and its replacement by more participatory development strategies on a much wider scale.

Secondly, groups offer to the deprived segments of the communities, especially women and youth, possible economic emancipation and social integration. In the Sahel, strong women's groups are active in income generating activities on a large scale, including such areas as vegetable gardening, sheep fattening, soap making, cloth dying, village shops, etc. Especially in the forest zones of central Africa, both mixed groups and groups with only female members have emerged, thereby enabling women to become active in all economic sectors including those traditionally reserved for men. In the Smallholder Development Project in the forest zones of Guinea, the swamp development groups comprise around 36% women participating in an activity which use to be predominantly male.

Furthermore, being a member of a group is quite often the only means of having access to credit and savings facilities provided by rural financial intermediaries. Also, the emergence of grassroots associations and pressure groups that are fully independent from politicians and the central administration encourage good governance and the restoration of democracy throughout the region.

Weaknesses of the Group Promotion Approach

Like any other tool for rural development, group promotion activities encounter a number of practical difficulties that limit their efficiency. *First*, experience shows that while the approach promotes genuine participation when applied seriously, it is not applied consistently throughout the project cycle. On one hand, it is applied mainly during implementation and entails encouraging the participation of the rural poor in sharing operating and maintenance costs of village infrastructures. However, this is rarely accompanied by genuine participation in either the management or the decision making processes of project activities. On the other hand, the approach is seldom used during the design and preparation stages. Consequently, the rural poor often find themselves confronted with fait accompli decisions made for them by others long before.

Second, during the actual implementation of the group promotion activities in the field a number of specific operational difficulties have been registered by IFAD-supported projects, including:

- the low level of training of staff, the abuse of confidence and vested power by group leaders, the emergence of conflicts over responsibilities between traditional village notaries and group leaders, and attempts at harnessing undue political influence by leaders;
- excessive charges without being provided adequate support services;
- farmers' organizations are considered by governments as an integral part of the project and, therefore, hierarchically under their responsibility;
- the legal frameworks under which the groups exist are often unclear, and they are neither legally protected nor recognized legally or administratively.

These difficulties can luckily be overcome by focusing on a few critical areas affecting the establishment and functioning of the village groups, namely the institutional frameworks and the implementation arrangements for development projects. *First*, on the institutional level, it is proposed to introduce appropriate modifications in the legislation in most countries concerning associative movements and the political and administrative frameworks. The changes in the legal frameworks should aim at decentralizing decisions and the funds to the local level where the groups will be operating. In particular, the administrative and budgetary procedures, as well as the funding mechanisms, should be adapted in such a manner as to allow groups to make their own investment decisions and to search for co-financing from whatever source is available.

Furthermore, the clarification of the demarcation line between municipal (or communal) and village community projects is becoming increasingly important following large scale decentralization in almost all countries in West and Central Africa. This would avoid conflicts and ambiguities in the mechanisms to be put in place to channel funds to the initiators of the projects. It is advisable to introduce separate channels for municipal and village community initiatives for a variety of reasons. Ownership and responsibility for project implementation will be different for one thing, as will be operational management and maintenance. The subordination of one type of financial channel to the other would be impractical.

Also, it will confuse public and private budgeting, accounting, contracting, and disbursement procedures. It may result in introducing cumbersome bureaucratic set-ups to manage funds at the national level that are meant for the villages and are better managed there.

Second, the implementation arrangements should be adapted to aim at simplifying the procedures for the design, organization and management of projects in such a way as to encourage the participation of the rural population, or their representatives, in decision making at all stages of the project cycle. This would entail essentially: the identification of the real needs and expectations of the targeted beneficiaries through participatory tools; the precise indications about the different roles and responsibilities of all partners and stakeholders; the introduction of participatory monitoring and evaluation tools; and finally the provision of technical and managerial training of group members.

These are the main characteristics of the IFAD-initiated Sahelian Areas Development Fund (SADeF) Program in Mali which is a precursor of the new type of Flexible Implementation Arrangement and Scheduling Project discussed below.

The SADeF Program in Mali: A Program Built on Mutual Trust between Villages, Government and IFAD

Key Elements of the Program

The convergence of different trends conjugated into the *Sahelian Areas Development Fund (SADeF) Program*, which can only be described as the most advanced, albeit daring, form of participatory development and flexible lending that IFAD has tested so far. Its main underlying rationale is that rural people, if (and when) given the possibility, have the ability and willpower to take their own investment initiatives. Also they will be willing to put forward their own equity contribution for such investments, in the form of cash and/or kind, according to their means and limitations.

The program represents a radical shift from the old ways when project designers and implementers tended to believe that the poor lack knowledge about investment opportunities in their own environments. They assumed the poor were incapable of making suitable investment decisions for themselves. Also, the government and donors pretended that since they provided the investment funds they had the exclusive right to decide how they were spent. In the end, the poor rural dwellers saw through the game. They learned how to receive so-called *development assistance* without taking responsibility for either its success or failure: the government's and the donor's money belonged to no one, after all! Despite all their good intentions, previous programs and projects continued to blindly spoon-feed the poor. They ignored local knowledge and know-how instead of using it as a springboard.

The SADeF Program is daring in comparison to other IFAD operations in Mali. For the first time, putting faith in the poor was accepted, controlling *a posteriori* what they did with IFAD and government provided funds instead of *a priori*. Also, it was agreed to give implementation responsibility for big operations to private associations run by grassroots village-based institutions, specially created for that purpose. Finally, IFAD accepted co-financing of investment ideas of the poor rather than coming up with its own detailed design of what must be done. All of these provisions make the program a gamble, albeit one that has long been overdue considering the evolutions among rural communities in Mali.

The ultimate goal of the program is to reduce the incidence of poverty among rural households in the Sahelian zones of Mali, including large sections of the four administrative regions in the center of the country, namely Kayes, Koulikoro, Mopti, and Ségou. This should be achieved through improved incomes and living conditions for rural dwellers in the poorest and most remote parts of the regions. The area is vast, extending over an area nearly the size of the United Kingdom or Ghana, and is inhabited by over 3.5 million people living in some 4,900 villages. It is generally characterized by mixed farming systems (under rainfed and flood recession farming), but with pastoralism the dominant system on the northern fringes. The main crops grown are rainfed millet, sorghum, and maize, plus paddy under flood recession or irrigated schemes. The program will initially focus its attention on the 1,700 poorest villages (with an average of 100 families each) in 81 communes. They were selected based on the fragility of their Sahelian ecology, the extent of their vulnerability to fluctuations in food production, and the presence or lack of local infrastructure. The population comprises sedentary crop farmers (mainly Bambara, Soninké, Kassonké and Dogon), nomadic herders (Fulani) and a few nomadic fishing communities (Bozos). Poverty is prevalent and about 11% are female-headed households, more than double the national average. Women represent 48% of the total labor force compared to 46% nationally.

The main focus of program activities will be on the empowerment of grassroots associations and village groups through a flexible mechanism that will allow them to identify, formulate and implement their individual priority investment demands. Therefore, it will offer rural communities its participation in their own initiatives through a number of services and funding arrangements that fall within the remit of IFAD's mandate. The activities to be supported have been divided into three broad menus (or components), which include: (i) village infrastructure development support, including the establishment of a *"Village Infrastructure Development Fund (VideF)"* for promoting community and group microprojects under the direct responsibility of the stakeholders, and the provision of required training and support services to identify, effectively manage and operate the microprojects; (ii) the promotion of decentralized financing services (SFDs) by local savings and credit banks (*"Caisses d'épargnes et de crédit* (CECs))" to meet the demand for individual productive investments; and (iii) program management by an appropriately sized private organization.

Decentralized Organization and Management by Private Grassroots Institutions

The unique opportunities offered by the convergence of views of all stakeholders made it possible for the management structure of the SADeF Program to meet several requirements for effective grassroots participation in decision making. First, it will operate under private law and establish a lean and flexible institutional set up. Second, it will have a single authority at the national level to facilitate the relations with the government and the external partners. Third, it will enjoy full autonomy in management at the regional level. And fourth and last, allowance has been made for village stakeholders to progressively take over the responsibility for program management. Another major consideration is that for a participatory and demand driven approach to be credible, it requires the shortest possible command line between stakeholders and decision makers and speedy procedures at the regional, national and external levels to be able to respond quickly to expressed valid demand of village communities.

Demand for microprojects must come from a village association or group of at least ten people from different households before it can be accepted. Each group will establish a management committee of three to four persons. A regional association, operating under private law, will be established in each region to oversee program activities there. They will comprise representatives of the federations of village associations and groups, NGOs operating in the region, the Regional Chamber of Agriculture, and local municipal councils. Their role will include approving microprojects and annual work programs and budgets. The overall responsibility for the program at the national level will be given to a national association, whose membership will include representatives of the regional associations, the coordination body of national and foreign NGOs, the Association of Consulting Engineers, and the Permanent Assembly of the Chambers of Agriculture. In line with the disengagement policy of the state, the Ministry in charge of Rural Development will delegate overall responsibility for program implementation to the national association. It will retain, however, full responsibility for overseeing and controlling the activities, in close consultation with IFAD and the village stakeholders.

Flexible Implementation Arrangements and Schedules

The rationale of the program dictated that maximum flexibility be given for the selection of activities to be supported and in scheduling their implementation. This built-in flexibility is crucial as a tool of adapting to the exigencies of a holistic approach towards poverty alleviation, which accepts the challenge that the needs of the poor are in fact wide-ranging and have variable time frames. Therefore, the satisfaction of those needs have to be provided in a manner that is necessarily appropriate and hence variable and adaptable.

Consequently, the SADeF Program is the first IFAD operation designed under the new Flexible Lending Mechanism (FLM) adopted in September 1998. Under its flexible design, efforts were made to refrain from deciding in great detail about all the activities that will be carried out during its 10-year implementation period. Nor was any firm decision taken as to when and where the activities will be carried out. Instead, the program intends to create a funding mechanism that will be used to co-finance investments initiated privately by individuals or groups within the target communities. For this to work, appropriate procedures and channels need to be in place for identifying suitable components of the investment portfolios of rural communities and/or of their individual members. Once identified for intervention, an investment portfolio will benefit from the necessary fine-tuning and retrofitting through a feasibility study in order to enhance its chances of success. Thus, the sense of ownership by the rural communities of their own investment decisions will be given the appropriate priority and symbolism. Under such circumstances, all stakeholders have a real interest in the success of the joint operation. The joint operation is also more likely to address a real need of at least one of the stakeholders (the rural dwellers) as opposed to some perceived needs of all stakeholders, which invariably end up being left by the wayside.

Furthermore, the Program's ten-year implementation period is divided into three cycles of three, four, and three years respectively. The main objective of the first cycle is to establish the institutions, mechanisms and procedures, to ensure their workability and to develop a limited number of investment activities. During the second, the investment activities will be expanded throughout the program area. The third and last cycle will focus on consolidating the achievements and executing an appropriate exit strategy to ensure the sustainability of the operations.

Out of the full envelope of financial assistance for the ten years, only those earmarked for the first cycle have been described in any detail. Even this was limited to the criteria for selecting geographical coverage and the broad listing of the menu of activities to be supported during the first three years.

Under the FLM approach a series of so called "*triggers*" or pre-conditions have been developed to decide whether or not to proceed with, delay or cancel the subsequent cycles of the Program. A joint review involving all stakeholders, especially those among the village communities, will be carried out at the end of the cycle in order to determine whether the pre-conditions have been met and to recommend an appropriate course of action. The decision to proceed to subsequent cycles will be taken jointly by all the stakeholders. In this exercise program level monitoring and evaluation will play a critical role in tracking progress in attaining the triggers. Therefore, the M&E system has to be sensitive and systematic. It will consist of close monitoring using specialists, complemented by participatory monitoring and evaluation by stakeholders, plus independent evaluations by outside institutions.

14 Akina Mama wa Afrika – a Women's Development Organization

Sandra Okoed

To achieve the goal of people-centered sustainable development, policies and programs should ensure secure livelihoods, adequate social measures including safety nets, strengthened support systems for families, and equal access to and control over financial and economic resources. Therefore, all economic policies, institutions, and resource allocation should adopt a gender perspective to ensure that development and dividends are shared equally.

Akina Mama wa Afrika's Role

Non-profit organizations such as Akina Mama wa Afrika, a women's development organization, are vehicles for the increased participation of women in the development process. Multifaceted, highly participatory approaches have enabled Akina Mama wa Afrika, to complement the process of poverty eradication by working as a link between women at the grassroots level and legislators at the national level.

In spite of progressive policy frameworks such as the Ugandan Constitution, which prohibits gender-based discrimination, and policy documents such as the National Plan to Eradicate Poverty, the Domestic Relations Bill and Land Coalitions (Akina Mama wa Afrika is a member of both coalitions) have had to pressure the government to address unjust laws and provisions which prevent women from co-owning land with their spouses.

Over 35 years of women's concerted efforts forced the Ugandan government to initiate a project on the reform of domestic relations laws through the Uganda Law Reform Commission in 1993. In 1964, a body popularly known as the Kalema Commission was established to make recommendations for family law. Although recommendations were made, few were adopted in the final legislation.

The Domestic Relations Law is crucial to defining the legal status of women. This discriminatory law affects women's access to and control over land, property rights, and inheritance rights. The activities of the Domestic Relations Bill Coalition focus on the key actors involved in the law reform process. To date this coalition has been able to accomplish the following:

1. Lobby key leaders of opinion such as the Minister of Justice and Speaker of Parliament.
2. Convene sensitization programs in the rural areas.
3. Spearhead a comprehensive multilingual media strategy.

Government Initiatives

The Agricultural Sector Modernization Plan of 1998 - 2001 is an innovative approach to poverty eradication. This approach aims at increasing agricultural production and at improving household food security, nutrition and incomes through extension linkages to farmers. Incomes have improved, but weather variations and prolonged droughts in many parts of Uganda have limited significant economic gain. More than 75% of the households in Uganda depend on agriculture as a main source of income. Women provide 80% of the labor involved in food production while owning only 7% of the land. In Uganda women usually live on their husbands' land. With fluctuating statuses and roles as sister and daughter on the one hand and wife on the other, a woman has an ambiguous position, and she is prevented from receiving land from either family trees.

Government interventions support programs which aim at promoting food security but do not address the issues of ownership, access to and control over resources. This gender based, culture specific subordination gives women little access to the components of production, such as land. Women are robbed of the opportunity to appropriate the rewards of their labor. This inequality has direct implications on the health, nutrition and financial status of women and their children.

Government ministries such as the Ministry of Tourism, Trade and Industry in collaboration with members of the Ugandan private sector have organized programs to develop the entrepreneurial capacity of women. Through the Uganda Women's Entrepreneurs Association, women are encouraged to participate in trade shows, trade fairs and exhibitions at the national and international levels. Although women are more visible in this arena, the commodities and services they market tend to account for less than those of their male counterparts. However, Akina Mama wa Afrika has been training members of the Uganda Women's Entrepreneurs' Association in leadership skills. As a result, members of the Association testify that they have improved their strategies for business ventures and interests.

The establishment of credit and financial services such as the Entadikwa Scheme, the Poverty Alleviation Project, and the Uganda Participatory Poverty Assessment Project is symbolic of the Uganda government's commitment to alleviating poverty. The Entadikwa Scheme for providing loans places great emphasis on targeting women for its programs. Women from northeastern Uganda were able to set up handicraft kiosks and food processing units. These programs, recognizing women's disadvantages, encourage them to engage in traditional projects that were formerly not lucrative, such as making handicrafts and pottery.

The Decentralization System

In order to redress the historical imbalances, the Ugandan government adopted a decentralized system of government. The Local Government Act of 1995 empowers locally elected members along with communities to plan for development. Members of the community are able to voice their views and to participate in the planning and implementation of local development activities.

In each of Uganda's forty-five districts, the local council structure is headed by the elected chairperson, who as part of the executive branch helps to form the executive committees at the respective levels. Members of the executive branch propose policies for their specific legislative bodies. The councils are comprised of the representatives of the people, while the civil service staff implement decisions. This structure is replicated at the various levels. Community issues are handled at the LC 1 level; at the LC 2 and LC 3 levels, extension officers and community development workers are allocated resources. At the LC 5 level, decisions are made about district development plans and district policies.

Successes

Decentralization has effected positive change, facilitating the growth of local development planning. With its adoption, the compiling of development plans became mandatory for all districts. Two success stories, which combine initiative and strategic thinking, are those of the Rakai District in western Uganda and of Mukono in central Uganda. In Rakai, officials produced a development plan through a highly participatory process. The long process involved consultations with members of the community and lower administrative units. In Rakai this process was facilitated by the Danish International Agency for Development, while in Mukono it was implemented without donor assistance.

In keeping with the 1995 Uganda Constitution, the decentralized system of government allocated one-third of the seats in the local council to women. The LC 1 and 2 levels have witnessed increased participation of women. Women at these levels articulate their concerns, but few have utilized available resources to establish gender sensitive development programs. Realizing the need for capacity building, Akina Mama wa Afrika has a program which trains women councilors at the LC 3 and 5 levels of government in leadership skills.

Decentralization has also dictated a new approach to the administering of foreign aid in Uganda. Previously, all donor assistance was channeled according to an agreement between foreign donors and the central government. Now, however, the donor agency maintains agreement with the central government, but also negotiates a parallel agreement with the district. This was the case in Rakai. This direct relationship has ensured efficient implementation of programs.

Challenges

In spite of all the gains, the decentralized system of government faces several challenges. There is a lack of knowledge about the process of decentralization at the grassroots level. Women interviewed during a study (Ahikire, 1999) on the effectiveness of the local government stated that despite having heard about the system, they lacked a clear understanding of how it works.

Instead of performing the tasks of an empowering mechanism working from the bottom upwards, the decentralization system has taken on an approach working from the top downwards and taking away powers. This is characterized by lack of rapport between LC 3 and 5 councilors with members of the community. Rapport between leaders at the LC 1 and 2 levels and the community is better.

The breakdown in communication between the councilors and members of the community poses a serious problem for the provision for and access to information. Meetings are called at short notice, and minutes of the meetings are rarely released in sufficient time. Radio programs announce council meetings at inconvenient times, while newspapers and memos often do not reach the appropriate people. This has resulted in mistrust between the council and members of the community.

The increased number of women in the decentralized system has not resulted in the implementation of more gender sensitive policies and decisions. Women councilors require training in leadership, lobbying and budgeting. Issues discussed at the LC 1 and 2 levels are more community based in nature; at the LC 5 level, the issues revolve around planning and linking problems from various parts of the district to the national level. Limited exposure hinders women from effectively participating in the local council system.

There is an increased deterioration of health and education services. This has resulted in higher school drop out rates and in an increased number of people resorting to herbal medicine and traditional doctors. This defeats the government's intention to increase the access and quality of services such as health, education, transport and environmental management.

All in all, the realization of people-centered, sustainable development involves the establishment of a progressive, gender sensitive, economic framework which ensures good governance and security. It is only through this multifaceted approach that the needs of the poor will be addressed and ultimately solved.

15 The Darewadi Watershed Development Project in Maharashtra, India

Robert D'Costa and Abraham Samuel

Four years ago, Darewadi, a remote and drought prone village in the rain shadow region of the state of Maharashtra, India, was a picture of despair. The village was isolated, without any assurance of drinking and irrigation water. Other natural resources were depleted, especially those necessary for rural livelihoods like biomass and soil. Villagers had to migrate to areas with resources to earn their livelihoods. They toiled in the sugar fields of the rich farmers or in the brick kilns of the contractors. Some of them took to herding sheep, which further depleted the already fragile ecosystem. The poor environment, coupled with the absence of any alternate sources of livelihood, had caused the community to lose its vibrancy. It became withdrawn and fatalistic. Agricultural production, even in a year of reasonably good rain, was not sufficient for even three or four months - labor opportunities were scarce, and there was little possibility of even basic education for the children. Women had to work hard, either in the places to which they had migrated or in their own village fetching water, fuel and other basic needs. In order to redeem this desperate situation and to transform despair into hope, the only solution possible was resource mobilization and judicious utilization of the conserved resources by the community through locally evolved practices, institutions and management structures.

At this point, the Watershed Organization Trust (WOTR), which supports non-governmental organizations (NGOs) and Village Self Help Groups (VSHGs) to undertake participatory natural resource management along watershed lines, came into the picture. The challenge was to win the confidence of the people and make them aware of the interrelationship between the environment, health and the quality of human life – the social, economic and cultural interplay. The generation of awareness was achieved through constant interaction, audio-visual aids and exposure visits to areas where people were conserving and mobilizing resources for the improvement of their lives. The next stage was to mobilize and to build the capacity of the entire community to undertake the responsibility of managing their own resources and lives. People agreed to contribute voluntary labor and follow the social fencing principles such as a ban on free grazing and tree felling. A simple but scientific and people-oriented technology was adopted for soil conservation, arresting the rainwater runoff and harvesting water, as well as greening the mountains and wastelands. A series of technical treatments (contour trenches, gully plugs, farm bunds, contour bunds, check dams, etc.) along with bioregeneration (plantation, grass seeding, etc.) were undertaken. The once degraded landscape was slowly transformed, providing adequate drinking and irrigation water with increased soil moisture for better crop production and sufficient (sometimes even surplus!) fodder and fuel.

This transformation of Darewadi would not have been possible without the emergence of effective local institutions. These were willing to discharge their responsibility according to the authority and legitimacy conferred on them by the village community. The formation of the Village Watershed Committee (VWC), giving representation and voice to all sections that had earlier been marginalized (women, landless, lower caste, etc.) became the backbone of the village development. They interfaced with the civil society organizations - WOTR - as well as government departments for the holistic development of their village. They planned, implemented and monitored all the activities and evolved systems and procedures for management and conflict resolution.

The Darewadi VWC, which is the official project holder, is a registered body, having 22 members (including six women) nominated by the Village Assembly (Gram Sabha). Once a month, they have their official meeting in which they plan, make decisions about project implementation, and provide ongoing monitoring. They also jointly manage the finances with the support organization (WOTR). There are eight women's Self Help Groups (SHGs) in Darewadi, and these groups have an Apex Body, the Samyukta Mahila Samiti (SMS), which is functioning well. The women are co-partners and active contributors to the management of their watershed and the integrated development of their village. They have undertaken a number of activities for drudgery reduction and enhancement of the quality of their lives, such as soak pits, kitchen gardens, improved cooking devices, water supply system, toilet construction, etc. A number of income generating activities such as dairy, nursery, homestead poultry, etc., have also been undertaken. Most important of all, they manage their savings and credit groups with internal lending, which provides immediate loans for their basic needs.

In all of these activities, in order to enhance the management skills of the VWC and women SHGs, WOTR provides support through training, exposure visits, farmer-to-farmer extension, experience-sharing workshops and gatherings. The enhanced capacity of the villagers and their determination to take charge of their own development has also had its effect on the other government developmental agencies, which have contributed inputs such as potable drinking water supply, treatment of government forest lands, private horticulture plantation, construction of latrines, roads, etc.

Through microfinance support provided by WOTR, the women's group has started a dairy. At present they own 48 improved cows and sell over 800 liters of milk every day to milk cooperatives in the city. Increased agricultural production due to increased availability of water and an enhanced soil moisture regime has created linkages with markets for selling the surpluses. Pearl-millet farmers have diversified with the cultivation of vegetables. Growing cotton, onions, improved cereals and pulses, the farmers can now sell their surpluses to big cities like Pune, Ahmednagar, Mumbai, etc. This is possible because the farmers appreciate the "minimal risk" to their projects because of the assured availability of water.

Environmental impact is at times difficult to quantify, but the observations of the people substantiate it. The cool breeze, the clean and filtered water in the stream, the survival and vigor of plant growth (with over 100,000 plants recorded), the grass carpet on the mountains and wastelands where previously only black rock and cactus existed, the chirping of the birds, the contented faces of women who formerly had to wait endlessly for water tankers or travel long distances to fetch water, these things all bear witness to the changes that have taken place.

The local organization, the VWC, and its members are the custodians and managers of their ecosystem. The members have clearly assigned responsibilities known as portfolios. This helps to better coordinate their effort, to utilize human resources in the best way and to allocate responsibilities with authority. Depending on the various portfolios, these subgroups have responsibilities for such matters as banning free grazing and tree felling, the judicious utilization of regenerated natural resources, voluntary labor contributions, financial management, maintenance of impact records, maintenance of assets accounts, organizing meetings of VWC and Gram Sabha, etc. With moral incentive being the guiding force, the VWC members discharge their duties in consultation with and with the participation of the villagers. The enhanced capacity of the VWC of Darewadi has enabled it to undertake watershed development in the neighboring village of Chaudariwadi. The Participatory Operational Pedagogy (POP), developed by WOTR and practiced by the people of Darewadi during the implementation of the watershed development work in their village, has proved very helpful.

Darewadi, which was once a remote and isolated village, is now a web of activity. Visitors come from far and wide to understand and share their success story. Many of them go back with the resolve to replicate this effort in their own areas. The villagers themselves explain the changes to the visitors, and some of the VWC and women's group members even work as resource persons to create awareness in other villages. Due to the "demonstration effect" of Darewadi, many villages in the vicinity have also taken up natural resource management along watershed lines.

In order to conserve the developed natural resources, the VWC has had to use some social fencing methods and make some difficult decisions, such as banning the direct lifting of water from storage structures, the digging of bore wells, and the cultivation of water-intensive crops such as sugar cane. Just as scarcity leads to competition, abundance leads to greed. Both can create conflicts. A firm resolve, the necessary institutional structures and an ownership of the project together with unity among all stakeholders - vertically as well as horizontally - are necessary to overcome this natural instinct. The people of Darewadi have shown how this can be achieved.

For further sustainability of the watersheds, the VWCs of the different areas have come together to form a federation, which will take up the responsibility of monitoring, evaluating, innovating and creating accountability structures. They will also interface and advocate with the local government institutions and development structures.

16 Grassroots Movements as a Means to Overcome Rural Financial Market Failures

Franz Heidhues

In many developing countries dynamic growth is accompanied by a growing income disparity between urban centers and rural areas. Rural areas often remain centers of poverty with underdeveloped infrastructure, poor school and health facilities and few diversified production activities. Regional development concepts need to be expanded to give proper priority to strengthening rural urban linkages and complements. These include:

- sustainable productivity increase in agriculture in rural areas (agricultural research for rural area problems, extension, rural credit);
- infrastructure and institutional development to facilitate access to agricultural inputs, equipment and services;
- processing and marketing facilities and their link-ups to urban facilities and markets;
- development of non-agricultural activities in trade, transport, tourism etc.

These measures and actions need to be geared towards reducing the income gap between rich and poor. This implies improving access of the poor to land, water, trees and other natural resources, to factor and output markets, to technological know-how and to physical and social infrastructures. In these efforts particular attention needs to be given to the landless, tenants and women, both in assuring access to resources, institutions and markets, and to provide fair and enforceable ways of legal process and conflict resolution. Of particular importance to them is access to education, training and health services.

In this respect, informal rural finance markets have to be especially considered as opportunities to reduce the income gap between rich and poor. In many developing countries rural finance markets (i.e. agricultural credit and savings institutions, insurance agencies) are not working well. In particular, formal credit institutions often fail to reach small farmers and other rural householders. Some have no interest in extending credit to small farmers, trades people and handicraft micro entrepreneurs; others ask for conditions that many of these poor people cannot fulfill. Thus formal rural credit institutions may ask for land as security against extending credit, but poor people, particularly women, often do not have land or other assets that could serve as collateral. Other credit organizations often lend only for crops which they themselves market, i.e. where they can deduct from the sales revenues the credited amount and in this way directly collect credit and interest. Often, the poor can resort only to moneylenders or traders, possibly at very high interest rates.

Table 16.1. Four Women Meeting once a Week, often at Market Day

	A	B	C	D	Σ	Recipient of Σ			
Week 1	50	50	50	50	200	A			
Week 2	50	50	50	50	200		B		
Week 3	50	50	50	50	200			C	
Week 4	50	50	50	50	200				D
Σ of payment by member	200	200	200	200		200	200	200	200

As a remedy for these deficient rural finance markets, informal savings and credit groups, also called ROSCAs (rotating savings and credit association), have formed grassroots organizations. In Cameroon and other countries in West Africa, they are also called Tontines. They function as described in Table 16.1.

These grassroots organizations cater to important needs of the poor. They provide:

- savings possibilities and credit at low cost,
- without collateral,
- without restrictions on the use of funds, i.e. can be used for high priority consumption needs (food, medicine, school fees, etc.),
- quick and without red tape (the sequence of who receives the collected amount may be modified according to need).

Apart from credit and savings facilities these institutions provide their members with insurance, a particularly important service in rural areas where often formal insurance services are not available. They also establish trust and encourage group formation for other services (e.g. women groups may organize baby/child care while the other women work the fields). The group helps to better articulate the needs of the poor and to bring pressure on local authorities to pay adequate attention to them. It has also been observed that they help in preventing or resolving conflicts, e.g. in disputes about land, water or tree use rights.

While these are extremely important and valuable functions, such grassroots institutions also have their limits. The above example of a credit/savings group normally provides only relatively short-term credit/savings possibilities; it is not suitable for financing long-term investments, such as buildings, farm equipment, animals etc. By their nature, they operate on a local/village level and cannot carry out intermediary functions at the regional or national level. Thus, they are not able to transfer savings surpluses from one village to other areas where there may be a need for funds. They also cannot cover risks that affect whole regions or larger areas, such as major flooding or drought occurrence. Also, it is not certain that all members of a village, particularly those most in need, have access to such groups. Complementing the services of local informal groups with the activities of formal institutions linked to nationwide networks has often proven to be a necessary and useful format.

While such grassroots organizations give valuable insights into the functioning of locally based organizations and into the needs of rural people, there is a danger in idealizing them. They are far from being a heaven of peace and harmony caring for the poor and helping each other. They often have strict norms that are enforced by severe sanctions including social pressures, intimidation and exclusion from local community life and social security nets. Threats of witchcraft are also part of village life. Where power and economic resources are unevenly distributed, local power structures often are reflected in the structure of grassroots organizations.

As a result, decentralization and delegating decision making to locally based organizations does not automatically lead to a better integration and participation of the poor. A careful, village specific analysis of the local institutional framework, the resource and income distribution and the political power structure is needed to find a suitable design for decision making in local development processes. Grassroots institutions may well be an important part of this framework, but they are not automatically the optimal solution.

17 The Barefoot Approach

Bunker Roy

By far the most crucial and immediate problem that needs to be addressed in the elusive search for a viable sustainable development model is the colossal and criminal waste of natural resources today. Most of the efforts with very good intentions have in fact led to serious problems of injustice, dependency, exploitation and discrimination. It is obvious that the disparity between the rich and the poor and between the urban and the rural is widening every year. The *Barefoot College in Tilonia*, a very small village in the middle of the Indian deserts of Rajasthan, was established in 1972 with the goal of tackling these fundamental problems in a new way. It has taken twenty-nine years of trial and error efforts and of successes and failures to develop some conceptual clarity about the "barefoot" approach to sustainable development. Implementing simple, radical and in fact futuristic ideas, approaches and methods so that the rural poor could respect and indeed "own" the project (because it was theirs after all) only confirmed our belief in the low cost replicability of the concept of the *Barefoot College (BC)*. Two hundred years ago, when there were no urban trained doctors, teachers, engineers, how did the rural population tackle their own problems? In rural areas all over the world where there is a high percentage of illiteracy, the oral tradition is very strong. Rural communities passed on their knowledge, skills and wisdom from one generation to the next, improving, adapting and improvising through the years. With "modern" development appearing on the scene, this valuable human resource is withering away and being wasted. With no value and respect for their skills, people are migrating and dying slowly in urban slums. It has been the firm belief of the *BC* that this knowledge, skill and wisdom can be revived, reserved and widely utilized for the poor's own development. It is the only long-term solution to reverse migration and bring some self-respect and dignity back to rural lives.

    The *BC* has a healthy disrespect for degrees handed out by formal educational systems certifying technical competence. The definition of a professional as accepted by the *BC* is a person who has a combination of competence, confidence and belief. Community acceptance, respect and demand for a person's skills is a more enduring and valuable certificate than any paper handed over by a college or university. Thus the need to recognize and respect the technical competencies of traditional midwives, bonesetters and healers (using indigenous medicines) is central to the beliefs of the *BC*. The success rate of water diviners in finding drinking water in deserts has been discovered be as good as any geologist or geophysicist. The water diviner is definitely less expensive and more environmentally friendly. Traditional communicators (street theaters, glove puppeteers, and street singers) have their own technical skills to reach the illiterate poor with vital social messages far more effectively than the mass media, the printed word or television. Information about the importance of safe drinking water, the legal minimum wages, the importance of sending children to school, and gender equality issues are only a few of the many messages conveyed in the evenings through theater and puppet shows. The technical knowledge skills of limiting waste and of recycling in rural communities is indeed astounding, and the *BC* has benefited from this collective knowledge.

People's lives in rural communities are centered around an inherent and deep respect for the natural resources around them. Sun, water, air, earth and fire are elements that they *never* abuse, over-exploit, waste or commercialize. They have internalized this as a part of their lifestyles. The *BC* has understood this deep respect for the environment, listened to the advice of the elders, and developed an alternative that has baffled the so-called "experts." Urban based, certified engineers are on the verge of losing their battle to provide safe drinking water to the rural poor around the globe. This is because they cannot look beyond conventional solutions such as hand pumps and piped water supply. Their work has resulted in a colossal waste of money in drilling rigs guzzling fossil fuels and in over-exploitation of groundwater; also there has been no change in urban lifestyles (flushing toilets, washing machines, watering lawns) which set examples of how to criminally waste water. For hundreds of years, people have collected rainwater to meet their basic needs. Since 1986, the *BC* has shown how effectively this low cost technology of collecting rainwater instead of allowing it to go to waste has changed the quality of life for the poor. A complete rainwater harvesting tank has been constructed for less than Rs 1.50/liter (0.5p/liter). This has provided direct employment in the village, encouraged a sense of ownership, given the people direct control of the distribution of available water and reduced dependency on urban engineers. In 1998 - 1999, the *BC* collected 12 million liters of water in 107 remote rural schools attended by over 3,000 shepherd boys and girls. Because of the water, the attendance of the girls increased significantly in these remote night schools. The *BC* is the *only college* in India which relies fully on solar energy. With 15 kws of solar panels, the *BC* provides lighting to a library (40,000 books and 100 magazines); a kitchen and a dining hall that feeds over 50 people; a soil and water testing laboratory; a solar electronic workshop that makes, installs, repairs and maintains complete solar lighting systems; and runs a 250 line telephone exchange, e-mail and 20 computers. Over 2,000 houses in the Himalayas, 15,000 ft up in Ladakh (Kashmir), have received solar electricity, which has been maintained by 20 barefoot solar engineers, some of them semi-literate, for the last 6 years. 400 solar lanterns have been produced by unemployed and unemployable rural youth who have barely received 5th standard primary school education. These lanterns give light to the 150 night schools (thus replacing kerosene lamps), so over 3,000 shepherd boys and girls are attending school for the first time.

In the *BC*, *nothing* is wasted. Agricultural waste is recycled into handicrafts such as mobiles and sold. Over 50 women earn an income from the sale of these mobiles, which are also exported. Old tires from tractors and jeeps are made into swings for children. Waste paper, unsolicited glossy reports, and newspapers are all recycled into making glove puppets, masks, envelopes and copybooks. Falling leaves are collected in biomass plants to make gas for the medical laboratories which carry out pathology tests. All the roofs (66,000 sq.ft) of the College are connected to an underground, 400,000 liter tank to collect rainwater.

Used matchboxes, pens, railway and bus tickets are utilized to make toys and for scientific experiments in the night schools. Leftover cotton is used to make rag rugs which are exported abroad. The *Barefoot College* offers a model that is easy to replicate anywhere in the world. Where old culture and tradition is alive, where there is respect for natural resources and where people are prepared to work on their development at their own pace with self respect and dignity, this is the *only* viable alternative model that will work in the twenty-first century.

18 IFOAM – the International Organic Grassroots Movement

Bernward Geier

Organic agriculture is an excellent showcase of a successful grassroots development and movement. Farmers, consumers, scientists and traders realized that the chemical intensive conventional or so-called modern agriculture is ultimately destructive not only to the environment but also to rural development and the livelihood in villages. Refusing to follow the dead end road of conventional agriculture, the organic movement has developed a holistic system of producing food and fiber, which is not only environmentally and economically sustainable but also socially sound.

Organic farming is nowadays practiced by hundreds of thousands of certified farmers all over the world. The *International Federation of Organic Agriculture Movements* (IFOAM) unites 770 member organizations in 107 countries. Leading European countries are Austria with 10%, then Sweden and Switzerland with 8% of the total farmland converted to organic farming. Especially during the last 10 years, Asia, Latin America and Africa have also seen a very rapid development of organic agriculture. The number one herbal tea in Egypt is certified organic. In projects in Senegal and Uganda respectively, over 3,500 and 7,000 farmers grow cotton as well as their staple food organically. In Mexico, cooperatives with up to 7,000 farmers have found a future in agriculture by combining fair trade and organic farming methods, which both pay premium prices. For the year 2000, the market for certified organic products has reached worldwide an estimated volume in the range of over 20 billion US dollars.

What do these few highlighted figures actually show us? They demonstrate that the trend to take value out of farming and villages can be reversed. What the villages and their people desperately need is value adding, which means both better prices for their agricultural products and job creation through local value adding to the primary production of food and fibers. This is exactly what organic farming offers. The demand for organic products is far greater than the supply. The appreciation of consumers for organic quality motivates them to pay a higher (actually reasonably cost covering and fair) price. Organic farming offers a lot of potentials to add value, for example, through innovative marketing approaches like farm gate sales, organic farmers' markets, box and delivery schemes, etc. Many organic farmers have successfully increased their incomes and created jobs by adding value to their primary products such as making milk into yoghurt or cheese, milling grain and baking bread or by direct delivery to consumers. Agro-tourism on organic farms is also an interesting aspect of value adding.

While discussing the matter of adding value, we have to ask if we still have our values straight. To quote the chairperson of the agricultural committee of the European Parliament Graefe zu Baringdorf (an organic farmer!): "*If butter is cheaper than shoe polish and milk cheaper than mineral water, we have a serious problem of value and priority setting in our society.*" (The cap on a beer bottle costs as much as the farmer receives for the barley used to brew the beer. For the flour in a bread roll, which costs 30 cents, the farmer receives only one cent).

We must change the devastating development that has led to the fact that money today is made mainly on the "backs" of farmers and not on the farms. What we need is a paradigm and reality shift in our consumption patterns to make sure that we keep our farmers in farming and therefore always have access to healthy and organic foods. The organic movement will continue to prosper as a grass roots and bottom up movement with a strategy to convert to organic farming field by field, village by village and region by region. With this strategy, the farmers and the villages indeed have a new outlook for the 21st century and beyond: Organic farming not only produces healthy crops but also provides more jobs and better livelihoods.

19 SEKEM - an Egyptian Cultural and Economic Initiative

Klaus Merckens and Ahmed Shalaby

The SEKEM[1] initiative is one of the models of integrated development projects in Egypt. Comprehensive development is the aim of SEKEM, where the economic, cultural and social spheres of life interact to form a sustainable way of development and build a model for the future.

In the *economic sphere*, based on a new management of the value adding chain from the farmer to the consumer, SEKEM promotes partnership and transparency to ensure high quality products and justified prices while taking into account the welfare of people and the earth. In a period of almost twenty years SEKEM contributed substantially to the growth of environmentally friendly agriculture in Egypt. Biodynamic methods which have been adapted for their effectiveness in arid zones were used for the first time in Egypt, converting the desert sands into fertile soil and a green, flowering landscape. Since then, a steadily increasing number of biodynamic farms in Egypt have served as the base for a modern development of Egyptian agriculture. More than 180 farms involving more than 800 small farmers cultivating about 3,000 hectares biodynamically and several specialized companies have been established to ensure the proper production and marketing of theirs products. Directly over 2,500 people are working in the different SEKEM Companies.

The *cultural sphere* is designed to complement working with learning and research in all fields of human life. SEKEM's cultural contribution enables people to determine and realize their own unique and culturally appropriate development path. The *social sphere* at SEKEM develops integrated social forms conscious of the collective responsibility for social uplifting based on a better awareness of the individual capacity to grow and develop.

The Egyptian Society for Cultural Development (SCD) is the basis for the cultural and social life of the members. SCD is a NGO striving to create culturally and socially legitimate forms of development that contribute to local, national, regional and international development. SCD thus serves as both a local and a global model of sustainable development. Its activities consist of service provisions including the Adult Education Program and the Vocational Training Center. These services are available to anyone from the local area. Furthermore, specialized development projects focus on the design and management of specific project activities (e.g., to increase health awareness and to establish a set of guidelines for organic agriculture in Egypt).

SCD believes that improvements in living standards must be accompanied by an enhancement of people's moral and cultural awareness if this goal is to be fully realized. So, for instance, farmers in association with the SEKEM initiative have

---

[1] SEKEM is a transcription of an hieroglyphic word that means: "vitality from the sun."

easy access to education. They are encouraged to enjoy music and to develop their capabilities in whatever areas they like. They share the sense of security with their families, to whom the different services provided by the SCD are also offered.

The commercial companies daringly promote a market strategy based on transparency and cooperation between people. This is rather unique in a world of market competition and *"survival of the fittest"* ideologies. The system of re-channeling part of the revenue received from the commercial companies of the initiative to the social and cultural NGOs proved to be very successful, sustaining the whole system.

II. SUSTAINABLE USE OF NATURAL RESOURCES

Agenda 21 Principle 7, Earth Summit, Rio 1992

Principle 7 States shall cooperate in a spirit of global partnership to conserve, protect and restore the health and integrity of the Earth's ecosystem...

Sustainable Energy Systems

Introductory Statement

Peter Hennicke

Sustainable energy systems present different and comparable challenges for the industrialized countries (ICs) and for the developing countries (DCs): In the ICs the *absolute* amount of non-renewable primary energy consumption per capita has to be decreased by about 50 to 75% by 2050 if climate protection and risk minimization are taken seriously. In most OECD countries a *relative* decoupling of GDP and energy consumption is already going on. Energy consumption is constant or decreasing but economic indicators, industrial production and the GDP are increasing. This trend is encouraging, but concerning the aim of sustainable development the increase in energy productivity must be fostered: Not only a relative but also an absolute decoupling of GDP growth and (non-renewable) energy consumption is necessary; per capita energy consumption levels in ICs are far too high, unsustainable and not transferable to a growing population in the DCs.

Of course, in the DCs the challenge facing sustainable energy systems cannot be an absolute decrease in energy consumption, but a *decrease in growth rates*. With further development and a growing population, energy consumption in DCs will inevitably increase. By giving highest priority to energy efficiency the *growth rates* of energy consumption could be slowed down and, at the same time, the necessary increase in *energy services* (like efficient heating, cooling, lighting and cooking) and in the standard of living could be accelerated. Therefore, the crucial questions concerning a worldwide sustainable energy system are:

- Is it possible to achieve the IPCC goal for reductions of worldwide CO_2-emissions in 2050 (e.g. -50% of the 1990 emissions)?
- Is a strategy "increasing wealth and decreasing energy consumption" technically feasible?
- Can the climate protection goals be combined with risk minimization (e.g. phase-out of nuclear power) and an economically reasonable path towards sustainability?

The recently published "Factor-Four" Scenario of the Wuppertal Institute has given some answers to these questions.[1] The analysis had to be based on a complete worldwide model of the energy system, representing 11 regions and a detailed database for 160 countries, similar to the model used by IIASA and the World Energy Council (WEC). The Wuppertal Scenario is a technology oriented "Bottom up" approach with a special focus on the demand side and energy efficiency. It explicitly takes such enduse technologies like "hypercars," "passive houses", energy efficient production processes and most efficient appliances into account. Following otherwise the general basic assumptions from the "ecologically driven" WEC/IIASA-scenarios, the more enduse oriented scenario analysis of "Factor Four":

- shows that climate protection (CO_2-reduction by 50%) and risk minimization (e.g. nuclear phase-out) by 2050 could be combined with a threefold increase in GDP and a reasonably increased standard of living in DCs;
- identifies the three "green pillars" of sustainable energy systems, namely giving priority for the rational use of energy (RUE), fostering the market introduction of combined heat/cold power production in industry and district heating (CH/CP), and increasing the share of renewables (REG);
- outlines possibilities and demands for technology transfer between industrialized and developing countries ("leapfrogging").

In short: Decoupling energy supply from energy services is technically feasible; to build up sustainable energy systems based on the above-mentioned three "green pillars" is a challenge for both the developing and industrialized countries. The problems are not primarily the technologies but rather the implementation and the fostering of more efficient use of energy. In particular, there is a need for capital and know how transfer from ICs to DCs, advanced instruments and more ambitious energy policies. Comprehensive bundles of instruments and a broad policy mix to foster sustainable energy systems are needed. There are many different ways to achieve more energy efficiency and to foster the use of renewable energies adapted to the local conditions. These include standards, labeling, capacity building and education, incentive programs (e.g. Demand Side-Management), intelligent combinations of market forces and regulation and acceleration of research and development for renewable energy and efficiency technologies.

While promoting rural development, it must be stressed that the problems of increasing energy supply and energy services cannot be solved without placing the highest priority on energy efficiency in both production and use of energy. In addition, greatest effort must be made to promote a broad mix of affordable renewable energies.

However, before discussing the improvement of energy efficiency in rural areas, a minimum of infrastructure must undergird their development. The

---

[1] Lovins, A. / Hennicke, P., *Voller Energie - Vision: Die globale Faktor Vier-Strategie für Klimaschutz und Atomausstieg*, Campus Verlag, Frankfurt, September 1999. "Factor Four" refers to the ideas of Ernst von Weizsäcker and Amory and Hunter Lovins.

importance of the worldwide electrification of rural areas seems to be obvious, especially when one considers that rural development is a prerequisite to agricultural development. The electrification of villages is a major component of the infrastructure needed to ensure their development. According to *Bartali*, the inadequacy of commercial energy becomes apparent because of the ongoing modernization of rural areas. The electrification of these areas must be achieved through the most appropriate techniques. As an example, the efforts made in Morocco to provide rural electrification are analyzed.

Deppe describes the route to sustainable energy systems in an industrialized country. Some possibilities for a substantial reduction of energy consumption (based on the concepts of Least Cost-Planning and Third Party Financing) are summarized. As a representative of an energy supply company, he outlines an innovative approach to the reduction of energy consumption that also benefits the companies involved.

Gupta gives insights from lessons learned in thirty years of intensive research on renewable energy technologies in Indian villages and describes renewable energy sources employed for community sustainability in rural settings. One major lesson is that a network of eco-communities linked physically with clean transportation and psychologically with communication technologies will bridge the existing urban rural divide. Here, the gridiron constraints of city-scale utilities will be replaced with local utilities based on green power.

Finally, *Mwakasonda* emphasizes the urgent need for most African countries to formulate their energy policies with the aim of immediately overcoming the total dependence on imported petroleum as well as minimizing the biomass energy use. This would reduce deforestation and its ecological and social consequences. Promoting renewable energy technologies for rural areas in Africa is an important component of national economic planning. Along with satisfying the basic needs of the rural population, it also reduces the degradation of natural resources.

Agenda 21 Principle 4, Earth Summit, Rio 1992

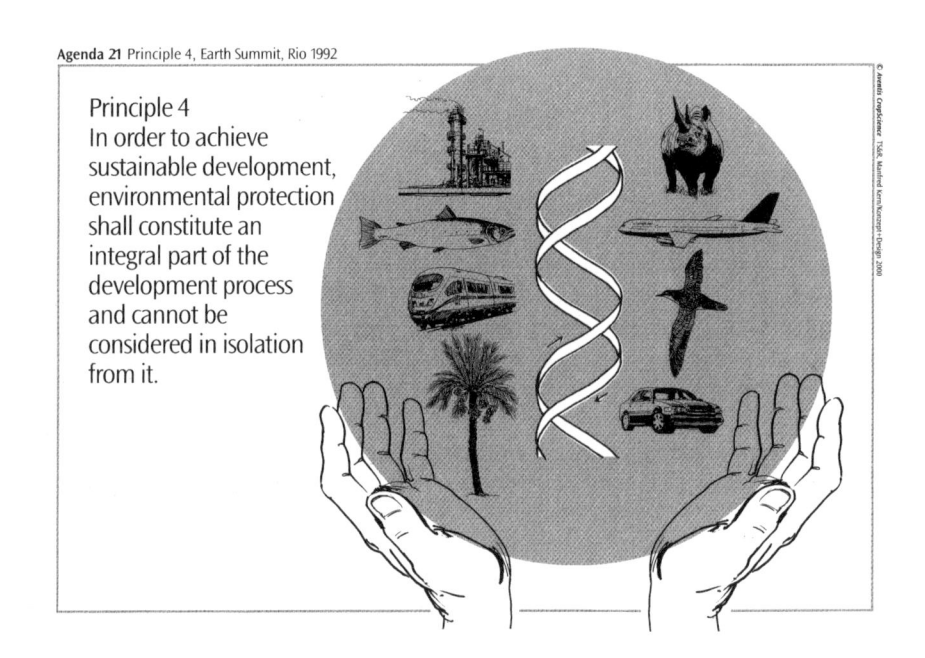

Principle 4
In order to achieve sustainable development, environmental protection shall constitute an integral part of the development process and cannot be considered in isolation from it.

20 Rural Development and Sustainable Energy Systems

El Houssine Bartali

Rural areas are essential components of the development process in each country. Furthermore, no development of rural zones can be achieved without adequate infrastructures (roads, electricity, water supply, schools, and health services). Many developing countries provide an insufficient electricity supply for their rural areas, for example only 5% in Senegal and 18% in Morocco. Energy supply stimulates the socioeconomic development of rural areas, especially in poor hilly regions. The supply of electricity to remote areas, inaccessible farms and decentralized consumers remains a worldwide challenge. Today, with the movement of rural areas toward modernization, the inadequacy of commercial energy becomes the key problem of integral development in rural areas. Electrical power is the first issue needing to be resolved. Global demand for oil, mainly from developing countries, is expected to grow from 22% to 34% during the next fifteen years. This may well put a strain on the world's remaining oil reserves.

Rural Development and Electrification

A global objective of agricultural policy is to maximize economic growth by reducing the disparities between rural and urban areas. A package of agricultural policies is to be designed and implemented to reduce commercial protection, rationalize pricing systems, protect farming activities and implement social policy in order to stabilize populations in the rural areas. A strategy for rural development is a prerequisite to the strategy for agricultural growth. Importantly, this growth is not only a question of yields but also of rural development. It is not agricultural growth which ensures rural development; it is rather rural development which allows agricultural growth. A major change affecting rural life is the development of small urban centers which are closer to the rural areas than to the cities. Key facilities and services are available in these centers. Infrastructures such as electricity, drinkable water, schools, training and health services are there. Such small urban centers are promoting economic and development activities in rural areas. They represent an important alternative to the rural exodus and a space for the diffusion of new technologies in agriculture.

In an era dominated by the development of electronics and biotechnologies, a major concern is bringing large groups of the rural population away from marginalization and providing them with the means of subsistence.

Rural areas in developing countries are suffering from the lack of electrification. It is believed that only one third of the estimated 2.5 billion people living in rural areas has access to grid based electricity (De Groot, 1996). About

300 Wh per person per day is the minimum amount of energy needed for the improvement of the basic living environment.

Efforts to improve the supply of electricity to the rural world face a number of difficulties, such as the disparity and low density of the population and low consumption levels. In addition energy losses, both technical and non-technical, must be expected. With this in mind, it is estimated that revenues from the sales of electricity can hardly cover 30% of the cost of production and distribution (Biermann et al., 1990).

Rural Electrification in Morocco

Around the mid-seventies, the government decided to improve the low percentage of rural villages supplied with electricity. However, an evaluation made 20 years later showed that the situation had not improved much. The average electrification rate per year remained low, around 10,000 households per year. The following criticisms were made about the approach followed and the measures taken:

- vision was limited to programs which were unable to ensure rural electrification of the whole country in a short period of time;
- institutional programs were restricted to grid based electrification and in any case were not able to meet the needs of rural households;
- finances were dispersed in many different programs; funds were not available and the system was incoherent because of the various financing modes.

Learning from these lessons, a new approach was developed, and a global rural electrification program was institutionalized in 1996. This was managed by the National Board of Electricity (ONE, 2000). The program aims at the following:

- to supply the entire territory with electricity within a short period of time;
- to integrate the various electrification techniques in order to meet the needs of each household within acceptable technological and economic conditions;
- to integrate all the financial resources that can be mobilized for rural electrification.

The program is designed to be participatory, and three partners are involved in its financing: local counties, beneficiaries and the national board of electricity. This board's contribution covers 55% of the cost of the energy supply. Appropriate modes of financing are adopted in order to encourage the participation of the rural population.

In order to achieve the electrification of as many rural houses as possible in a short time, the criterion used to select the most suitable method is the minimum cost per household. The program target is to provide electricity to 6,000 villages between 1996 and 2000. For the majority of these villages, the cost per household is less than US $ 1,000. However, for the development of the general electrification scheme, it was found that the criterion of minimum cost while necessary is insufficient. Therefore, other criteria must be considered:

- In order to improve the equilibrium between regions, the national board bears the financial burden of the necessary infrastructure needed for the electrification of inaccessible rural areas.
- A study has identified villages with highly dispersed populations. These villages can neither be electrified from an interconnected grid nor from a mini local grid. They are eligible for a supply through individual photovoltaic kits.
- Analyzing the internal economic return rate allows the comparison of various methods and helps select the appropriate one for each village.

These analyses help to shape the re-orientation of the plan so that it can benefit whole regions within the country; they also help to clearly define the most appropriate electrification methods and adapt them to each site. An evaluation of electrification programs made at the end of 1998 showed that the number of electrified villages steadily increased: 557 in 1996, 1,044 in 1997 and 1,227 in 1998 (ONE, 2000). It should be emphasized that the global rural electrification program has a number of major advantages, such as:

- The costs of rural electrification are reduced in comparison to previous programs.
- International donors are willing to support the programs of the present plan of electrification. This shows that the financing of rural electrification in Morocco will be long lasting.
- There is an increasing demand for electricity supply from the population, its representatives and local authorities. This will undoubtedly lead to a greater acceleration of the implementation of the global rural electrification program.
- Participation of the large majority of villages in the country in establishing a master scheme for electrification helps develop a clear vision of the needs and efforts to be undertaken.
- Integration of new and renewable energies within the rural electrification program promotes understanding about these techniques at the national level. It also promotes better organization of contractors working in this field.
- The creation of a market for rural electrification enables the introduction of large international contractors. As a consequence of this, additional employment opportunities are created as is a better sense of competition between local contractors in this field.
- A master scheme was developed using GIS techniques. This system integrates data to make the final computations of the electrification costs per household in villages. To establish this scheme, data from more than 90% of the population of the country were used. The analysis of the components of the technical and economical evaluation has addressed the following parameters: population disparity, regional balance, cost per household, and internal return rate.

For the purpose of improving the balance between regions to be electrified, the intervention of the Board of Electricity is sometimes necessary. Using its own funds, it must construct the infrastructure needed to supply isolated and inaccessible villages.

In Morocco, the economic planning for the next 5 years, which emphasizes the development of decentralized electrification, aims at supplying nearly 550,000 households with electricity. Of this number, 450,000 will have grid based electricity. The electrification rate of the country is expected to reach 60% in the year 2000 and could become 65% by the year 2003 (ONE, 2000).

Alternatives

The agricultural sector needs to be provided with small but crucial supplies of commercial and non-commercial energy which are required for food production. Many nations are faced with important questions to answer (Stout, 1990): Is it important to determine what options and strategies are available? Is it possible to maintain a balance between energy and food supplies as the world population continues to grow? What are the feasible alternatives to petroleum? Will renewable alternative energy resources be cost effective? The continuously rising costs of energy resources, the continuously deteriorating ecological environment and the technical advancement and modernization of equipment led to the development of some appropriate energy systems.

In early 1998 a strategic vision called *Plant/crop based renewable resources 2020* was unveiled in the United States. There is growing concern about fossil fuel emissions which contribute to global warming. New uses for renewable resources could increase recycling given that many products will be biodegradable (Douglas et al., 1999). In order to meet energy needs, two options must be considered:

- the increase of energy supply through the development of renewable energy alternatives, and
- the decrease of the energy demand through a more efficient management of energy systems and the design of reduced-energy farming alternatives.

Renewable Energies

The technologies of renewable energies are progressing through every stage of development from research to being fully commercial. On the one hand, energy crops are emerging from extended research on variety and harvesting machinery. On the other, hydro schemes for generating electricity have been operating commercially for decades (Hunter, 1996). Energy crops, even the woody crops, have a very low energy yield after conversion to electricity when compared to the yield of wind or photovoltaic conversion systems. However, the potential for growing energy crops is very large.

The opportunity for economic gain from harvesting renewable energy resources certainly exists for many farms. Technologies are advancing rapidly in this field; costs are coming down; and a number of governments are establishing programs to take advantage of this. Fragile rural economies of less favored areas have an urgent need for rural development while at the same time they have some resources in abundant supply. Energy crops can become an important opportunity when the market for supplying fuel to power stations starts to develop further, probably in the next few years.

When villages are of a certain size and the population is too dispersed, national grid based electrification is feasible, and the cost per household is reasonable. For other villages, it may be necessary to consider *decentralized electrification* with one of the main sources being electrification with solar energy, the photovoltaic.

The advantage of *decentralized photovoltaic systems* is that they can supply, in any place and for a long period of time, very small quantities of energy which are suitable for rural populations. However they represent an unaffordable investment for some villagers. Such costs can be reduced by 30% provided there is market development (Legendre, 1997). Complete systems of investment management in the sector of decentralized energy are to be designed to manage small quantities of energy. In order to achieve their goals, they should be integrated into a global strategy:

- Promotion of the product at the level of thousands of economic operators settled permanently in the rural areas: volumes of capital must be mobilized and managed within the context of advanced decentralization; this requires the maximum mobilization of local operators who take on the responsibility of collecting the finances.
- The development of synergies between the small investors involved by means of a modern approach to the network.

In addition to improving living conditions in the villages, such a strategy of rural electrification also positively impacts the local economy in the following ways:

- Many small and medium sized enterprises established in the rural areas are strengthened.
- The implementation of such programs stimulates the development of markets of supplies and services for the realization of electrical installations.
- The strategy lays the ground work for the long-term development of the energy market in the rural areas.
- Local operators will establish a significant field of reference and acquire training capacities allowing them to manage various services in the rural areas.
- The possibilities of technology transfer are abundant and will translate into a high local value added and a better control of the implementation of decentralized electrification.

In Morocco, the global rural electrification program presented earlier includes the development of solar energy in its strategy to supply rural areas with electricity (ONE, 2000). The PV kit is a suitable solution for the electrification of dispersed villages (one kit per household). It is especially appropriate because of its low cost and its ability to meet the needs of the population in rural areas. The strategy of developing the use of PV kits in the case of Morocco has set a target number of 200,000 households to be equipped with PV kits.

The estimated cost for this operation is about 190 million Euros. Although this objective may seem easy to reach, its implementation is not. The National Board of Electricity does not have a monopoly and therefore does not control the management of this mode of electrification. Furthermore, several projects are implemented in various parts of the country by different operators: associations, NGOs and public entities. In addition, PV kits are considered domestic equipment; thousands of households have acquired solar panels which were either purchased locally from the market or imported. The best strategy adopted to develop the use of PV kits is based on a partnership involving the private sector, counties and associations of villagers. This partnership has two dimensions, finances and mode of implementation. This approach may take a long time, but it is important to involve:

- the private sector, including the main operators in the field: dealers, equipment makers, boards of engineers and service providers;
- NGOs, associations and federations of associations providing assistance to rural households;
- counties and local authorities.

The identification of the appropriate marketing formulas of the PV kits, which are adaptable to diverse situations, requires a research development effort. The research center for the development of renewable energies, the private sector and any other entities concerned with rural electrification are collaborating on this effort.

The sensitization of village associations is important in order to successfully achieve the electrification of villages. An increasing number of villages have been supplied with PV kits. A project underway, funded by KFW and started at the beginning of this year, aims at supplying some 70,000 rural households over a period of three years. Another project is being negotiated for co-funding by the European Union.

In conclusion, decentralized rural electrification will keep developing during the next two years and is expected to reach a rate of 20,000 households per year. In order to assess the importance of this figure, it should be considered that the total number of households equipped with PV kits all over the country prior to the year 2000 did not exceed 20,000. Over the period from 1999 to 2003, the goal is to equip 70,000 households (ONE, 2000).

Hybrid plants, utilizing indigenous renewable energy sources including biomass, wind and solar radiation, represent an emerging effective environmentally compatible solution. Such plants combine various types of energy conversion and storage units. The *electrification through mini grid* makes it possible to use energy potentials that are available locally. Their expansion and connection to conventional grids can be easily achieved. Small power hybrid systems (1 KW range) can be used to supply electricity to rural locations that are cut off from interconnected grids. The systems are also suitable to supply energy to health centers, schools, water supplies and cottage industries (Heier et al., 1997). Research development projects are being conducted in order to develop and implement rural electrification by hybrid systems (wind, solar and generators) or by biomass power stations. The results of such projects will make it possible to analyze the possibilities of using such techniques for the electrification of villages. Furthermore a survey study of adequate sites to receive electricity from micro hydraulic power stations is under way.

Small hydropower is clean and widely used in rural areas all over the world for economic development. New techniques have been used for the development of SHP stations, thus reducing their construction period and cost. SHP becomes a high efficiency and low cost power energy for the rural areas, particularly the remote ones (Gaorong, 1998). Many SHP stations were built in combination with the construction of farmland irrigation projects. These formulated an important and comprehensive rural economic background and promoted the development of SHP.

References

Biermann, E., F. Corvinus, T.C. Hergberg and H. Hofling (1992): Basic Electrification for Rural Households. GTZ, Eschborn.

De Groot, P. (1996): A Photovoltaic Project in Rural Africa: A Case Study. Proceedings of the World Renewable Energy Congress, p. 163.

Douglas, L., J. Faulkner, S. McLaren and B. Mustell (1999): Renewaable Ressources 2020. In: Ressource Magazine. March 1999, ASAE, St. Joseph, Mi, USA, pp.11-12.

Gaorong, L. (1998): Small Hydropower and Rural Electrification in China. Proceedings of the XIIIth International Congress of CIGR, vol 5, pp. 1-7. Editor: Prof. H. Bartali, ANAFID, Rabat Morocc.

Heier, S., W. Keleinkauf and F. Raptis (1997): Rural Electrification with Hybrid Plants. Proceedings of the 20th International Conference of the IV CIGR Section on Rural Electrification and Rational Use of Energy in Agriculture, 23-25 April, 1997, Agadir Morocco.

Hunter, Alastair G.M. (1996): Renewable Energy Opportunities for Agriculture. Proceedings of the International Conference on Agricultural Engineering, Madrid 23/26 September, 1996, pp. 679-680.

Legendre, B. (1997): Mandosol. Un nouveau concept de gestion de l'energie pour la mise en oeuvre de programmes d'electrification rurale a grande echelle. Proceedings of the 20th International Conference of the IV CIGR Section on Rural Electrification and Rational Use of Energy in Agriculture, 23-25 April, 1997, Agadir Morocco.

Office National d'Electricité (ONE) (2000): Presentation du Plan d'Electrification Rurale Global. Janvier 2000.

Stout, B. A. (1990): Handbook of Energy for World Agriculture. Elsevier. Applied Science. Elsevier Science Publisher LTD, New York.

21 The Route to Sustainable Energy Systems in Germany

Erich Deppe

To begin with, there are two important points for consideration for everyone working in Germany in the area of energy systems for sustainable use:

- Germany has a 100% level of electrification, and the energy-intensive industries have already been fully developed. This statement may appear banal, but it shows that there are only a few isolated cases where the development of infrastructure can be influenced by considerations of the economical use of energy.
- Energy consumption in Germany and its negative effects on the environment, as in most industrialized countries, lie considerably above the level acceptable for a sustainable economy. Therefore, the route to sustainable systems includes a substantial reduction in the level of energy consumption.

Although the reduction of energy consumption is not the goal of a business-oriented energy supply company, there are still possibilities for integrating supplier's goals with those of a sustainable economy. They are briefly described in the following.

If it could become possible to supply the amount of energy desired by customers in an efficient manner – that is, with a minimum use of primary energy - then the business interests of an energy supply company and a positive effect on the environment would complement each other. Certainly, the most important points here are the use of modern, efficient power stations, the use of fuels with a low specific production of CO_2 (i.e. mainly natural gas), and finally the combination of energy and heating, which compared with the separate productions of heating and electricity can lead to significant savings in the use of primary sources of energy.

What is commonly described as "end energy" is often not usable for the customers. Electrical energy is certainly described as end energy, but for customers it is only then sensibly used if, for example, areas can be illuminated with it during periods of darkness or if buildings can be provided with electrically powered air conditioning. This means that in conventional relationships, the customers of energy supply companies actually only form a part of the value-added chain. In doing so they often, because of lack of knowledge or financial problems, use the energy ineffectively. This presents an opportunity for customers, energy supply customers and the environment. If the energy supply company takes over the last part of the value-added chain, then it also thereby extends its own value-added chain. The actually desired uses are still available to the customer, and the environment is burdened with less CO_2. With lighting contracts, this functions for example through the signing of a lighting contract with the customer. The *Stadtwerke Hannover AG* modernizes the lighting installation and sets up an efficient lighting system. The customer then has better lighting for prices equal to or somewhat less than the previous costs of electricity. Since the new lighting system consumes less electricity than the former one, the energy supplier saves on primary energy and thus also on the costs of primary energy. Thus it can be in the supplier's own interest to help contribute to lower energy consumption.

The issue of lighting also played a role in the context of the *Stadtwerke Hannover AG's* Least-Cost Planning Program (LCP). A scientific case study carried out between 1992 and 1995 on LCP and on Integrated Resource Planning (IRP) has shown that a significant part of Hannover's electricity consumption can be reduced both technically and economically. The accompanying profit to the local economy which would accompany this would certainly represent an even greater business deficit for the *Stadtwerke Hannover AG*. It was therefore decided to use some of the resulting cost-saving benefits to customers for the financing by moderately increasing the prices.

In the framework of test programs carried out between 1996 and 1997 and an implementation phase run in 1998 and 1999, some fourteen different sponsoring proposals for private and business customers in Hannover were issued. The results of the test program alone led to a reduction in power station output of 3.2 MW. This result was 18% greater than expected. Viewing the situation in early 2000, these efforts have cut back annual electricity consumption by ca. 19,000 MWh, which corresponds to the annual needs of around 9,000 households in Hannover.

Another theme also deserves closer examination. Up until two years ago, there were many organizations and companies in the greater Hannover region involved with the protection of the environment and climate. Unfortunately, because of the groups' divergent aims, the public was left with a confusing picture. As a result, it was almost impossible to show the effectiveness of measures which had been employed for climatic protection. In addition, there was a high level of administration with little systematic selection of the supported measures. With the establishment of *proKlima*, it is now possible to combine the activities of the city of Hannover and its surrounding towns and communities with those of the *Stadtwerke Hannover AG*. The resulting combination covers financial means, decision making procedures, application procedures, etc. An evaluation procedure has been agreed upon for reviewing sponsorship requests. One important parameter which has to be studied is the achievable level of CO_2 avoidance, calculated over the period of effectiveness of a single particular measure. Further criteria are also considered, e.g., the degree of innovation of the suggested measures. Because of these efforts, the flow of a large part of the expenditures intended for climatic protection within the greater Hannover area through *proKlima* is ensured, as is a guarantee that these resources will be effectively used. This also means that less effective measures will not be supported.

The ideas of the "LCP" and "*proKlima*" also led to changes in the attitudes and behavior of energy customers. More importantly, customers were confronted with the concept of ecological electricity. This "eco-electricity" is energy gained from regenerative sources. Since September 1999, the *Stadtwerke Hannover AG* has offered customers electricity gained from regenerative sources of energy under the brand name "enercity and cares". This offer is extended to customers both inside and outside its own network area. Clients who are prepared to pay an additional charge of 8 Pf/kWh can receive electricity produced only from regenerative sources. Because of this surcharge, it is just as profitable for the *Stadtwerke Hannover AG* to sell its customers eco-energy as "normal electricity". Allowing customers to choose different types of energy from the same company also keeps customers from changing over to ecologically oriented competitors.

The following points about how to proceed can be derived from these experiences:

- It is important to have a clear understanding of the situation, i.e., of relationships, modes of action and possible starting points.
- The strengths of all interested groups and possible supporters should be combined.
- Structures have to be created or modified to facilitate the ability of all participants to act based on the recognized relationships and modes of action.

22 Renewable Energy Sources for Community Sustainability in India

Chaman Lal Gupta

In twenty-five years, we have learned many lessons from intensive research in developing, creating and fine tuning renewable energy technologies relevant to Indian villages and their current life styles; from incursions into the field with vastly varying degrees of intensity, scale and impact; and from sporadic monitoring. Some of these lessons are summarized in the following section.

Lessons Learned (1975 - 2000)

Welfare enduses such as electric lights and clean cooking and drinking water, high on the agendas of donors and governments, are desirable for healthy and decent living. However, they do not make a visible impact on the acceptance of sustainable technology choices except in remote or difficult terrains. For reasons of status, renewable energy technologies are considered poor cousins of the LPG and grid power.

Renewable energy is achieving acceptance in cities only because of utility failures. This means that energy security rather than sustainability is the determining factor for the urban sector. This will also be true for villages in the twenty-first century.

Income-generating projects requiring irrigation water, grinding of cereals, decorticators, paddy hullers, oil presses, etc., have a greater impact if renewable energy systems are employed. These systems must be reliable, convenient, and no more costly to run than captive diesel gensets. The price of energy is competitive if a 40 to 60% plant load can be assured by proper maintenance and load management. All financial juggling aside, most such systems do not pay back capital costs. Pollution abatement, reduced carbon loading and energy security are desired as additional benefits. The only convincing determinant for choosing these options is value-adding enduses on a long-term basis.

For evaluating sustainability, annualized life cycle costs per sustainable livelihood created can take into account efficiency per unit of capital investments, equity, proper use of natural resources (particularly those available locally), and utilization of local skills including their upgrading. Use of local resources and skill-based enterprises are naturally low in capital costs and intensive in manpower, thereby creating livelihoods at low costs. To be sustainable, however, efficient options on an equitable basis must be chosen.

The goals of the promoters of technology interventions must be the same as those of the users. Transparency, empathy and dedication are at least as important prerequisites for a project as competence, correct choice of site and appropriate choice of technology for success. Psychological determinants are critically important in countries like India. These are best served by local participation in all decisions from the beginning, including whether or not a technology intervention is needed at all. This empowerment may take time, but it is essential. It is far better to be slow and steady, laying a good foundation for a process, than to leave a string of failures in the wake of fast, overly ambitious races to targets. Post-mortems do not help development.

Some Essential Energy Systems

Cooking systems: Currently, cooking is responsible for 90% of domestic rural energy consumption and about 63% of total rural energy consumption at 15 GJ/capita per year. It is one of the most regionally specific functions. Even normally reasonable people can be unreasonable about changes in cooking methods because of deep-seated traditions and preferred tastes. Therefore, it is not easy to change some methods of cooking or the fuels used for them. The fact that the fuelwood required for cooking causes deforestation is quite controversial, but a combination of available possibilities (fuel-efficient cook stoves, solar cookers and cooking by biogas generated from animal wastes as well as availability of RDF and biocoal derived from agrowastes/ garbage) could mitigate the problem. I do not see ethanol being used for cooking at least in rural Asia at any time in the foreseeable future.

Rural energization: The key to rural stabilization is enhancing affordability by creating long-term jobs at minimal capital investment and income addition through value-added projects generated in times of underemployment. This would require energy. Onsite rural energization via locally available feedstock for green power through biomass seems to be the best current option both ecologically and economically. Capital costs are less than grid stations, and energy tariffs are comparable, provided that the plant load factor is around 60% and that feedstock and loads are available locally.

Rural electrification: Rural electricity is a symbol of status as well as convenience. SPV systems for less than 100 watt connected loads per family are the preferred choice because of the similarity to grid operation. Families are readjusting their budgets to meet such expenditures in non-grid areas. In grid areas, building integrated SPV and tail end voltage boosting could be the possible enduses. Fuel cells working on microbial hydrogen or photolytic hydrogen will be options in about 10 years.

Sustainable Energy Systems for Villages

The confluence of economic and ecological stability integrated with the psychological perceptions of the villagers is absolutely essential for any real sustainability. This translates into a proper culture of development and heartsets (as opposed to mindsets) which can give rise to appropriate strategies and systems that are intelligently formulated. The mind alone is incapable of solving the problems of life. Once this is accepted, certain axiomatic guidelines become self-evident:

Priorities, targets and choices are determined by participants in the project area from the beginning. Local maintenance teams are imperative. Necessary skills, if not available, have to be taught.

Hybrid energy systems are usually called for as opposed to single systems. Constituent subsystems should be chosen on the basis of minimum cost per unit efficiency rather than minimum cost or maximum efficiency. This allows for hands-on innovation at the local level, which is crucial for involvement and economic empowerment as well as for the running of systems.

Analysis of projected loads and fuels or ambient resources available should be done professionally. Only enduse efficient appliances and processes should be used on the demand side. System configuration should satisfy the sustainability index defined by the least annualized cost per unit of livelihood created on a long-term basis.

To have credibility, efficiency and costs must be secondary in pilot trials when compared to effective work making the best of available environmental conditions. Countless technologies have failed because of confusion between effectiveness and efficiency.

The Auroville Experience

In 1968 Auroville was founded for the express purpose of providing "*a site of material and spiritual research for a living embodiment of an actual Human Unity*". By definition it had to be a sustainable community. Over the last thirty years, reliable data and know-how have been generated for technical, financial, social and operational aspects needed as the "hardware" for community sustainability. The result of this effort should be the birth of properly conceived, sustainable "ecocommunities", unconstrained by the rigidities of utility grids and the falsity of the rural-urban divide.

The first experiment for integrating renewable energy sources with buildings was carried out in the mid-70s. This included a climate-conscious design integrating solar cooking and water heating, rooftop rainfall harvesting, a biogas plant for mixed wastes and a roof mounted aerogenerator. The concept was well ahead of its time and was not fully successful, because the technologies had not yet matured. However, it provided the first hard data on costs and assessment of the integration problems including the human ones. Since then, many energy autonomous houses have appeared in Auroville but mostly retrofitted with devices. There are also one or two communities of about 20 persons each which are off-grid and have wind powered water supply and bio-filter waste recycling systems. The primary motivation has been philosophic commitment. A scientific basis for eco-community design is under study and should hopefully lead to a manual for planning *rurban communities*. This is probably one of Auroville's best offerings to India as it glides past its half century of freedom; it is also a gift from India to the developing world for the next century.

Concluding Remarks

Defining sustainable energy systems for villages in the twenty-first century is a bit like gazing in a crystal ball. Images are sufficiently clear, however, to allow for observations. The realization of sustainable energy systems assumes a sustained and patient labor of love and does not yield any dramatic successes or allow for wishful thinking. There has to be a total transformation of the rural-urban divide from both sides to form a network of eco-communities linked physically with clean transport and psychologically with communication technologies. The gridiron constraints of city scale utilities will be replaced by local utilities. These will be based on green power, rainfall harvesting and closed loop waste recycling systems as primary features of a new paradigm of development for the twenty-first century. Auroville is a hands-on, ground-based experimental community steadily moving towards this goal.

23 Sustainable Energy Systems in Tanzania

Stanford A.J. Mwakasonda

In most African countries, the major indigenous energy sources are hydropower, coal, natural gas, woodfuel, and solar and wind energy. In most cases, the extent to which these sources are tapped is very low. Therefore, although endowed with abundant indigenous energy sources, the availability of energy still poses an obstacle to development for most countries in Africa.

The energy problem is compounded by the fact that these countries in some cases have to use up to 50% of their export earnings for importing petroleum. In light of this, for most of these countries the energy policy goals must focus on overcoming problems caused by the total dependence on imported petroleum. Another immediate goal of the national energy policies is the minimization of biomass energy use in order to reduce deforestation with its ecological and social consequences. Although not articulated in a concise form, the development of the energy sector in Tanzania, like many other African countries over the past three decades, has followed a consistent pattern based on the following principles:

- the need to exploit the abundant hydroelectric energy sources;
- the need to develop and utilize other indigenous energy sources such as coal and natural gas;
- the need to step up petroleum exploration activities;
- the need to arrest the depletion of wood through the evolution of more appropriate land management practices, afforestation and use of more efficient woodfuel technologies;
- the need to develop the human resources potential;
- the need to develop an indigenous capacity for research and development in energy systems technologies.

New and Renewable Energies in Africa

New and renewable energy technologies have been investigated for promotion and use in the African countries by different institutions and organizations. The policy needs for renewable energy focus on providing substitutes for woodfuel, petroleum and energy technologies applicable to the needs of people in the rural areas. In advocating the spread of the use of renewable energies, however, it is important to learn from the experiences of renewable energy popularization in other African or developing countries, in particular the constraints associated with this endeavor. In an Energy Research Project conducted in Tanzania by the University of Dar es Salaam, a number of common problems were identified, a few of which are highlighted here:

- *Lack of funds*: a number of windmills have stopped operating due to lack of funds for spare parts.
- *Lack of routine maintenance*: this has led to the breakdown of a number of windmills and biogas units.
- *High cost of investment*: in most cases, successful demonstration has not been followed by successful extension because of the high investment cost of units such as windmills, biogas plants and photovoltaics.
- *Social problems*: there have been reports of biogas plants stopping to work due to lack of biomass or to stolen cows.
- *Technical assessment problems*: in some cases, water pumping systems stopped working because of the lack of underground water.
- *Perception problems*: there have been cases where there has been a classical problem of differences in the perceptions of the energy problem and the preferences of energy sources by consumers. While technically sound and working, some biogas plants have been abandoned because electricity was introduced into the areas.
- *Lack of people's participation* from initiation to implementation of projects has been a problem.
- *Negative attitudes*: in some cases sponsors of projects have been reluctant to teach beneficiaries how to operate plants, because they do not believe the beneficiaries can grasp "sophisticated" technology.

In spite of the above problems found in most African countries, the geographical and climatic conditions of these countries give them a very good potential for most renewable energy technologies. However, the utilization of renewable energy in Africa is still in its infancy.

Review of Rural and Renewable Energy in Tanzania

Biomass energy accounts for more than 90% of Tanzania's energy consumption, while the rural sector accounts for almost 85% of the energy consumption. The per capita consumption of commercial energy such as petroleum and electricity is quite low at about 0.04 toe, which is significantly lower than the average of 0.7 toe for low income sub-Saharan Africa.

Renewable Energy

Biomass (predominantly firewood) in rural areas is mainly used for cooking. 90% of the households in these areas use the traditional three stone fireplaces for cooking and kerosene for lighting. On the other hand, Tanzania possesses an immense amount of renewable energy sources which could be utilized instead of oil and fuelwood, thus reducing deforestation and supplementing the overall national energy balance. Important renewable energy sources include:

- *Biomass fuel sources*: Tanzania has vast resources of agricultural, forestry and animal wastes in the form of biomass. In the past decade the yearly agricultural wastes amounted to more than 8.0 million tons or 11.0 million cbm of wood equivalent. These included 12,500 tons of cashew nuts residue, 8,400 tons of coffee husks, more than 150,000 tons of sugar residue (bagasse and molasses), more than 900,000 tons of cotton residue and more than 5 million tons of maize residue. More than 40% of Tanzania's land is made up of forest woodlands. The domestic animal population is estimated at 20 million. The amount of animal waste is substantial. Most of these biomass resources can be used efficiently in biomass conversion technologies, such as direct combustion, pyrolysis, briqueting, biogas generation, gasification and ethanol production.
- *Solar energy resources potential*: Tanzania has long sunshine periods with an average daily solar radiation of between 4.0 - 4.5 kWh/m^2. Seasonal variations in insolation levels occur with a peak in December and January.
- *Small hydropower*: It is estimated that Tanzania has more than 80 small hydro potential sites of up to 2 MW of installed capacity. The total potential capacity is about 40 MW. The small hydro resources are distributed in such a way that 77% of these sites have the potential capacity of less than 1 MW and 23% of between 1 and 2 MW.
- *Wind energy potential*: Tanzania has an average wind speed of 3 m/s. Higher wind speeds are encountered along the coast, on the islands, on the highlands and around the great lakes.
- *Geothermal potential*: Geothermal energy is based on the extraction of steam or hot water from underground natural reservoirs. Research work is still underway to determine the geothermal potential in Tanzania. Several hot water bodies that could be utilized for geothermal power production have been discovered.

Energy Efficiency Systems

There is a lot of room for increased energy efficiency in most energy use areas. To date energy efficiency efforts have mainly been limited to promotion of improved stoves and improved charcoal production techniques. Conversion losses involved in charcoal production from wood are high, with 12 cbm (10 tons) of solid wood required to produce 1 ton of charcoal because of the low efficiency of the traditional earth mound kilns used in the production.

Crop processing and industrial use of fuelwood is also inefficient. 50 cbm of wood or more is required to cure 450 kg of tobacco as compared to 15 cbm of wood for 450 kg of tobacco for improved curing bans. The government has initiated various efforts aimed at reducing these inefficiencies.

In the sixties, the Ministry of Natural Resources introduced steel kilns in several forests around the country. Although these kilns were of much higher efficiency than the traditional kilns, the project did not last because of the high cost of the kilns and the associated management problems. Other efficient methods of charcoal production were introduced.

It was noted that one of the biggest constraints in the introduced technologies for efficient production of charcoal was the nature of the kilns, which had to be constructed at one particular location. Traditional charcoal producers are always moving from one place to another, depending on the availability of the appropriate trees for charcoal production. They therefore find any production method that requires them to be stationed at one particular place unsuitable.

Sustainable Energy Systems and the Environment

Dams and reservoirs required to generate *hydropower* can provide additional benefits in terms of better water management, development of irrigation systems to increase agricultural land, impoundment beneficial for the abatement of orchocerciasis transmitted by simulium damnosum which breed in the rapid sections of streams and rivers. They are also beneficial in the clean production of power needed to accelerate industrialization and socioeconomic development, tourism development and recreation, as opposed to production of electricity by thermal generators that emit greenhouse gases.

Some environmental benefits of *solar energy* are:

- reduced use of fossil fuels, and therefore reduced bulk of pollutants emitted by such fuels, and conservation of non-renewable energy sources;
- reduced use of wood energy and other non-commercial energy sources;
- is relatively "neutral" as far as excess heat rejection is concerned;
- no land is required for water and space heating and cooling; rooftop solar collectors are used;
- no emissions or wastes to be disposed of since solar thermal systems are closed systems.

Conclusion

Use of renewable energies, which is also closely associated with rural electrification, improves the quality of life by greatly facilitating lighting, cooking and heating. People in rural areas can then devote more time to child care, education and income-generating ventures instead of collecting firewood. Health facilities can also be improved by the use of renewable energies. Planning for the dissemination of the use of renewable energy technologies is an important component of national economic planning. Apart from satisfying the basic needs of the rural population, such an initiative is an indirect factor in checking deforestation. It can also create job opportunities. While it is recognized that the use of renewable energies in Africa is still facing a difficult situation created by the lack of funds, the high initial costs and the overdependence on donors, there is still room for appropriate policies that will provide for the extended use of renewable energies. Such policies could emphasize the importance of the use of renewable technologies, not as services delivering benefits to society but as part of a rural development package. Such policies could place greater emphasis on rural community involvement, greater allocation of domestic resources and creating a good environment for the adoption of renewable technologies.

References

Ministry of Energy and Minerals and AF Swedish Management Group (1999): Proceedings on the Review of the 1992 National Energy Policy.

Mwandosya, M.J. and M.L. Luhanga (1990): Energy, Research, Development and Extension Projects in Tanzania. IDRC, Canada.

Mwandosya, M.J. and M.L. Luhanga (1991): Proceedings of the Seminar on the National Energy Policy for Tanzania. Stockholm Environment Institute.

Parke, M. E. (1974): The Use of Windmills in Tanzania. University of Dar es Salaam.

Ranganathan, V. (1992): Rural Electrification in Africa. African Energy Policy Research Network, AFREPREN.

Agenda 21 §14.34 Earth Summit, Rio 1992

Land use

1992	2000	2025
5.4 billion people	6.25 billion people	8.3 billion people

Sustainable Soil Management

Introductory Statement

Helmut Eger

Soil, like air and water, is essential to support life on earth. Over 90% of human food and livestock feed is produced on the land, on soils which vary in quality and extent. Although the world's current development and progress is characterized by urbanization and globalization, rural areas will remain the major production area for the world's food demand, which is steadily increasing due to rapid population growth, especially in developing countries. But most of the world's soils, the fragile basis for agricultural production, are threatened by human mismanagement, unsustainable land-use practices, severe degradation and desertification.

Soil degradation has already affected one-third of the world's agricultural soils, particularly soils less suitable for cultivation which are nevertheless used for agriculture, grazing and other purposes. The first global overview of the current status of soil degradation in all its forms, known as GLASOD, was produced in the late 1980s and showed that the dominant processes of soil degradation are erosion by wind and water, chemical contamination and a diminution of nutrients (organic matter). These degradation processes usually lead to a reduced soil fertility and as a consequence considerably diminish the productivity of the land.

The significance and the essential functions of the soil (biodiversity, carbon sequestration, waterhousehold, food production, etc.) for human existence justify an intensive commitment to soil conservation, erosion control and sustainable land management. The international community has long recognized the vulnerability of the soil-system and the complexity of the problem. So far, a wide range of measures to support sustainable soil management and conservation of natural resources for future generations have been applied, but the success of projects and programs has often been limited.

In recent years, institutions and organizations concerned with soil conservation have therefore emphasized the importance of legally binding instruments like the UNCCD and an intensified international discourse between farmers, practitioners, politicians, scientists and decision makers from business, government and civil society. In order to find appropriate solutions to the problems of soil degradation and to ensure sustainable development in the affected regions, the dialogue between all the stakeholders involved has to be deepened and extended. It simply has to become "global". The following section focuses therefore not only on looking at the major causes of problems in rural areas and their relation to soil degradation but also on exchanging experiences and learning from examples.

The contributions foster transnational sharing of insights and facilitate an open debate on conflicting issues. They stress the importance of coherent and transparent land-use planning for degraded landscapes as well as the necessity to implement and realize erosion control programs (*Vlek et al.*). Balanced organic and mineral fertilization is identified as a remedy to improve the N:P:K ratio, which is often critical to plant growth and crop yield (*Krauss*).

Furthermore, emphasize is put on the role of active participation of the local population and the need to consider traditional and indigenous knowledge of land-use practices in the planning and implementation of soil conservation projects (*Diallo*). Another contribution discusses an approach to attaining the goal of building the fertility of tropical African soils to levels never before attained, stressing the importance of the phosphorus capital in the soils (*Mokwunye*).

The final article confronts the reader with seven widely believed myths about sustainable soil management in sub-Saharan Africa and the more complex reality behind them (*Breman*).

Additionally, useful political and institutional frame conditions have to be established in order to favor sustainable land management, to secure land rights and to strengthen the role of farmers and among them especially of women, who have been neglected for the longest time. Partnership, instead of competition, can lead to an improved awareness of the critical situation and is also a way to establish practical and cost-effective synergies between the different efforts in the struggle against soil-degradation, which will definitely be one of the big challenges in the new century.

24 Balanced Fertilization: Integral Part of Sustainable Soil Management

Adolf Krauss

Population Growth and Urbanization Call for More, Better and More Varied Food

Global population increases annually by about 80 million. Although the growth rate will decline, it is expected that the global population will reach the 8 billion mark within the next 20 years. Most of the population growth will occur in developing countries, where between now and the year 2020, the population will increase by almost 40%. In contrast, during the same period the increase in industrialized countries will be under 4%. There is also a considerable shift in the ratio of people living in rural areas and in towns. Urbanization in developed countries is currently about 74% and in developing countries 40%. However, within the next 20 years half the population in developing countries, searching for jobs and food, will be living in towns.

More people need more food, and urbanization changes the diet towards more animal protein, wheat, rice, high quality vegetables and fruits. In China, for instance, town people eat more meat (27 kg red meat) and less rice (68 kg) than their rural counterparts (17 kg meat and 103 kg cereals) (Rozelle and Jikun Huang, 1999). Correspondingly, the per capita demand for meat will increase in developing countries by 60% but in developed countries only by 5% (Rosegrant et al., 1995). Furthermore, in response to changing demand with urbanization, the acreage under vegetable and fruit cultivation in China increased by 4 and 8 times respectively over the past 30 years. The area under cereal cultivation even declined slightly.

On the global scale, it is expected that the total demand for cereals as food and feed will increase by 2.4% annually, reaching about 3.4 billion tons in 2020. However, the projected cereal production of 2.7 billion tons in 2020 would leave a gap in supply of 700 million tons unless crop yields, especially in developing countries, can be increased substantially.

Land and Water Are Becoming Scarce - The Existing Land Must Produce More

The global availability of arable land is decreasing and will further decline from currently 0.24 ha per capita to 0.17 ha in 2020. The situation in Asia is most striking. Twenty years from now only 800 m² per person will be available for crop production there. A similar trend is expected in India. The ratio will drop from the current 0.14 ha to 0.10 ha by 2025. Moreover, "... *the quality of land (in India) likely to remain available for agriculture due to severe competition from urbanization, industrialization and civic needs, will be poor...*" (Kanwar and Sekhon, 1998). In this context, Hanson (1992) reports that only about 12% of the tropical soils has no inherent constraints. Of the remaining land, 9% has limited nutrient retention capacity, 23% aluminium toxicity, 15% high P fixation and 26% low potassium reserves.

What has been said for land availability also applies for water. Withdrawal of water in developing countries will increase by 43% between now and the year 2020, in developed countries by 22%. But, in developing countries, the demand for water for domestic and industrial uses will double, reducing the supply for agriculture (Pinstrup-Andersen et al., 1997).

Fig. 24.1. Regional Cereal Yield, Current Development and Projected Demand

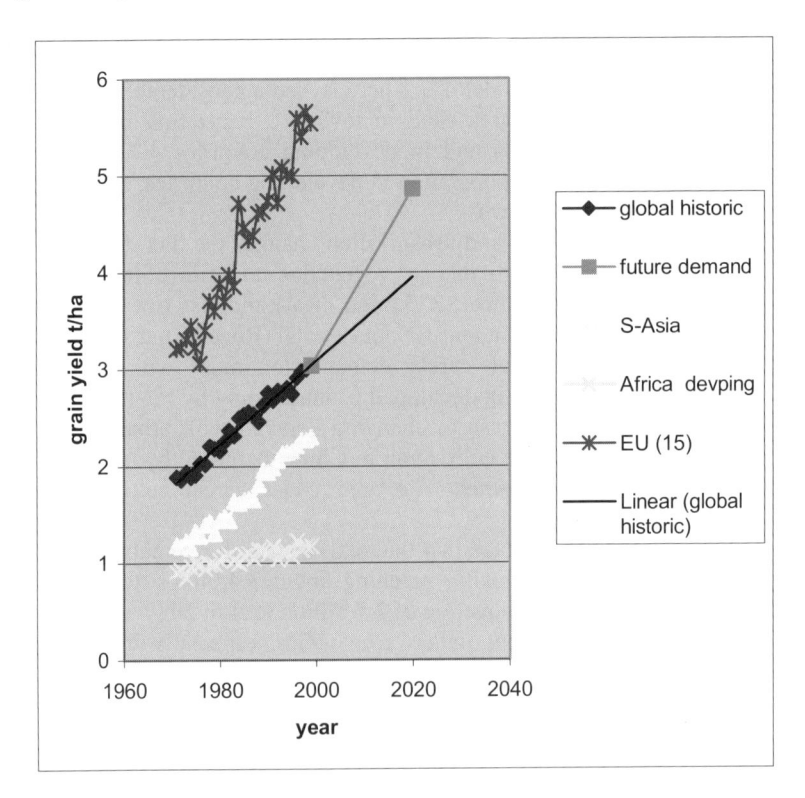

Source: FAOSTAT, 1998

In consequence, horizontal expansion in food production is hardly possible unless further deforestation and use of marginal land are accepted. The necessary increase in production has to come through higher yields and denser cropping sequences, i.e. through higher productivity of the remaining land and water. To match the expected cereal demand of 3.4 billion tons in 2020, cereal yields must be increased from currently 2.9 to almost 4.9 t/ha and rice yields by 60 to 70%. The current rate of increase as shown in Figure 24.1 is far too low to meet the future demand. Had India, for example, relied on area expansion for increased grain production, it would have had to more than double the area currently under cultivation. But India, like the other Asian countries, does not have spare land.

The Higher Yields Required Remove More Nutrients which Must Be Replenished

The need for higher soil productivity to feed future generations is evident. Higher yields remove more nutrients from the field, and these must be replenished. The global mean yield of 3 t/ha cereals together with the straw withdraws about 81 kg nitrogen, 15 kg phosphorus and 75 kg potassium from the field, the equivalent of 3½ bags of urea, 1½ bags of DAP and 1½ bags of potash fertilizers. Other nutrients are lost by leaching, erosion, fixation, etc. as shown schematically in Figure 24.2. These must also be replaced.

Native soil fertility cannot support the necessary yield increase. Nutrient inputs through depositions, sedimentation and biological nitrogen fixation are far from sufficient to satisfy the nutrient demand of high yielding varieties. Mineral fertilizer is the primary source of nutrients and usually contributes 35 to 50% to yield increases. Numerous field trials conducted by FAO and the private sector have shown that one kg of mineral fertilizer can achieve under farmer's conditions about 10 kg additional yield of cereals. In this context, a rather close relationship between use of mineral fertilizers and cereal production can be seen, taking sub-Saharan Africa and developing Asia as examples. Fertilizer use in Asia grew over the past 40 years from virtually nil to more than 20 kg per capita. During the same period, per capita cereal production increased from less than 200 kg to more than 300 kg. In contrast, use of fertilizers in sub-Saharan Africa fluctuated around a rather minute level of 2 – 3 kg; the corresponding cereal production declined from around 150 kg per capita during the sixties to around 130 kg indicating loss in soil fertility.

Recycled plant residues, green manure or farmyard manure are important components in soil fertility management, at least minimizing the loss caused by export of food and thus nutrients into urban centers or across national boundaries. However, in principle, use of organic manure or recycling of plant residues cannot compensate for all losses unless additional nutrients enter the farm from outside such as in the form of concentrate feed. In this context, Belgian dairy farms have a positive nutrient balance although fertilizer usage has declined because imports of concentrate feed for animals have increased (Michiels et al., 1997).

Fig. 24.2. Schematic Nutrient Flow in Agroecosystems

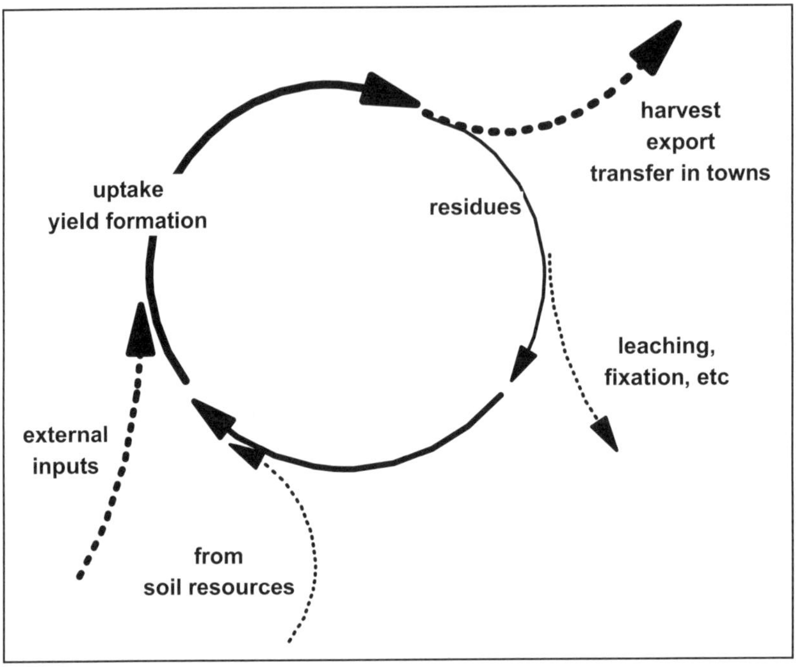

[Figure: Schematic diagram showing nutrient cycle with arrows connecting: uptake/yield formation, residues, harvest/export/transfer in towns, leaching/fixation/etc, from soil resources, external inputs]

Source: after van Noordwijk, 1999

The situation in developing countries is different. In China for instance, although use of organic manure is still increasing, the share of nutrients there from total nutrient input has declined steadily from 100% in the fifties to currently 30% (Xie et al., 2000). In Pakistan, the use of farmyard manure is inversely related to farm size. Small farms of up to 1 ha use on average almost 5 t/ha, whereas larger farms of 10 to 20 ha apply less than 100 kg/ha (NFDC, 2000). Lack of affordable labor and misuse of crop residues as fuel and building material are the reasons.

Use of Mineral Fertilizer to Support Soil Fertility Has Become Common Practice, but Fertilizer Use Is out of Balance

Global consumption of mineral fertilizers increased steadily after World War II up until the late eighties, when economic constraints (especially in Eastern Europe and the FSU), ecological considerations, set-a-side programs, crop prices, etc, caused a substantial setback in fertilizer use. Subsequently, the use of nitrogen fertilizer recovered fairly well whereas the use of phosphate and especially of potash fertilizers are still below the level achieved during the late eighties although crop output continued to increase further. This resulted in fertilizer usage becoming unbalanced in two respects:

- the nutrient ratio, especially of N to K,
- the ratio of fertilizer nutrient to nutrient removed by crops, i.e. the input/output ratio.

The NK ratio in fertilizer use declined in the past 20 years from a roughly balanced value of 1:0.4 to currently 1:0.26. This ratio contrasts sharply with the ratio in which plants absorb N and K. Cereals for instance take up nitrogen and potassium in almost equal quantities; vegetables and root/tuber crops absorb even more potassium than nitrogen.

There are substantial regional differences in the N:K ratio in fertilizer use ranging from 1:0.43 in West Europe or North America to 1:0.1 in India or even 1:0.06 in West Asia and North Africa (the WANA Region). Developing countries in general consume roughly 65% of the global N use but only 45% of K use. Some of the reasons for the regional differences in the nutrient ratio are:

- belief in unrestricted K supply of cultivated soils;
- spectacular crop response to nitrogen when mineral fertilizers are introduced, so relating yield mainly to N fertilizer usage;
- rather discreet crop response to potash fertilizers in terms of quality, disease and stress resistance;
- misinterpretation of soil test results;
- simply lack of knowledge;
- unawareness, unavailability of potash;
- inadequate promotion of balanced fertilization in developing countries.

Concerning the nutrient output/input ratio or the balance between nutrient removal by crops and fertilizer use, fertilizer imbalance is especially evident in developing countries.

As shown in Figure 24.3, use of nitrogenous fertilizers in developing countries has reached the level of N removed by crops. Increasing use of phosphate fertilizers is beginning to close the gap. However, potassium removal by crops exceeds by far the use of potash fertilizers; the negative balance in potassium is steadily growing. India for instance has currently a deficit of about 7 million tons K_2O whereas N use more than covers N removal by crop. China's use of potash lags more than 11 million tons K_2O behind potash removal; but N use exceeds N removal in crops by nearly 7 million tons. The situation in sub-Saharan Africa is even worse because all 3 major nutrients are in deficit. IFDC estimates the actual crop nutrient requirement to be 3.9 million t N, about 2 million t P_2O_5 and 2.9 million t K_2O. The current fertilizer use is about 0.7 million t of N, 0.4 million t P_2O_5 and 0.23 million t K_2O, i.e. use of mineral fertilizers in sub-Saharan Africa covers around 20% of N and P requirements and less than 10% of potassium requirements (Henao and Baanate, 1999).

Fig. 24.3. Fertilizer Use in Relation to Nutrient Removal by Crops in Developing Countries

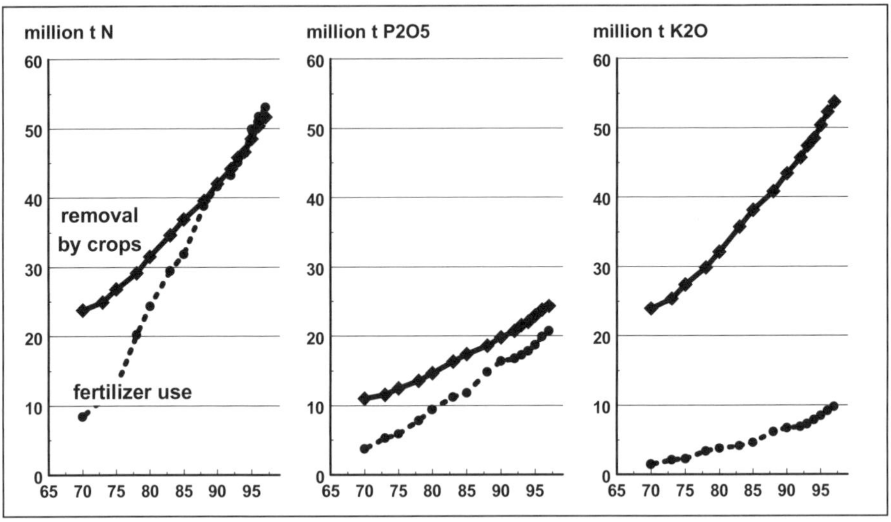

Source: FAOweb, 2000

The nutrient balance in developed countries evolved rather differently than that of developing countries. Until the late eighties, in developed countries fertilizer application exceeded nutrient removal by crops, thus substantially building up the soil fertility. This applied to all three major nutrients. During the early nineties, fertilizer use in developed countries decreased sharply, mostly in response to uncertainties in land titles, lack of credits, etc., during the economic reforms in East Europe and the FSU. As indicated earlier, economic constraints, ecological considerations and set-aside programs in Western countries also caused some reduction in fertilizer use. N and P fertilizer usage is close to the nutrient withdrawal by crops. K deficiency is also evident in some developed countries.

Imbalance in Crop Nutrition Leads to Soil Nutrient Mining and Consequently to Soil Nutrient Depletion

Extended soil nutrient mining as observed in Africa not only depletes the nutrient reserves, it also affects chemical soil characteristics such as pH, base saturation and the content of free aluminium (Table 24.1). The latter is toxic to the roots of most cultivated plants. Root length is drastically reduced, which restricts exploration for soil nutrients and water. Not only is yield much reduced, but the plants are also very susceptible to drought.

Table 24.1. Effect of Soil Nutrient Mining on Soil Properties (Tanga Region, Tanzania)

	Virgin soil	Cultivated soil
Soil pH	6.2	5.2
Org. C [%]	2.1	1.7
Exch K [mmol/kg]	5	1
Exch Ca [mmol/kg]	68	13
Exch Mg [mmol/kg]	26	5
Base saturation [%]	80	21
Exch Al [mmol/kg]	0	9
Al-saturation [%]	0	42

Source: Hartemink, 1997.

Detailed information is available on the impact of soil K mining on soil fertility. As readily available K is depleted, K supply to the soil solution depends more on K release from the non-exchangeable or slowly available fraction, the K reserve. As shown by Mingfang et al. (1999) on soils in North China, the rate of release of K from the exchangeable fraction is much higher (5 - 9 mg K/kg/min) than that from the non-exchangeable fraction (0.1 - 0.5 mg K/kg/min). Consequently, with both the depletion of soil K and the corresponding increase of the contribution from non-exchangeable K, the yield declines because the release intensity cannot cope with the demand of high yielding crops.

Therefore, so long as the yield level was rather low, there was understandably little or no response to K on the alluvial soils of the Nile, Indus and Ganges systems. On these heavy textured soils, the rate of release of K from the reserves was high enough to meet the K demand of a small crop. But with transition from subsistence to market oriented agriculture, the introduction of high yield varieties, double cropping and intensive use of nitrogen, the demand for soil K increased to the extent that the native K supply of these soils could not cope with it. With declining native K reserves, crop response to potash increases as observed in India. Evaluation of long-term experiments revealed that during the last 25 years crop response to N has decreased, whereas the response to P increased by 20% and to K even by 160%.

With progressive exhaustion of soil K reserves, applied potassium is more strongly fixed. Because applied K is trapped in the interlayer spaces of clay minerals, crops may fail to respond to normal K dressings. Unfortunately, the occurrence of K fixation and the resultant absence of crop response often misleads the fertilizer advisor, and this worsens the situation. At the same time, K depleted soils tend also to fix NH_4, thus lowering the availability of fertilizer N.

The Soil's Potential to Cover Nutrient Demand and to Handle Stress Situations

Results from 650 field trials in Germany showed that, at the lowest level of exchangeable K, cereals lost 18% and leafy crops like potato or sugar beet almost

40% of their yield potential. In long-term field trials at Rothamsted, UK, barley receiving 144 kg/ha N yielded less than 2 t/ha on exhausted soil with 3 ppm P and 68 ppm K but 5 t/ha on fertile soil with 140 ppm P and 329 ppm K (Johnston, 1994). Lack of sufficient P and K impaired utilization of natural resources and reduced the efficiency of inputs like N fertilizer, causing a burden to the environment, because the content of residual N in soil is much higher following a poor crop than after a high yielding crop.

The value of lost yield opportunity due to unbalanced fertilization can be considerable. For instance, Bulgaria lost around 135,000 t of wheat, 28,000 t of maize and 53,000 t of sunflower, or the equivalent of almost US $ 30 million due to unbalanced fertilization (Nikolova and Samalieva, 1998).

Apart from reducing the yield, unbalanced fertilization also impairs the quality of the crop. This is also true for nutritional, hygienic, organoleptic and functional properties as well as to the environmental compatibility in production. Numerous field trials conducted by IPI in India, Egypt, Hungary or China showed higher contents of protein in wheat, of oil in groundnuts, of rapeseed or soybean, of sugar in cane or beet, and of more aromatic components in tea if balanced fertilization was employed.

Tolerance to climatic and biotic stress is also much affected by soil fertility. Early wilting in drought is a typical indication of K deficiency. Water stressed plants have significantly reduced photosynthesis with ultimately lower yield. Yield of triticale was more than halved by drought on soils inadequately supplied with potassium (Wyrwa et al., 1998). Use of K alleviated the effect of drought stress and increased the yield to about 83% of unstressed plants, because potassium, as an osmotically active ion, improves the water economy of the plant and regulates stomatal movement.

Frost hardiness is also improved by balanced fertilization. Finally, adequate potash use in relation to N decreased incidence of fungal diseases in 70% of the investigated cases. There was a similar improvement in resistance to bacterial (69%) and virus (41%) diseases and to insects and mites (63%) (Perrenoud, 1990).

At the same time, the farmer lowers input costs by applying fewer agrochemicals to more tolerant plants, a contribution to a more environmentally compatible crop production.

Balanced Fertilization Provides the Basis for Sustainable Crop Production; Who Are the Beneficiaries?

Numerous long-term field trials prove that the initial level of soil fertility and thus crop yield can be maintained provided the nutrient balance is in equilibrium. As indicated earlier, the introduction of modern high yielding varieties affects the nutrient balance by removing more nutrients due to a higher biomass. Of course, there are genotypical differences in nutrient use efficiency. At the same level of soil fertility, the yield of modern wheat varieties is 50% higher than that of less advanced land races (Karpenstein et al., 1986). A better spatial exploitation of the root system and the release of acidifying and/or chelating substances can be explained by the genotypical differences in nutrient uptake efficiency.

The present trend in plant breeding, especially in the development of transgenic lines, focuses more on specific traits such as resistance to a particular insect or virus or on tolerance increase to a particular herbicide. Development of transgenic lines for high-yielding crops has a lower priority because so many different genes are thought to be responsible for yield performance. Of course, as limiting factors like disease susceptibility are eliminated, thus increasing yield, an indirect effect may be to increase the nutrient requirement. As far as potassium is concerned, the future could see the development of improved efficiency in K uptake systems in order to utilize low external solute concentration or to improve the competitiveness of K uptake in presence of excessive Na under saline conditions.

Who Benefits from Balanced Fertilization?

The farmer reaps higher yields of better quality. Production costs are lower, and profits are higher. Also, competitiveness on the market is greatly improved. Increasing the stress tolerance of crops also improves yield potential under adverse conditions, which gives the farmer confidence to reinvest in agriculture. Table 24.2 summarizes the economics of potash use achieved in on-farm trials in India. By investing one rupee in potash the farmer gets returns of between 5 and 20 rupees.

The nation: increased food production coupled with better quality improves food security; it lowers import requirements and increases export opportunities. Higher yields increase purchasing power in the rural area, attracting other business; they create jobs and reduce migration. In this way, balanced fertilization contributes to social security in the rural area.

The natural resources: by raising yield, balanced fertilization improves the efficiency of land, water and energy. In so doing, it reduces pressure on land and water resources, prevents deforestation and protects marginal land. Concerning the energy gain achievable with balanced fertilization, IPI on-farm trials with wheat in India showed an output/input ratio of 41 to 136 for K. Although the yield increment and thus the energy gained with potash is lower than usually obtained with nitrogen, the energy output/input ratio is considerably higher than that shown for N use on wheat in Germany (Biermann et al., 1999). Potash has a lower energy equivalent (3 MJ/kg K_2O) than nitrogen (35 MJ/kg N).

The environment: improved fertilizer use efficiency leaves smaller nutrient residues in the soil. This protects the groundwater and reduces losses of N by volatilization. In China, after harvesting cabbage, under farmers' practice, i.e., primarily N and P, some 140 kg/ha NO_3-N were left in the soil. It declined to less than 40 kg/ha when N was well balanced with K. Vigorous plant growth as resulting from balanced fertilization covers the ground more quickly and stabilizes the soil surface. This helps to reduce soil erosion and runoff, the major sources of nutrient losses for instance in sub-Saharan Africa.

Table 24.2. Economic Response to Potash (IPI On-Farm Trials in India, 1993-1998)

Location	Crop	Treatment $N-P_2O_5-K_2O-S$	Value / Cost Ratio potash
W-Bengal	jute	40-20-30-20	15.8
W-Bengal	mustard	80-40-40-20	11.2
W-Bengal	potato	120-100-100-20	19.4
W-Bengal	rice	80-30-60	6.3
W-Bengal	rice	60-30-30-20	18.9
Orissa	groundnut	0-40-60	9.2
Orissa	rice	60-30-30	5.4
Orissa	rice	60-30-30-20	10.1
H.P.	potato	60-100-75	5.3
M.P.	soybean	30-80-25/25	8.9
M.P.	wheat	100-50-25/25	12.9

Conclusion

"*Fertilizers feed the world*" is the slogan of the international fertilizer industry. Global transfers of food and thus nutrients, and especially the transfer from the arable land into towns, means that farmers have to rely increasingly on external sources of nutrients to replace those lost from their fields. Whether the external source is a bag of mineral fertilizer or the nutrient in concentrate feed *via* FYM is irrelevant as long as the deficit in the nutrient budget is closed. Full respect is paid to those who advocate organic farming, but systems based only on recycling, green manure, etc, will fail in the long run because of inadequate compensation for exported or otherwise lost nutrients.

Balanced fertilization involves considering the whole spectrum of nutrients. The focus on nitrogen and phosphorus during the initial phase of fertilizer use is understandable because of the spectacular response of the crop to N. The demand for K emerges as a result of increased yields produced by N or N P in the farmer's first experience with fertilizer. A demand for S and, sooner or later, Mg and the trace nutrients will follow. There is a growing need for site specific fertilizer recommendations according to the crop type, yield level and soil conditions. Balanced fertilization should include the use of organic manure as an integral part of programs designed to secure sustainable soil fertility.

The soil nutrient capital is not an inexhaustible resource and must be replenished according to the nutrient withdrawal. With the need for intensification of crop production, the demand of crops for readily available soil nutrient increases. Soil-borne nutrients may suffice for small crops in subsistence farming but are insufficient for a crop destined for the market.

References

Biermann, S., G.W. Rathke, K.J. Hülsbergen and W. Diepenbrock (1999): Energy Recovery by Crops in Dependence on the Input of Mineral Fertilizer. Martin-Luther-Universität Halle-Wittenberg, Germany, EFMA, Brussels, Belgium.

FAO (Food and Agriculture Organisation): Production Yearbooks, Fertilizer Yearbooks, several issues, website. FAO, Rome.

FAOSTAT (1998): FAO Statistical Database. FAO, Rome.

Hanson, R.G. (1992): Optimum Phosphate Fertilizer Products and Practices for Tropical Climate Agriculture. In: Proceedings of Int. Workshop on Phosphate Fertilizers and the Environment. International Fertilizer Development Center, Muscle Shoals, Alabama, USA, pp. 65-75.

Hartemink, A.E. (1997): Soil Fertility Decline in some Major Soil Groupings under Permanent Cropping in Tanga Region, Tanzania. In: Geoderma 75, pp. 215-229.

Henao, J. and C. Baanante (1999): Estimating Rates of Nutrient Depletion in Soils of Agricultural Lands of Africa. IFDC, Muscle Shoals, USA.

Johnston, A.E. (1994): The Rothamsted Classical Experiments. In: Leigh, R.A. and A.E. Johnston (eds): Long Term Experiments in Agriculture and Ecological Sciences, CAB International, pp. 9-37.

Kanwar, J.S. and G.S. Sekhon (1998): Nutrient Management for Sustainable Intensive Agriculture. Fertiliser News 43 (2), pp. 33-40.

Karpenstein, M., K. Scheffer and R. Stülpnagel (1986): Anbauvergleich zwischen alten und neuen Winterweizensorten bei unterschiedlicher Anbauintensität. In: Kali-Briefe 18 (3), pp. 219-226.

Michiels, J., I. Verbruggen, L. Carlier and E. van Bockstaele (1997): In- and Output of Minerals in Flemish Dairy Farming: The Mineral Balance. In: Proceedings of 11th Intern. World Fertilizer Congress of CIEC on Fertilization for Sustainable Plant Production and Soil Fertility, September 7-13, 1997, Gent, Belgium, pp. 695-702.

Mingfang, C., J. Jiyun and H. Shaowen (1999): Release of Native and Non-Exchangeable Soil Potassium and Adsorption in Selected Soils of North China. In: Better Crops International, vol 13 (2), pp. 3-5.

NFDC (2000): Fertilizer Use at Farm Level in Pakistan. Survey Report No. 4/2000, National Fertilizer Development Center, Islamabad, Pakistan.

Nikolova, M. and A. Samalieva (1998): Economic Potential of Fertilizer Use in Bulgaria with Particular Reference to Potassium. In: Proceedings of the 11th Int. Symposium on Codes of Good Fertilizer Practice and Balanced Fertilization, September 27-29, 1998, Pulawy, Poland, pp. 416-422.

van Noordwijk, M. (1999): Nutrient Cycling in Ecosystems versus Nutrient Budgets of Agricultural Systems. In: Smaling, E.M.A. et al. (eds): Nutrient Disequilibria in Agroecosystems. CABI Publishing.

Perrenoud, S. (1990): Potassium and Plant Health. IPI-Research Topics No. 3, International Potash Institute, Basel, Switzerland.

Pinstrup-Andersen, P., R. Pandya-Lorch and M.W. Rosegrant (1997): The World Food Situation: Recent Developments, Emerging Issues, and Long-Term Prospects. Food Policy Report. IFPRI, Washington, D.C.

Rosegrant, M.W., M. Agcaoili-Sombilla and N.D. Perez (1995): Global Food Projections to 2020: Implications for Investment. Food, Agriculture, and the Environment. Discussion Paper 5. IFPRI, Washington, D.C.

Rozelle, S. and J. Huang (1999): Supply, Demand and Trade of Agricultural Commodities in China. Marketing Opportunities: World Trade Competition. Agricultural Outlook Forum, February 23, 1999.

Wyrwa, P., J.B. Diatta and W. Grzebisz (1998): Spring Triticale Reaction to Simulated Drought and Potassium Fertilization. In: Proceedings of 11th Int. Symposium on Codes of Good Fertilizer Practice and Balanced Fertilization, Pulawy, Poland, September 27-29, pp. 255-259.

Xie Jian-Chang, J.-M. Zhou and R. Haerdter (2000): Potassium in Chinese Agriculture. Chinese Academy of Science, International Potash Institute Basel, Switzerland, Hohai University Press, pp. 380.

25 Land Degradation: Causes and Prevention

Paul L.G. Vlek, Daniel Hillel and Sonya Teimann

A prevalent notion is that human abuse of the environment is a rather new phenomenon, mostly a consequence of the recent population explosion and of our expansive technological economy. Ancient societies are presumed to have been more prudent than ours in the way they treated the basic resources of land and water. Research has shown this notion to be in the realm of romantic fiction. In fact, manipulation and modification of the environment was a characteristic of many societies from their very inception. Long before the advent of earth-moving machines and of toxic chemicals, even before the advent of agriculture, humans began to affect their environment in ways that tended to destabilize natural ecosystems. Our ancestors did this by using fire to clear tracts of forested land in order to encourage the growth of shrubs and grasses and thus to attract grazing animals that could be hunted or trapped. Clearing trees also served to promote the growth of plants that produced edible products for direct human consumption. The collection of such products and the trapping of grazing animals eventually led to domestication and to the advent of agriculture.

While pre-agricultural interference with the environment was limited in extent and severity, the practice of agriculture greatly intensified that interference. Once it began to spread, agriculture quickly became the chief agent of environmental transformation.

As long as agriculture was confined to small enclaves while the greater continental area remained practically undisturbed, the earth's environment as a whole was not severely threatened. Degraded land could be abandoned and thus allowed to recover gradually, while new tracts were cleared, in succession. But, with the growth of population brought about by the very success of agriculture in improving food security, the cultivated enclaves also grew, until entire extensive regions were put under continuous cultivation. To compensate for the loss of natural fertility and to achieve ever-higher yields, farmers have applied increasing quantities chemical fertilizers and pesticides. In arid regions, irrigation has been introduced to overcome drought. Erosion, waterlogging, salination, pollution, and the eradication of numerous species – such were the unforeseen but now global consequences of humanity's expansive and often injudicious management of soil and water resources.

According to global studies by IFPRI:

- 1.5 billion additional people will be on the planet by 2020, almost all in poorer developing countries;
- it is increasingly difficult to find productive new land to expand the agricultural base; and
- the natural fertility of agricultural soils is generally declining.

In short, an increasing population will have to be fed from a decreasing area of agricultural land. Of the nearly 5 billion hectares of land found in the tropics, the FAO estimates that 57% has a high potential for agricultural production. Only a small part (1.46 billion ha) of the total agricultural land in the world is in use as cropland (Katyal and Vlek, 2000). In 1998, 80% of the world's population lived in developing countries with only 58% of the total land area and 54% of the cropland area. Suitable new land for crop cultivation in the tropics has virtually disappeared, resulting in a doubling in the pressure on land. Per capita cropland declines as the population increases. To date only 0.24 ha of cropland is available per person worldwide (WRI, 1998), about half of what it was half a century ago. This value also holds for Africa, whereas in South America cropland availability is nearly twice as high and in Asia it is only half.

"*Thou shall not transgress the carrying capacity*" Hardin (1999). This fundamental ecological principal highlights the limits that nature imposes on us when it comes to the use of a natural resource such as land. Ignoring that physical boundaries of land cannot be stretched sets in motion a vicious cycle of events leading to land degradation. By its very nature, agriculture is an intrusion upon and a disruption of the natural environment, as it replaces a natural ecosystem with an artificial one, established and maintained by humans. Encroachment into new, often marginal, lands or into ecosystems with bio-regulatory functions such as tropical rainforests, steep- or wet-lands often comes with high on-site or off-site costs such as soil degradation, desertification, down-stream flooding, or loss in biodiversity.

The moment a farmer delineates a tract of land, separating it from the contiguous area by arbitrary boundaries and turning it into a cropped field, the farmer is in effect declaring war on the native species, now treated as weeds or pests to be eradicated by all possible means. Moreover, the act of cultivation and the subsequent removal of the crop tends to deplete the organic-matter and the nutrient reserves of the soil by hastening decomposition, reducing replenishment by plant and animal residues, and causing erosion of topsoil. Soil organic matter is also readily lost when organic matter inputs are reduced upon cultivation and more so upon intensification. The productivity of agricultural lands is to a large extent determined by the fertility of the soil, which in turn is largely determined by its organic matter content and stored nutrients.

The pressure to grow more food forces cropping in quick succession, thus shortening the fallow period. This robs the land of the opportunity to rejuvenate, as was possible when lands were vast and population was limited. Thus, soil organic matter disappears and loss of soil structure follows. Structure-poor soils easily fall prey to erosion, the principal soil-degrading factor. The denudation of fertile topsoil by erosion imperils sustainability of agriculture. As productivity falls, the land's carrying capacity shrinks. This makes a soil fragile and exposes it further to forces of degradation.

People encroach upon cropland to serve the requirements of a growing non-agricultural sector. Generally, the better quality cropland is the victim, forcing farmers onto land of lower quality. On pastoral lands, people realize their expanding needs by maintaining large herds of domestic animals – a characteristic penchant shared by developing-world herders. Exceeding the number of animals dictated by carrying capacity degrades pastures, leaving behind plant life of no or low significant economic/fodder value. So it is no wonder that whether cropland or rangeland, three out of every four hectares of degraded land worldwide are located in developing countries.

The insidious progress of land degradation leads to irreversible degradation and land-flight. Were reversible degradation serviced in time, the spread of irreversible degradation would be far less than currently occurs. In our opinion, enduring reversible degradation should receive constant monitoring and timely attention. It is caused primarily by exploitation of natural resources beyond their carrying capacity. Solutions to combat land degradation lie in the management of its causes. However, the intricate web of human actions and nature's responses that cause the spiral of land degradation suggests that there are no easy options to combat it. While managing demographic pressure should receive priority, the solutions to combat land degradation require local answers based on land use and climatic variations. They have to be in harmony with local needs and people's expectations. Integrated data on land and soil degradation and on the socioeconomic environment is the basis to formulate strategies on sustainable land use. Our knowledge base and monitoring capabilities are currently woefully inadequate for that task (Wood et al., 2000).

References

Hardin, G. (1999): The Ostrich Factor: Our Population Myopia. Oxford University Press, Oxford, UK.

Katyal, J.C. and P.L.G. Vlek (2000): Desertification – Concepts, Causes and Amelioration. ZEF – Discussion Paper of Development Policy, Nr. 33. Center for Development Research, Bonn.

Wood, S., K. Sebastian and S.J. Scherr (2000): Pilot Analysis of Global Ecosystems: Agroecosystems. WIR and IFPRI, Washington. D.C.

WRI (1998): World Resources. A Guide to the Global Environment 1998-99. WRI/UNEP/UNDP/World Bank, New York.

Agenda 21 §14.44 Earth Summit, Rio 1992

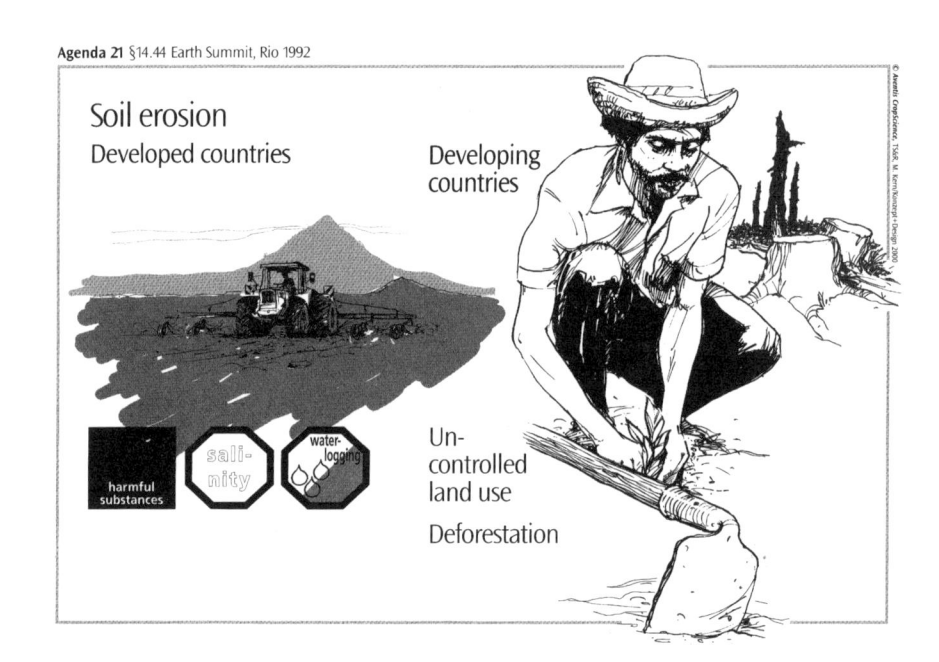

Soil erosion
Developed countries

Developing countries

harmful substances

sali-nity

water-logging

Un-controlled land use

Deforestation

26 The United Nations Convention to Combat Desertification

Hama Arba Diallo

Inappropriate use of soil has manifested itself in several forms, including soil erosion, reduced water catchment, deforestation, drought and desertification. More than any other environment-related phenomenon, the deterioration of land resources in the world's dryland areas is one of the grave problems that has contributed to the impoverishment of people living in these areas. Achieving sustainable soil management has to involve long-term integrated strategies that focus on improved productivity of land, rehabilitation, conservation and sustainable management of land and water resources. In doing so, it is recommended to give proper attention to involve and promote cooperation among all relevant actors at all levels. This will establish a better understanding of the nature and value of land resources as well as integrate the issue of sustainable soil management in the overall development policy. The process of rehabilitation of degraded land requires the enactment of an appropriate policy and supporting institutions as well as an enabling environment that ensures the participation of pastoralists, farmers and the other land users. The most important consideration is to place the cooperation of the inhabitants at the forefront of all the initiatives geared towards sustainable soil management. In the context of the United Nations Convention to Combat Desertification (UNCCD), the majority of inhabitants of dryland areas derive their livelihood from dryland agriculture and pastoral livestock production. The linkage between these two production systems has been necessitated by the need to minimize risks in these fragile ecosystems.

The important issue of promoting sustainable soil management cannot be dealt with adequately without a strong commitment of the national states. Decision makers at the national level ultimately will have to set the rules concerning land use and land use tenure. In addition, policies designed for sustainable soil management will necessarily have to rely on market forces and provide real opportunities and incentives to vulnerable land users. In African and Asian regions for instance, extensive consultative processes have been launched and their strategies for the development of National and Regional Action Programs to combat desertification include priority activities such as integrating sustainable soil management through the use of early warning systems and desertification monitoring and assessment to support their agricultural production systems; integrating local and traditional knowledge and techniques into modern scientific research; and integrating research findings into the farming production systems. The relationship between environment and trade along with harmonizing and popularizing incentive measures within national and regional policies are also taken into account.

Successful soil management, particularly in the arid, semi-arid and sub-humid areas, would improve the living conditions of local communities, alleviate poverty and stem the other associated problems such as migrations that result in environmental refugees.

27 Do African Soils only Sustain Subsistence Agriculture?

Uzo Mokwunye

Africa enters the twenty-first century as the continent with the most daunting development challenges. The inability to feed its population, projected to reach one billion by 2025, is one of these challenges. Data for 1995 from the World Bank reveal that less than one-half of the countries in Africa were able to ensure the sufficient food energy of 2,200 calories per day. This compares with 3,400 calories per person per day for Western Europe and 3,600 for the United States. Tropical Africa is the only region where the absolute number and percentage of chronically malnourished population have continued to grow. The cause of this dire situation is low agricultural output. Cereal yields currently average 1.0 t/ha. This compares poorly with average yields of 4.8 t/ha for China, 2.2 t/ha for India and 5.7 t/ha for the USA. Furthermore, there is a gap between the potential yield of cereals in research stations and the actual yields obtained by farmers. There is no shortage of suggestions for African policy makers. However, it is apparent to many scientists that no matter how effectively other conditions (policies, input/output markets, infrastructure, research and extension, etc.) are remedied, the imbalance between soil, food and population must first be addressed. Only 9 million skm of land, or 29% of Africa, has soils that could be classified as having little or no major constraints to agricultural production (Eswaran et al., 1997). This very limited land area supports about 400 million people, or 45% of the present population in Africa. Food insecurity stems from the fact that subsistence agriculture is the current mode of crop production. The question arises: "*Are the soils of tropical Africa capable of sustaining more than subsistence agriculture?*" To answer this question, we must first understand the nature of the soils, their limitations, appropriate methods of management and what it would take to improve the productive capacity.

The Environment of Tropical Africa

The climax vegetation of tropical Africa varies from the tropical rainforest, where rainfall exceeds 1,500 mm in 5 to 10 months of the year, to the tropical desert vegetation, where annual rainfall of less than 200 mm occurs in less than one month. The different ecological zones and the important soil associations are presented in Table 27.1. The soils were formed from Precambrian parent materials. Aided by high temperatures and, in most cases, high moisture regimes, weathering of these very old rocks has been extreme. Under such conditions, the soils that have developed are almost devoid of basic cations. The Oxisols, Ultisols, Alfisols, Entisols and Inceptisols (Mokwunye, 1994) are fragile in both their chemical and physical characteristics. The main constraints of all the soils are summarized below:

Table 27.1. Agroecological Zones of Tropical African and the Main Soils

Ecosystems	Main Soils	Countries included
Miombo acid savannas	Ultisols	Zambia, Malawi, Southern Zaire, Angola
	Oxisols	Mozambique, Tanzania and Madagascar
Tropical rainforest	Ultisols	Liberia, Sierra Leone, Eastern Nigeria,
	Oxisols	Cameroon,
		Republique Central Africaine, Gabon, Congo, Zaire,
		Madagascar
Guinea savanna	Alfisols	Senegal, Guinea, Côte d'Ivoire, Togo, Benin, Ghana, Nigeria, Cameroon
Eutrophic East	Alfisols	Kenya, Tanzania, Uganda, Mozambique
African Savannas	Vertisols	
Sahel	Sandy	Mauritania, Mali, Burkina, Nigeria, Chad, Sudan
Sudan savanna	Alfisols	Senegal, Mali, Burkina, Nigeria, Ghana, Cameroon, Chad, Sudan
Highlands	Oxisols	Kenya, Uganda, Rwanda, Burundi, E. Zaire,
	Andisols	Ethiopia
Wetlands	Aquapts	Mali, Chad, Zaire, Zambia, Gambia, Sudan
High Veldt savannas	Alfisols	S. Zambia, Zimbabwe
Low Veldt savannas	Alfisols	Zimbabwe, Botswana
	Entisols	
Kalahari	Entisols	Angola, Namibia, Botswana

Source: Sanchez et al., 1991.

Low levels of soil organic matter: The importance of the organic component of the soil can be best described by making an analogy to the economy of a nation. In this respect, the organic matter content of the soil can be likened to a nation's central bank. Ideally, the central bank monitors the economic health of the nation. The importance of organic matter in soils of tropical origin is inestimable. Here are a few of the soil properties controlled by organic matter:

- the cation exchange capacity of the soil,
- the amount of nitrogen in the soil,
- the amount of sulphur in the soil,
- the amount of phosphorus in the soil,
- the soil's aggregate stability, and
- the buffering capacity of the soil.

In tropical African soils dominated by Oxisols, Ultisols and Alfisols, the organic matter content ranges from a high of 10% (for soils derived from volcanic ash and basic amphibolites) to a low of less than 0.1% (very sandy soils of the Sahel). Intensive cultivation, deforestation, bush burning and the removal of crop residue after harvest leave the soil poorer in organic matter. The rate of decline depends on soil texture, being faster in sandy soils. Reduced levels of organic matter result in poor stability of micro-aggregates making the soils more susceptible to erosion, surface sealing and reduced moisture-holding capacity. Because they have lost most of their bases as a result of leaching, most of the Oxisols and Ultisols are acid in character. In addition to acidity, these soils, which are poor in soil organic matter, have low buffering capacities. A low organic matter content reduces the soil's ability to complex the iron and aluminium oxides and hydroxides that affect the availability of such nutrients as phosphorus.

Low nutrient holding capacities: The Alfisols, Oxisols and Ultisols are characterized by the abundance of low-activity clays (primarily of the 1:1 lattice variety). These clays have variable charges and therefore changes in the pH of the soil solution result in drastic changes in the capacity of the soils to hold on to the exchangeable cations. Leaching of soluble nutrients below the root zone is thus a major problem. The combined effect of the leaching of nutrients, losses of nutrients through soil erosion and nutrient removal by crops leave the soils with a negative nutrient balance. This situation is exacerbated by continuous cropping without the addition of nutrient sources in the form of organic or inorganic nutrient sources. Over the last 30 years, estimates are that in about 200 million hectares of cultivated land in tropical Africa, an average loss of about 600 kg of nitrogen (N), 75 kg of phosphorus (P) and 450 kg of potassium (K) per hectare (Buresh et al., 1997) has occurred through these processes. It is further estimated that the current NPK depletion levels in tropical Africa have resulted in an average decrease of 7% in potential Agricultural Gross Domestic Product, whereby the loss varies between countries (e.g., Cameroon with 3% and Niger with 22%) (SWNMP, 1999).

Aluminium toxicity: In Africa south of the Sahara and north of the Limpopo, approximately 550 million hectares suffer from acidity AND the presence of free oxides of aluminium (Sanchez et al., 1991). These soils are known for their capacity to immobilize soluble phosphorus (P fixation). Immobilization of soluble phosphorus results in P deficiency symptoms such as poor development of root systems, stunting and late maturing of crops. In the drier parts of tropical Africa where rainfall reliability is uncertain, P deficiency resulting from aluminium toxicity can result in total crop failure.

Soil acidity: 19% or 481 million hectares of soils in tropical Africa suffer from acidity even when they do not have toxic levels of oxides of aluminium and iron. These soils characteristically are short of bases and only specially adapted crops can grow. Leguminous crops such as groundnut and cowpea (which are excellent sources of protein) cannot perform well in these soils as the acid conditions prevent the legumes from fixing atmospheric nitrogen because of poor nodule formation.

Poor infiltration and moisture retention: Shallow soils and soils full of gravel make up more than 27% (645 million hectares) of the soils of tropical Africa. These conditions result in poor infiltration and moisture retention in soils. In such soils, moisture stress at critical periods in the life of the crops is common. It has been noted that poor crop yields are often related more to poor soil moisture management than to absolute insufficiency of rainfall (World Bank, 1996). Low contents of organic matter aggravate the situation. But it is a vicious cycle since lack of moisture reduces biomass production that, in turn, lowers the amount of organic matter. In the absence of adequate moisture supply and organic matter level, biological activity is impaired.

Where Do We Go from Here?

The characteristics enumerated above present a formidable array of problems that seem to suggest that only subsistence agriculture can thrive in tropical Africa. As already noted, the population of Africa will top the one billion mark by the year 2025. Unless adequate measures are taken to improve the productive capacity of the soils, projections are that tropical Africa will be able to adequately feed just 40% of the projected 1 billion people (UNU/World Bank, 1999). If continued reliance on subsistence agriculture is inappropriate, what must be done?

Since the major constraints appear to relate to the quality of the soils, effective land management systems and technologies are crucial to sustaining economic production of crops from tropical African uplands. This translates, as we have already noted, to the maintenance of favorable levels of organic matter and soil physical properties, preventing soil erosion and replenishing plant nutrients leached out of the root zone or removed by crops (Lal, 1987). The intriguing thing is that, to a large extent, the technologies to achieve these goals are available for different ecological regions

Revisiting Tried and Tested Territories

If tropical Africa is to feed its population some 25 years from now, it must meet the challenge of building the fertility of the soils to levels never before attained. A tried and tested option for replenishing plant nutrients leached beyond the root zone or removed by crops is the application of inorganic fertilizers. Unfortunately, there are several well-meaning people outside Africa who neither comprehend the magnitude of the problem nor understand Africa's history and geology who would cry foul at the mention of inorganic fertilizers. Incidentally, while for the past 30 years 200 million hectares of cultivated land in tropical Africa have lost an average of 1,125 kg NPK per hectare, over the same period, commercial farms in North America increased their nutrient capital by 3,700 kg NPK per hectare (Pieri, 1989). The use of inorganic fertilizers does not imply the abandonment of organic materials for reasons already noted. However, the use of organic materials will not enable us to achieve our goal of building up the soil's nutrient capital to levels never before attained. This is simply because there will not be enough organic materials and, besides, the organic matter has been produced in soils that are already deficient in nutrients. As was aptly noted by Greenland (1992), *"Recycling existing poverty levels of nutrients can only condemn most of Africa to continuing poverty"*. Having said this, however, it is important to emphasize that the proper management of farm by-products such as crop residue is a vital aspect of maintaining soil fertility. A substantial body of evidence has accumulated to demonstrate that inorganic fertilizers alone are not sufficient to sustain high yields in the dominant soils of tropical Africa (Mokwunye, 1994). In a region where manure from livestock is scarce, crop residue is an important source of organic materials for soil fertility maintenance. Data from an ongoing trial (Bationo and Mokwunye, 1991a) in sandy soil in Niger where the performance of millet crop residue (CR) alone was compared with the performance of mineral fertilizers alone or a combination of mineral fertilizers and crop residue (CRF) showed:

- Millet yields from CR plots were initially inferior to those from plots where mineral fertilizers were applied. This difference lasted only until the fourth year as the yields from the CR plots gradually increased over time.
- The combination of millet residue with mineral fertilizers (CRF) always produced the highest yields. These yields were more or less additive.

One major obstacle to the exclusive use of crop residue for soil fertility improvement lies in its multiple values for the household (Baanante and Thompson, 1992). However, scientists (Bationo, personal communication) have shown that under farmers' conditions, the use of inorganic fertilizers increases both grain yields and stover yields. The increased stover production enables the family to meet its requirements for fuel, roofing, fencing etc. while leaving substantial quantities in the fields to act as mulch during the intense dry season and as manure during the subsequent rainy season. This is definitely a win-win situation.

A Bold New Initiative

For the past fifty years, soil scientists in tropical Africa have developed technologies to promote the effective and efficient use of inorganic fertilizers. African farmers today know the value of NPK fertilizers. However, at less than 11 kg/ha, fertilizer use in tropical Africa is the lowest of any region in the world. The major obstacle is the high cost of fertilizers for both governments (who used to subsidize them) and farmers (who have to pay the full cost today). For example, in the landlocked country of Zambia, approximately 230,000 metric tons of fertilizers are imported annually (K. Munyinda, personal communication). The import bill for fertilizers is second only to the import bill for petroleum products. The challenge lies in substituting appropriate indigenous sources of nutrients for the imported sources.

The most logical approach to attaining the goal of building the fertility of tropical African soils to levels never before attained is to increase the phosphorus capital in the soils. Why phosphorus? About 80% of the soils of tropical Africa have inadequate amounts of this critical nutrient element. In some of the soils, the native supply of the nutrient is so low that seedlings die when the phosphorus supply in the seed is used up (Bationo et al., 1986). As important as nitrogen is to crop production in tropical Africa, Mughogho et al. (1986) concluded that profitable returns to nitrogen application in the drier regions are possible only in the presence of phosphorus. To take care of phosphorus deficiency, imported phosphate fertilizers have been used throughout tropical Africa. In the Zambian example above, 65% of fertilizer imports or 150,000 metric tons constitute P fertilizers.

There is an exact congruence between the concepts of capital stocks and service flows in economics and that of nutrient pools and fluxes in soil science (Sanchez et al., 1977). Addition of nutrients using either inorganic or organic sources increases the nutrient pools. Nutrient fluxes during the growing season are analogous to service flows. Such fluxes subtract from the nutrient capital and are thus analogous to the concept of capital depreciation. The phosphorus pools and fluxes in the soil system can be summarized as follows: Agricultural P is immediately available to the crop. It can thus be analogous to the liquid capital in the bank. The P in the organic and inorganic pool is not immediately available to the crop. However, it can be called upon when a withdrawal is made from the "liquid capital". The P in this pool is therefore analogous to the "capital stock" (Capital P). Removal of P from the Agricultural P pool results in movement of P from the capital stock to the liquid capital (depreciation of the capital stock). The third phosphorus pool in the soil system is the Inert P. This is P that is part of the parent rock or P that has been irreversibly fixed by the soil constituents. This fraction is relatively unimportant for food crops. Both Agricultural P and Capital P can be enriched in the soil system. The addition of soluble fertilizers, for example, increases the Agricultural P as a first step. With time however, the soil clay fraction reacts with the soluble P and transforms it into the relatively unavailable Capital P. What is most interesting to the farmer is that both the liquid capital and capital stock can be increased (built-up). P sources such as commercial superphosphates are routinely used to improve Agricultural P. But of even greater significance for the African farmer is that some indigenous phosphate rocks can perform a similar function. Better still, the phosphate rocks can also be used to build up Capital P.

Tropical Africa has abundant supplies of phosphate rock. These range from the relatively highly reactive sedimentary deposits in the Tilemsi Valley in Mali and Minjingu in Tanzania to the relatively "unreactive" igneous deposits in Chilembwe (Zambia) and Dorowa (Zimbabwe) (McClellan and Notholt, 1986). Literature abounds on the agronomic performance of Africa's phosphate rocks (Bationo et al., 1986). Most of these materials have been found to be suitable for direct application to annual crops while others are only efficient for annual crops when partially acidulated, applied in combination with soluble sources or mixed with organic inputs (Bationo and Mokwunye, 1991b). One quality all sources have is their ability to continue to supply P to crops beyond the initial year of application. This "residual effectiveness" of P fertilizers in general and of phosphate rocks in particular is a significant property which enables phosphate rock to be used for the long-term improvement of the productive capacity of Africa's soils. This approach calls for the use of indigenous phosphate rock to increase the P capital in the soil systems. The capacity of phosphate rock to increase the capital stock of P is independent of the quality of the phosphate rock.

Because the P in the phosphate rock can be released to the crop over a long period,[1] the application of phosphate rock can be considered an investment in natural resource capital. The strategy may involve the application of a one-time dose of P or annual applications of smaller amounts of P using local phosphate rock. Whatever approach is adopted, the principle is to ensure that the service flows are maximized for a given crop or cropping system.

Will tropical Africa be able to feed its more than one billion people by 2025? We have taken note of the adverse soil fertility conditions in tropical Africa. We are convinced that subsistence agriculture is not the answer. We have also noted that there are technologies that can replenish nutrients lost to the soil/crop system. More importantly, we have suggested that local phosphate rocks can be used as an investment in natural resources capital to improve the long-term productivity of the soils. It should however be noted that nutrient capital build-up based on the use of local phosphate rocks is not a cure-all. The investment is worthless unless the following accompanying technologies are also adopted.

Unless the phosphate rock is highly reactive, its use whether in small or large doses will not immediately increase the level of Agricultural P. Therefore, a farmer who wants immediate returns from the investment will be highly disappointed. To avoid this disappointment, farmers should be encouraged to continue to use their traditional amounts of commercial fertilizers. Here is the punch line. The application of the phosphate rock increases Capital P. This reduces the outflows from the Agricultural P pool to the Capital P pool thereby improving plant uptake of P in the commercial fertilizer. In other words, P added by the farmer in the form of inorganic fertilizers remains longer as Agricultural P and can therefore be more efficiently used by the crop. Currently, African farmers apply much less than the required amount of NPK for optimum crop growth. A combination of the less than adequate dose of inorganic fertilizers that the farmer currently uses and the large dose of phosphate rock ensures that the service flows are maximized for the given crop or cropping system.

Runoff and erosion are the main mechanisms of loss of phosphorus from the crop/soil system. If the investment on P capital build-up is to be meaningful, soil erosion control technologies must be in place. It should be borne in mind that application of large doses of phosphate rock without ensuring that appropriate soil erosion control measures are in place does more harm than good as erosion promotes the pollution of rivers, lakes and groundwater. Fortunately, there are well-proven biological methods of erosion control such as growing leguminous hedges or vegetative filter strips along the contour (Kiepe and Rao, 1994; Garrity, 1996). These biological methods provide useful products for the household and are thus attractive to adopt.

Adoption of sound agronomic practices is essential to the success of the investment program. Such practices involve the use of appropriate techniques of land clearing, the adoption of tillage practices that take into consideration the

---

[1] Residual effect of Tennessee phosphate rock applied to a soil in Illinois was observed more then 100 years after application—M.P. Cescas, Ph.D Thesis, University of Illinois, Urbana-Champaign, 1969.

fragile nature of many of the soils of tropical Africa, the use of crop mixtures or rotations that include leguminous species and the use of improved crop varieties that can take advantage of the improved soil fertility.

Sharing Responsibilities

Agriculture is a way of life for the vast majority of the people of tropical Africa. Sustainable agricultural development will depend on strong partnerships between governments, research and extension workers and farmers. Governments must put in place appropriate reform measures to ensure open, competitive markets for both farm inputs and farm products. The technologies that ensure adequate replenishment of plant nutrients already discussed are not being used by the farmers, because they have not participated either in their development or in their testing. In tropical Africa, a poorly defined land tenure system provides little incentive for farmers to invest their meager resources on soil improvement. It is widely accepted that subsidies on fertilizers constitute unnecessary drain on national financial resources. The use of local phosphate rock to "re-capitalize" the fertility of the soils of tropical Africa is one initiative that must be financed by national governments. The task of improving the long-term productivity of the soils of tropical Africa through investment in natural resource capital cannot be left to those members of our society who are least able to afford it—the limited-resource farmers. African governments must develop national plans that would:

- affirm the commitment of national governments to tackle the issue of poor soil fertility,
- articulate the views and concerns of principal stakeholders,
- identify gaps in knowledge and information and outline processes to fill them,
- create incentives for continued generation and adoption of soil fertility management technologies, and
- spell out concrete strategies to mobilize internal and external resources to execute such soil improvement schemes as the use of phosphate rock to improve soil nutrient capital.

References

Baanante, C.A. and T.P. Thompson (1992): Shifting Cultivation, Population Growth and the Economic and Environmental Benefits of Fertilizer Use. IFDC Special Publications. Muscle Shoals Al. 35660.

Bationo, A., S.K. Mughogho and U. Mokwunye (1986): Agronomic Evaluation of Phosphate Fertilizers in Tropical Africa. In: Mokwunye, A. Uzo and Paul L.G. Vlek (eds): Management of Nitrogen and Phosphorus Fertilizers in sub-Saharan Africa. Martinus Nijhoff, Publishers, pp. 283-318.

Bationo, A. and A.U. Mokwunye (1991a): Role of Manures and Crop Residue in Alleviating Constraints to Crop Production: With Special Reference to the Sahelian and Sudanian Zones of West Africa. In: Mokwunye, A.U. (ed): Alleviating Soil Fertility Constraints to Increased Crop Production in West Africa. Kluwer Academic Press, Publishers, pp. 217-225.

Bationo, A. and A.U. Mokwunye (1991b): Alleviating Soil Fertility Constraints to Increased Food Production in West Africa: The Experience in the Sahel. In: Mokwunye, A.U. (ed): Alleviating Soil Fertility Constraints to Crop Production in West Africa. Kluwer Academic Publishers, Netherlands, pp. 195-215.

Buresh, R.J., P.C. Smithson and D.T. Hellums (1997): Building Soil Phosphorus Capital in Africa. In: Buresh, R.J., P.A. Sanchez and F. Calhoun (eds): Replenishing Soil Fertility in Africa. Soil Science Society of America Special Publications 51. Madison, Wis. 111-149.

Eswaran, H., R. Almaraz, P. Reich and P. Zdruli (1997): Soil Quality and Soil Productivity in Africa. Journal of Sustainable Agriculture 10 (4), pp. 75-94.

Garrity, D.P. (1996): Tree-Soil-Crop Intetractions on Slopes. In: Ong, C.K. and P.A. Hexley (eds): Tree Crop Interactions. A Physiological Approach. CAB International. Wallingford, England, pp. 299-318.

Greenland, D.J. (1992): African Soils-Fertility and Degradation. An Overview. CAB International Wallingford, England.

Kiepe, P. and M.R. Rao (1994): Management of Agroforestry for the Conservation and Utilization of Land and Water Resources. Outlook Agric. 23 (1), pp. 17-25.

Lal, R. (1987): Managing the Soils of sub-Saharan Africa. Science 236, pp. 1069-1076.

McClellan, G.H. and A.J.G. Notholt (1986): Phosphate Deposits of Tropical sub-Saharan Africa. In: Mokwunye, A.Uzo and Paul L.G. Vlek (eds): Management of Nitrogen and Phosphorus Fertilizers in sub-Saharan Africa. Marinus Nijhoff, Publishers, Netherlands, pp. 173-223.

Mokwunye, U. (1994): Greening sub-Saharan Africa: Meeting the Challenges of Desertification. Transactions of 15th World Congress of Soil Science. 9 (Supplement), pp. 331-343.

Mughogho, S.K., A. Bationo, B. Christianson and P.L.G. Vlek (1986): Management of Nitrogen Fertilizers for Tropical African Soils. In: Mokwunye, A.Uzo and Paul L.G. Vlek (eds): Management of Nitrogen and Phosphorus Fertilizers in sub-Saharan Africa. Martinus Nijhoff, Publishers, Netherlands, pp. 117-172.

Pieri, C.J.M.G. (1989): Fertility of Soils. A Future for Farming in the West African Savanna. Springer, Berlin.

Sanchez, P.A., K.D. Shepherd, M.J. Soule et al. (1997): Soil Fertility Replenishment in Africa: An Investment in Natural Resource Capital. In: Buresh, R.J., P.A. Sanchez and F. Calhoun (eds): Replenishing Soil Fertility in Africa. Soil Science Society of America. Special Publication 51, Madison, Wis., pp. 1-46.

Sanchez, P.A., M. Swift, S.W. Buol et al. (1991): Soil Research in Africa. A long-term Strategy. Report to the Rockefeller Foundation.

SWNMP (The Soil, Water and Nutrient Management Program) (1999): Progress Report to Donors. CGIAR Donor Meeting, China May 25, 1999.

United Nations University/World Bank (1999): World Food Day October 16, 1999. Press Release.

World Bank (1996): Natural Resource Degradation in sub-Saharan Africa: Restoration of Soil Fertility. Africa Region. World Bank, Washington, D.C.

Agenda 21 §14.44 Earth Summit, Rio 1992

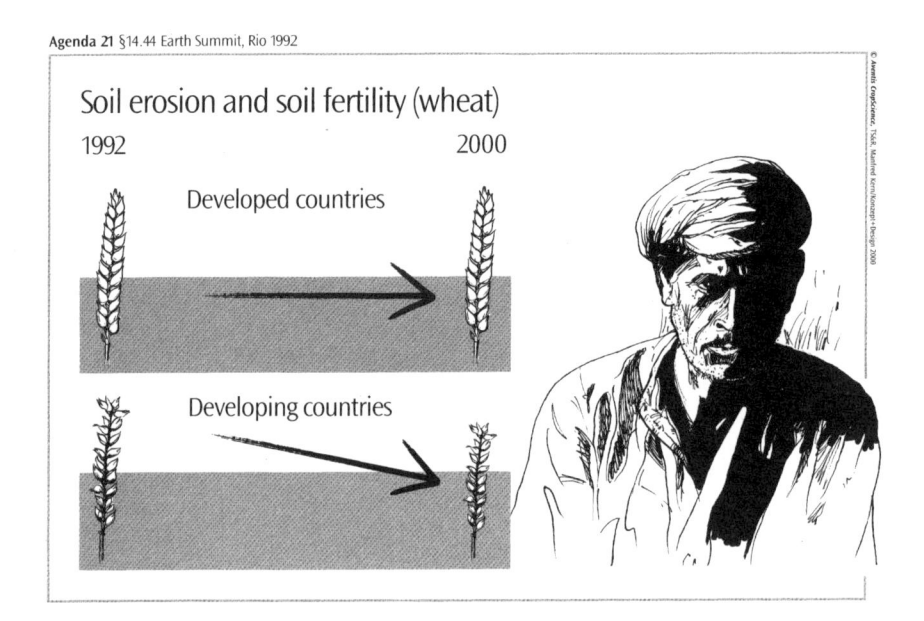

Soil erosion and soil fertility (wheat)

1992   2000

Developed countries

Developing countries

28 Natural Resource Use in Agriculture in Sub-Saharan Africa: Myths and Realities

Henk Breman

Efforts to stimulate agricultural and rural development in sub-Saharan Africa have been rather ineffective. Among the reasons are misconceptions of decision makers at all levels when considering farmers' capacities and strategies as well as overly optimistic estimations of the natural resources that sustain agricultural production. Seven widely believed myths are confronted with a more complex reality below:

1. "*African farmers are rather irrational and backward in exploiting their natural resources*": Both socioeconomic and agro-climatic conditions force farmers to overexploit their resources. Their understanding of these resources is such that in spite of the low quality, relatively high yields are reached. Sahelian pastoralists, for example, produce 2 to 10 times more animal protein per sq. km than animal production systems in semi-arid Australia and the United States (Penning de Vries and Djitèye, 1982; Van Keulen and Breman, 1991).

2. "*African farmers mine the soil of their fields of plant nutrients, and crop yields are decreasing*": Nutrient balances of production systems as a whole are indeed negative (Baanante and Henao, 1999). Nevertheless, average crop yields of staple food increase steadily at a rate of about 10 kg/ha/year, despite growing population density. Farmers have adapted their strategies to higher population densities; nutrients from ranges and wastelands are progressively concentrated on arable land because of the integration of crop and livestock and by direct transport of organic matter to the villages. Degrading rangeland and decreased animal production are measurable signals of overexploitation of the natural resources (Breman et al., in press).

3. "*Water is the most limiting factor; irrigation is a crucial tool in increasing crop production*": Even in the Sahel, nutrients are more limiting than water, both for arable farming and animal husbandry. Plant production in the Sahel can increase 3 - 5 times at average rainfall conditions by improving the soil fertility. Cereal yields of 2 - 6 t/ha instead of the actual 300 – 1,000 kg/ha are possible under improved socioeconomic conditions. Soil fertility improvement is a cheaper and more general solution than irrigation (Penning de Vries and Djitèye, 1982; Mokwunye et al., 1996; Breman and Sissoko, 1998).

4. "*Animal manure and other forms of organic matter manure are suitable alternatives for inorganic fertilizers*": Both the availability and the quality of different forms of organic matter are insufficient to permit African farmers to neglect inorganic fertilizer. An integrated – and largely complementary – use of organic and inorganic fertilizer aiming at soil quality improvement and increased efficiency of inorganic fertilizer has much better chances. (Mokwunye et al., 1996; Breman and Sissoko, 1998; Breman et al., in press).

5. *"Thanks to biological nitrogen fixation, the use of legumes can be an alternative for expensive nitrogen fertilizer"*: For generations, African farmers have used leguminous species as an element of their production system. The phosphorus deficiency of soils is one of the reasons that only a small fraction of the nitrogen requirements is realized through this option. But even when the use of phosphorus fertilizer increases, the human needs for carbohydrates prevent the fraction of land occupied by legumes from being high enough to cover the total nitrogen needs. Also, soil quality improvement is a prerequisite to making the use of the required nitrogen and phosphate fertilizers economically feasible, whereas legumes are far from the most suitable source of soil improvement (Penning de Vries and Djitèye, 1982; Breman and Van Reuler, in press).
6. *"Structural adjustment, privatization and liberalization will automatically lead to improved accessibility of external agricultural inputs"*: Since the structural adjustment programs in the eighties, the consumption of inorganic fertilizer has decreased in Africa as a whole. However, in about half of the African countries the reverse is true. But the increased demand is strongly linked to high-value cash crops. Experiences inside and outside of Africa have shown the importance of "enabling" policies to facilitate the transfer of the agricultural from the public to the private sector, i.e. policies that create an enabling environment for private investments fostering input and output market development. Moreover, given the very poor resource base of African agriculture in large parts of the continent, there seems to be an urgent need for well-targeted short- and medium-term public investments in soil quality improvement (Breman, 1998; IFDC, 1999).
7. *"African farmers should accept more responsibility for their own behavior, adopting agricultural practices that allow sustainable agriculture"*: It is up to governments to provide access for farmers to technical options for soil fertility improvement and agricultural intensification. Indeed, enabling policies that aim to improve socioeconomic conditions should be based on farmers' realities. They should therefore be accepted as the main stakeholders and be involved in the whole decision making process leading to those conditions (Anon., 1997; Sissoko, 1998).

References

Anonymous (1997): Framework for National Soil Fertility Improvement Action Plans. Prepared by IFDC in Collaboration with the SFI of the World Bank with Funding from the Netherlands Environmental Trust Fund. World Bank, Washington / IFDC, Muscle Shoals.

Breman, H. (1998): Soil Fertility Improvement in Africa, a Tool for or a By-Product of Sustainable Production? Special Issue on Soil Fertility/African Fertilizer Market, vol 11, No. 5, pp. 2-10.

Breman, H. and H. Van Reuler (2000): in press. Legumes, when and where an Option? (No panacea for poor African soils and expensive fertilizers). Paper presented at International Symposium on Balanced Nutrient Management Systems for the Moist

Savannah and Humid Forest Zones of Africa, Cotonou, Republic of Benin October 9 – 12, 2000. CABI, London.

Breman, H. and K. Sissoko (eds) (1998): L'intensification agricole au Sahel. Karthala, Paris.

Breman, H., S.K. Debrah, A. Maatman and H. Van Reuler (2000): in press. Factors affecting the Practicability of Soil Improvement through the Synergy of Integrated Use of Soil Amendments and Mineral Fertiliser. Adapted Farming in West Africa: 15 Years of research Results Offering New Perspectives for the Future, July 2000. University of Hohenheim, Hohenheim.

Henao, J. and C. Baanante (1999): Estimation Rates of Nutrient Depletion in Soils of Agricultural Lands of Africa. IFDC, Muscle Shoals.

IFDC (2000): A Strategic Framework for African Agricultural Input Supply System Development (CD-Rom, English and French version included). IFDC, Muscle Shoals.

Keulen, H. and H. Breman (1990): Agricultural Development in the West African Sahelian Region: A Cure against Land Hunger? Agric. Ecosyst. and Environment 32, pp. 177 - 197.

Mokwunye et al. (1996): Restoring and Maintaining the Productivity of West African Soils: Key to Sustainable development. Miscellaneous Fertilizers Studies No. 14, IFDC, Muscle shoals.

Penning de Vries, F.W.T. and M.A. Djitèye (eds) (1982): La productivité des pâturages sahéliens. Une étude des sols, des végétations et de l'exploitation de cette ressource naturelle. Agric. Res. Rep. 918, PUDOC, Wageningen.

Agenda 21 §18.47 Earth Summit, Rio 1992

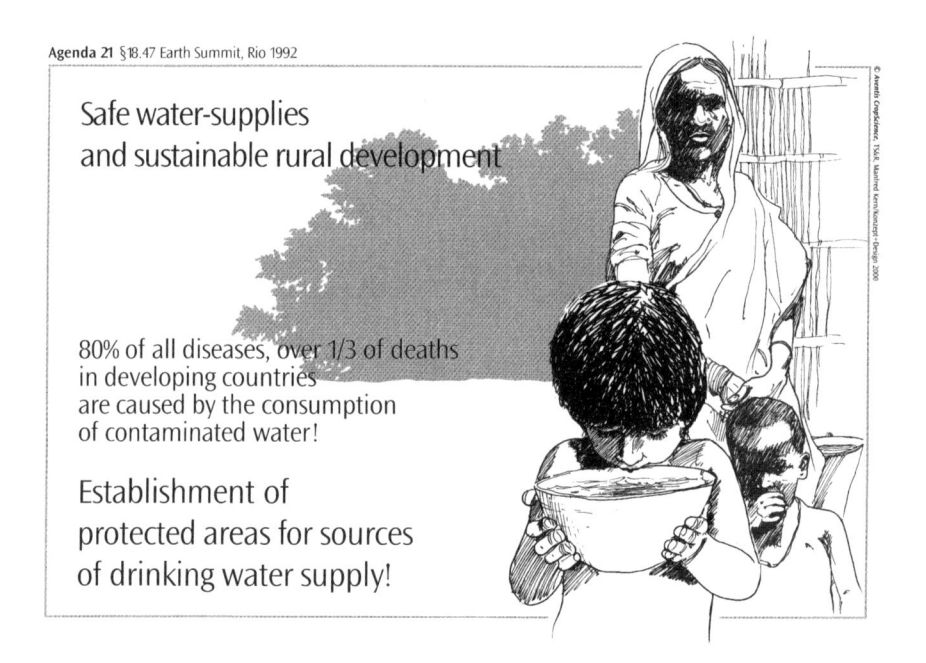

Safe water-supplies and sustainable rural development

80% of all diseases, over 1/3 of deaths
in developing countries
are caused by the consumption
of contaminated water!

Establishment of protected areas for sources of drinking water supply!

Enough Water for All

Introductory Statement

Winfried Polte

Our "blue planet" seems to have enough water for everyone. More than 70% of the surface of the earth consists of water. But almost all of it, namely 98%, is salt or brackish water. As a consequence, less than 1% of the existing water on earth can be used directly for drinking purposes.

Due to increasing consumption and utilization on the one hand and high population growth on the other, drinking water is becoming more and more scarce in many countries of the world. The debate over the issue of water has become a prominent international topic during the last decade. Let me explain in what way the water problem is global by referring to three important challenges for the international community in this context:

1. International cooperation on transboundary water resources: Rivers, lakes and groundwater resources do not observe national borders. That means that in most cases at least two countries are affected by a deterioration in water quality. This has very often been a source of conflict even for countries without water scarcity. For water-stressed countries, it is mostly the distribution of water rights that has led to high political tensions between neighboring countries. But at the same time the highly interdependent nature of transboundary water resources can also have a substantial integrative potential that can lead to a stable situation due to mutual dependencies. It is therefore a political, economic and ecological necessity for the international community to find regulations and agreements between different countries that assure the sustainable utilization of water resources for the mutual benefit of all countries involved.
2. International trade with water and water technology: The existence of water is a precondition for every form of life. Because of this life-sustaining value, in many countries water is regarded as a human right and not as a tradable commodity. But at the same time water is a scarce economic resource; an adequate price for this good could avoid wastage and excessive use. Only with an adequate price for water could the economic mechanism of supply and demand help to allocate the existing limited resources in an efficient way. One possibility offered by international trade is to import highly water-consuming crops and to reduce the amount of water used for agricultural irrigation at home. Another form is the direct import of water through pipelines. Under the precondition that the tariff system is structured so that the poor are able to afford the price of water, economic cooperation and trade between countries in

the area of water supply can help to reduce the burden for highly water-stressed countries.
3. Improvement of the living conditions for the poor: Almost 1.5 billion people living in developing countries have no access to clean drinking water. This is the case especially on the village level in the countryside, causing a heavy burden for women and children. As a result, water-related diseases are still one of the most important health problems in the world. In many water-stressed countries, policy failure and misallocation of resources worsen the inadequate access to drinking water, especially for the poor people. Therefore, access to clean water and hygienic wastewater treatment are important elements of a strategy to improve the fulfilment of basic needs and have also become a very important sector for international financial and technical cooperation between industrialized and developing countries.

Besides these international issues, which play a vital role at the global level, it is essential for the development of rural areas to concentrate on institutional issues of water conservation and management on the regional and local levels.

Merrey's paper focuses on the process of creating institutional arrangements for managing river basins, and it contrasts government-dominated approaches to form river basin management institutions with attempts to create stakeholder-based institutions. Two cases of the latter, in Mexico and South Africa, are compared to extract lessons for other countries.

Concentrating on agriculture, *Aeppli's* plea for irrigation shows that this sector has been neglected by international cooperation over the past two decades. Not only the importance of irrigation is emphasized, but also some important principles to be considered when planning and implementing irrigation projects are elaborated.

Zwarteveen questions the accessibility, democracy and participation by all users in user-based irrigation institutions. The gender issue is highlighted as one of the most important factors for exclusion, thus bringing into question the assumptions about the functioning of the "public sphere" and democracy on which the current model of participatory management is based.

Finally, according to *Dzikus*, there has been a paradigm shift in the context of water management from a pure supply focus to water conservation and demand management. This inspires hope that rural areas will be able to tackle their specific water problems. The philosophy and instruments of water conservation and demand management are described.

29 Creating Institutional Arrangements for Managing Water-Scarce River Basins

Douglas J. Merrey[1]

The World Water Forum held in The Hague in March 2000 highlighted the growing global concern about future water supplies and the complexity of the issues that must be faced if developing countries are to meet future demands for water. As its contribution to this forum, the International Water Management Institute (IWMI) developed and applied an interactive policy planning software to generate scenarios about future water supply and demand and its likely impact on food production. Some of the key findings include:

1. Nearly one-third of the population of developing countries in 2025, some 2.7 billion people, will live in regions of such severe water scarcity that they will have to reduce the amount of water used in irrigation to meet other water demands. This includes one-third of the populations of India and China which will live in regions facing absolute water scarcity. Nearly every country in sub-Saharan Africa will find the investment requirements to develop sufficient water supplies to meet their needs are far beyond their financial capacities.
2. The world's primary water supply will need to increase by 22% to meet the needs of all sectors in 2025. Only 17% more water will be needed for the world to feed itself if significant improvements can be made in irrigation productivity. However, at current levels of irrigation effectiveness, a 34% increase in water for agriculture will be required.
3. The people affected by growing water scarcity will continue to be the poor, especially the rural poor; and among poor people, women and children will suffer most from "water deprivation" (van Koppen, 2000). If the world fails to invest in finding and implementing solutions, the health, livelihoods and incomes of millions of poor people will deteriorate further. If large areas of India and China were forced by water scarcity to reduce irrigation, this would have enormous consequences for the livelihoods and food security of poor people in these regions.

The IWMI's researchers believe that despite these findings, the global community can meet future food requirements; make sufficient water available for domestic and industrial uses; increase food security and rural incomes in countries where poor people depend on agriculture for their livelihoods; and achieve higher water quality and environmental objectives. But this will require substantially increased productivity of water resources and the development of new water supplies.

---

[1] This paper reports "work in progress;" it is by no means complete and has not reached firm conclusions as yet. This research has been largely supported by a grant to IWMI from the German Federal Ministry for Economic Cooperation and Development (BMZ) through a contract administered by GTZ.

Two prerequisites for achieving these objectives are: 1) substantial, well directed financial investments, and 2) radical changes in policies, institutions and management systems in the water resources and agricultural sectors (IWMI, 2000; Vermillion and Merrey, 1998).

Creating River Basin Management Institutions

During the past twenty-five years or so, much attention has been directed to creating or strengthening local-level water users associations in both the irrigation and domestic water sub-sectors and to transferring financial and management responsibilities for water services to these associations. In the irrigation sector the record is mixed: some countries have done very well in transferring management of irrigation schemes to farmer-based organizations, but others have not. Even where "irrigation management transfer" (IMT) policies are judged to be "successful", IWMI's work shows that it is rare to find dramatic changes in productivity (Samad and Vermillion, 1999; Vermillion et al., 1999; Kloezen et al., 1997). IMT remains official policy but still unfinished business in most developing countries.

Nevertheless, with the rising awareness of the need to manage scarce water resources in an integrated manner, many countries have turned their attention to river basin management. Policy makers, researchers, and water managers now recognize that sustainable management of water resources must be done at the level of river basins. We cannot continue to work within specific sub-sectors like irrigation or domestic water supply; we must manage the allocations and interactions among these various uses. This has led many countries to embark on a new round of water reforms, focused on national level policies, and on creating new institutions for managing river basins.

Reforms at this level are even more complex and problematic than at the local level. The few successful cases to date are in rich countries with strong institutional capacities and well-educated publics. For countries where implementing even local level reforms strains the financial and implementation capacities, trying to implement river basin institutional reforms is difficult indeed. The political economy of such reforms is daunting, with strong vested interests at higher levels and weak local-level institutions affecting the capacity of the rural poor and small-scale farmers to have a voice.

Some countries in which IWMI is doing research, for example Nepal, are at early stages of thinking about what kind of river basin institutions might be desirable. Others, like Australia, have had long-term experience in reforming and building institutions to manage water-scarce river basins. The two basic patterns that have emerged are: 1) continuation, even strengthening, of government-dominated river basin management; and 2) promotion of broad stakeholder-based management institutions.

The first pattern, government-dominated river basin management, involves either one government agency taking a dominant role in managing the water resources and major infrastructure on a river basin or, in a few cases, a council that brings together the major government institutions involved in water management but excludes non-government entities. Such a council may often be dominated by a single powerful ministry. Examples of the pattern in which one government agency dominates are Turkey and the Mahaweli Authority of Sri Lanka. An example of government-dominated councils is found in Morocco. In these countries, strong government agencies continue to dominate water management. These agencies have either performed well enough to minimize pressures for reform, or they have been successful in resisting efforts at restructuring usually rising from strong political support for the status quo.

The second pattern, promotion of stakeholder participation in river basin management, is found to varying degrees in a number of countries. In Sri Lanka, for instance, a new water law is proposed which would facilitate the process of stakeholder mobilization for improved management of river basins. Sri Lanka seeks to avoid creating large new bureaucracies on its remaining relatively small rivers but recognizes the need to manage each basin more effectively.

Creating Stakeholder-Based River Basin Institutions: Two Cases

Mexico and South Africa are of interest for several reasons. Both river basins being studied by IWMI and our national partners are roughly similar in area (about 55,000 km^2); both are increasingly stressed in terms of demand versus available supply; both are characterized by competition among domestic, industrial and agricultural uses of water; both face serious environmental issues; both cross provincial boundaries (and in the case of the Olifants in South Africa, an international boundary); both contain both poor and relatively prosperous groups of people; and both are in the early stages of interesting, serious and fundamental institutional reform at the local and river basin levels.

Mexico: Lerma-Chapala River Basin[2]

The Lerma-Chapala River Basin in central Mexico contains about 900,000 ha of irrigated land and accounts for 79% of all water presently used in the basin. Irrigation is generally well-managed and productive but achieves low overall economic returns. About 35% of Mexico's GNP is produced in the basin, where about 15 million people live. From a water perspective, the basin is in crisis because water demand exceeds supply in all but the wettest years. About 109% of

---

[2] This section draws on forthcoming papers produced by IWMI researchers in Mexico; see especially Wester, Melville and Osorio 2000. Rap et al. (1999) provide an interesting analysis of the sources of the Mexican water reforms, while Mestre R (1997) provides a more "official" overview of the Mexican approach to river basin management reform with special reference to the Lerma-Chapala Basin.

the available water is developed and used, which shows the degree of over-commitment. In the terminology used by IWMI, this basin is "closed", i.e., there are no utilizable outflows and water depletion (use of water that makes it unavailable for further use) equals or exceeds the available supply.[3] As a consequence, the groundwater is being mined, and Lake Chapala, into which the river flows, is gradually drying up. This lake is the largest in Mexico, giving it a high symbolic value. It also generates significant tourism revenues.

Mexico is well-known for its important water sector reforms initiated in the early 1990s. Briefly, these included the creation of a new national water agency in 1989 (the National Water Commission, CNA), the promulgation of a new water law in 1992, the transfer of government irrigation districts to user organizations, and cautious introduction of water markets. The Lerma-Chapala River Basin Council is the first such council established in Mexico (1993), building on an informal Consultative Council initiated in 1989. More recently, separate "aquifer management councils" have begun to be established (from 1998). These reforms at the basin level are a direct response to the deterioration of the basin's water resources.

Mexico has proceeded very quickly in establishing new water management institutions at both irrigation district and river basin levels. In Asia, countries implementing IMT policies have done so gradually with some investment in animating and training local water users' associations. However, Mexico has followed what is sometimes called a "big bang" approach. Research carried out by IWMI (Kloezen et al., 1997) and others shows that the new irrigation associations have been effective in improving the provision of services and recovering costs from users, though the impact on agricultural performance is minimal. More recent work in one district in the Lerma-Chapala Basin raises questions about the long-term sustainability of these associations (Kloezen, 1999).

The Lerma-Chapala River Basin Council was also established quickly and began work almost immediately. The council includes representatives of the five states in the basin, various federal and state agencies, and representatives of various water use sectors. It quickly formed a Technical Working Group to analyze water data and prepare proposals for water allocation. Wester, Melville and Osorio (2000) argue that the council has been successful in agreeing upon and implementing a water allocation plan involving reductions of water supply to irrigation. However, the plan, while strictly enforced, has not stopped the continued reduction of the level of Lake Chapala. The council is therefore considering new measures.

A notable failure so far is in finding ways to reduce the mining of the groundwater. But a recent innovation is the promotion of "Technical Committees for Groundwater" by the federal water agency (CNA) and "Technical Water Councils" by the state of Guanajuato's Water Commission (Wester, Marañon-Pimentel and Scott, 1999). The former are intended as forums in which aquifer users, government agencies and civil society will interact under CNA's auspices, but they are not intended to have any legal status or authority. The Guanajuato

---

[3] See Keller, Keller and Seckler (1996) for a discussion of "open" and "closed" river basins.

Councils on the other hand are intended to be fully empowered user organizations for reaching agreement on aquifer management. It is too early to evaluate their impact, but the separation of the basin and aquifer councils seems to be a major weakness.

Mestre (1997) emphasizes that the river basin council is intended to be "*an open and plural forum*". The role of "Society" is seen as paramount and "*comprises non-governmental organizations, private sector organisms and individuals, academic and scientific actors, as well as a myriad of other social groups who participate in a regional water scenario*" (1997:142). He notes that society is "*commonly organized through diverse groups.*" This assumption that society is already organized and ready to participate in the new councils is important, explaining why Mexico has not felt it necessary to invest in social mobilization for either irrigation management transfer or for the establishment of river basin councils.

While this rapid approach has the advantage that the Lerma-Chapala Council has been able to get down to business quickly, two of the weaknesses identified by Kloezen (1999) as threatening the existence of irrigation associations may apply to the council as well:

- accountability mechanisms to users and information flows between users and their representatives are likely to be weak;
- dependence on the same leaders over time invites the possibility of rent-seeking, favoritism, and nepotism.

An additional potential weakness is the assumption that all important interests are organized and able to articulate their views effectively. There is no attempt to consider, for example, whether or not there are significant numbers of rural poor who are voiceless and facing "water deprivation" either in terms of basic human needs or in terms of access to water for productive uses. IWMI has been examining the extent and importance of less formalized "farmer-managed" irrigation in the basin, but the results are not yet available. It is quite possible that this small-scale sector is far more important than officially recognized and is not adequately represented.

The Mexican case can be seen as being characterized by a combination of continued government dominance and attempts to include and empower already-organized stakeholders in the river basin decision making process. South Africa is placing greater emphasis on the social mobilization stage, leading to a slower implementation process.

South Africa: Olifants River Basin[4]

South Africa is the only sub-Saharan country that is projected in IWMI's scenarios to face "absolute water scarcity" by the year 2000; countries so characterized will be forced to make difficult water reallocation decisions in the near future (IWMI, 2000). It is also important to note that about 60% of the country's water resources are shared with its neighbors.

The Olifants River Basin is the same size in area as the Lerma-Chapala, but its mean annual runoff at about two million cubic meters per year is a fraction of the 9.7 million cubic meters in the Lerma-Chapala. The river traverses two provinces and a major national park before crossing the border into Mozambique. The basin contains about 3.4 million people, and commercial and small-scale irrigation uses about 100,000 ha. Irrigation is the largest single user of water (48%). In the upper reaches thermal power plants generate almost 55% of the country's power, using coal from over 50 mines. Some water is imported into the basin to satisfy the power plants' requirements, but this is not a significant percentage of the total available water. Very small amounts are also exported from the basin to the cities. Pollution, largely from the mines, is a serious problem. All together, there are over 200 active mines in the basin for gold, platinum, tin, etc.; these are expected to expand significantly over the next decade.

Over half of the Olifants flow enters the river below its mid-section, making the middle area, where much of the irrigation is located, particularly short of water. About 65% of the total available water in the basin is already used. Much of the remaining water is in the lower tributaries and is difficult to develop for use in South Africa, though this may be seen as an opportunity for Mozambique in the future. In some years there is no flow at all into the national park, and continued development of the upper catchment is likely to prolong these low- or no-flow periods in the future. Although the basin is not as stressed as the Lerma-Chapala, it is also a "closing" water-scarce basin under increasing pressure.

An important feature of this river basin is that large areas, particularly in the middle portion, pass through former "homelands" set up under the previous regime. These areas probably account for more than half of the population, which is desperately poor after having been forced into marginal areas and provided with few basic services and little infrastructure. Since 1994 the new democratic government has devoted enormous effort to re-structuring the constitution, legal system, and policies and institutions to overcome the legacy of the apartheid system. Its reforms in the water sector must be seen in this context.

The new water management policies were developed through a long detailed process of public consultation and commissioned studies; it culminated in the National Water Act (No. 36 of 1998) and its companion Water Services Act (No.

---

[4] This section is based on research that is currently underway and not yet systematically reported, or reported in forthcoming papers (Stimie et al., 2001; Thompson et al. 2001). Figures are drawn from a section of the draft proposal currently under preparation for forming a river basin management agency (BKS 2000). See also Blank, de Lange, and Stimie (1999).

108 of 1997). The new policy adopts integrated water resources management at the "catchment," i.e., river basin in South African usage, level. Local water services are to be provided through "water users associations", while river basin management will be provided through *"Catchment Management Agencies"* (CMAs).

The policy embodies the following principles: equity in access to water resources, benefits and services; sustainability; optimal beneficial use; redress of past racial and gender discrimination; participation by stakeholders in decision making about water resources; "representivity" to ensure consideration of all stakeholder needs, interests and values; subsidiarity, i.e., devolution of responsibility to the lowest appropriate level; integration of water resources management functions; alignment of water resources management with other related departments' functions, and transparency to foster cooperation and encourage stakeholder support for decisions (DWAF n.d.).

The Department of Water Affairs and Forestry (DWAF) is the lead agency in implementing the new policy. The National Water Act makes the government responsible for overall water resources management as a public trustee, and it provides for licensing of water uses. But it also provides for reservation of minimum flows for environmental purposes and basic human needs, allowing any person to use water for "reasonable" domestic use, gardening, stock watering and recreation. The act also includes a specific "good neighbor" provision applicable to its internationally shared rivers.

Currently, DWAF is developing *"Catchment Management Agency Proposals"* on three of the 19 designated "water management areas" (defined as a large river basin, or several adjacent smaller basins), including the Olifants River. IWMI and its local partners are carrying out research in this basin on two related topics: small-scale irrigation among poor rural people, mostly in the former homelands, and basin-level hydrology and institutional reform. There are a large number of small-scale irrigation schemes in the basin, many of which were originally built and managed by the previous government. Most of these are not performing well. The government has adopted a policy of transferring ownership and management of these schemes to the users as part of a broader rural and agricultural development policy.

DWAF uses consulting firms to lead the process of developing a *"catchment management agency proposal"*. The proposal is intended to be developed through consultation with stakeholders and in its final form should lay out the broad scope and shape of the proposed *"catchment management agency"* (CMA). After a period of public comment on the draft proposal, the final version goes to the minister for approval. To date, there are no approved CMA proposals as the process has only recently begun. The Olifants proposal is to be sent to the minister before the end of 2000. Proposals are to be accompanied by independent reviews of the processes of developing the proposals and their provisions, assessing whether they meet the requirements of the policy and the National Water Act. IWMI has accepted the responsibility of playing this role for the Olifants proposal.

An enormous effort is being devoted to developing the CMA proposal. It will include the proposed name and defined water management area of the CMA; a description of the existing water resources and their management (drawing on existing, but somewhat out-of-date and incomplete previous studies and an ongoing separate study on the environmental reserve); proposed functions and institutional structure of the CMA; the feasibility of the CMA in terms of technical, financial and administrative matters; and a description of the consultation process followed.

So far, IWMI and its research partners have focused on the last issue, the consultation process.[5] The mining and industrial sectors, the suppliers of water to larger towns, and, most important, the commercial farmers are well-organized to represent and articulate their interests. The commercial farms are large modern farms, using the latest irrigation technologies and producing citrus and other high-value products for export. The government is seeking to balance the commercial farmers' need to have a reasonable and secure water supply with its policy to redress previous inequities. All of these interests are not only well organized but speak the language and come from the same culture of the consultants and DWAF officials.

On the other hand, the millions of rural poor in the former homelands are not well organized to participate effectively in a consultation process on water. Currently, one finds both "traditional" tribal chiefs, many of whom emerged in the apartheid era as a means of social control, and elected local councils which have little financial or managerial capacity. Neither of these are effective representatives of the local communities. The government has a major investment program to supply domestic water to these areas, but its approach has emphasized rapid construction of infrastructure to take care of a huge backlog of some 12 million people with no access to safe drinking water. Therefore, there has as yet been insufficient attention paid to strengthening local domestic water entities. Similarly, the small-scale irrigation sector is still unorganized and in most cases not profitable; the government is still pilot testing approaches to assist this sector. In fact, DWAF has not yet approved even one water users association under the new legislation.

A study carried out by IWMI's partners in a major tributary basin to the Olifants found that rural communities are unaware of the provisions of the new water law and of the CMA process despite efforts to inform people and offer them opportunities for expressing their views. Small-scale farmers had not heard about the CMA, and among municipalities and mining companies some knew; some did not. The irrigation boards providing water to large commercial farmers were however participating actively in the process (Stimie et al., 2001).

IWMI's initial observations of the public consultation process have raised many important issues. In short, the effectiveness of the process in the poor rural areas is doubtful. Two reasons for this seem most important. First, the consultants do not speak the local languages and indeed do not understand the local cultures of the rural poor. They have sought to overcome this by using facilitators who speak the local languages with only partial success. The minutes of the meetings

---

[5] An important partner in this endeavor is the South African office of IUCN. I am grateful to Saliem Fakir, IUCN Country Programme Coordinator, on whose reports to IWMI I have drawn here.

demonstrate that local people raise issues of immediate concern to them such as the lack of drinking water; however, the consultants are focused on higher level issues and make no attempt to relate the solutions to local problems. This is an issue of cross-cultural communication, or lack thereof, and can be addressed as such. One fear is that the well-organized sectors may yet monopolize access to most of the water, depriving the poor rural communities in spite of the strong political commitment to redress these inequities.

Second, the consultants and some DWAF officials clearly see developing the CMA as a largely technical process and do not recognize that in reality it is a quintessentially political process. Water is a political issue, especially when it is a scarce and valuable good and when access is so skewed. There are many conflicting views – and real conflicts – among stakeholders over water issues. Yet so far these are not being addressed or even articulated by the consultants. Experience from developing the first CMA proposal on the Inkomati Basin, where disagreements of some stakeholders with the proposal have delayed its finalization, suggests that not addressing or at least identifying these conflicts may lead to similar problems in the Olifants. Again, the political power of well-organized sectors, as well as possibly of local non-representative entities in the former homeland areas, may lead to continuing inequity in access to water.

Finally, it is important to note that IWMI and others have provided feedback to DWAF officials and to the consultants themselves about these concerns.[6] This is leading to the re-thinking of the process and consideration of ways to enhance the effectiveness of social mobilization efforts at the grassroots level.

Conclusion

In this short paper it is not possible to provide a complete analysis of the complex issues that arise when countries seek to implement new policies and create new institutional arrangements for river basin management. Indeed the processes are on-going, as is IWMI's research. But several general observations emerge from this overview.

First, there are important contrasts among developing countries in how they go about developing new policies and implementation arrangements. On one extreme, one finds a top-down almost entirely bureaucratic approach, driven by government agencies as the major stakeholders. In these cases, the process is essentially driven by a combination of technical concerns and inter-agency politics. It is important to acknowledge that some of these cases, such as in Turkey and Morocco, are characterized by relatively high performance in terms of productivity of agriculture for example. Nevertheless, there is no room in such approaches for less well-organized, "informal" interests, especially poor people, to participate and gain access to water. In countries characterized by large groups of

---

[6] IWMI's researchers have had similar close working relationships with the Lerma-Chapala Basin Council.

voiceless poor people, such an approach is unlikely to lead to overcoming water deprivation as a central element of poverty.

Second, the Mexico case exhibits a combination of a top-down largely government agency-driven process with inclusion of representatives of the organized users. An important result in Mexico is that the council has been able to begin addressing the serious water issues quickly; and including representatives of organized users lends the council legitimacy. This approach is entirely appropriate in conditions where the major stakeholders are organized, as the Mexicans assume to be the case, or where rapid economic growth is providing opportunities for poor people to improve their lives through other means. However, it is questionable whether many developing countries are characterized by these conditions. Therefore, following such an approach, while ensuring key organized sectors are represented and enabling rapid attention to problem-solving, also presents the danger of excluding large numbers of poor water users. As water becomes more scarce, this will amplify the degree of water deprivation among poor people.

The South African approach should therefore be of special interest to developing countries considering how to design new policies and institutional arrangements for river basin management. A clear disadvantage is the time it takes before the basin institution is able to address water resources management problems. In South Africa, there are independent processes underway to respond to demands for water from new mines, for example; decisions will either be postponed at potentially considerable cost in terms of economic development and job creation or will be made by DWAF with little involvement of stakeholders. On the other hand, successful empowerment of poor rural stakeholders could enable them to gain access to significant water rights, which are likely to be valuable assets. The water can be used directly for productive uses as well as for bargaining with mines and other commercial users needing additional water.

Carrying out public consultations in a manner that empowers local communities is extremely difficult when the population is not well-educated, not well-connected to urban centers and to mass media, and not well-organized about water. The time and effort required should not be underestimated. Nevertheless, if done well, there will be a greater likelihood that water deprivation problems characterizing poor communities will be addressed effectively and equitably.

References

BKS (2000): Development of a Proposal for the Olifants River CMA: Technical Situation Assessment for the Olifants River Catchment. Draft Chapter of forthcoming CMA proposal. BKS (Pvt) Ltd, Pretoria.

Blank, H., G. Marna de Lange and C. Stimie (1999): Olifants River Basin Water Accounting. Draft Paper. IWMI, Colombo.

DWAF (Department of Water Affairs) n.d. (1999): Implementation of Catchment Management in South Africa: The National Policy. Draft. DWAF, Pretoria. Website: www.dwaf.pwv.gov.za.

International Water Management Institute (IWMI) (2000): Research Contributions to the World Water Vision. IWMI, Colombo.

Keller, A., J. Keller and D. Seckler (1996): Integrated Water Resource Systems: Theory and Policy Implications. Research Report 3. IWMI, Colombo.

Kloezen, W.H. (1999): The Viability of Institutional Arrangements for Irrigation after Management Transfer in the Alto Río Irrigation District, Mexico. Draft Report. IWMI, Colombo. [Available in Spanish in IWMI's Latin America Series.]

Kloezen, W.H., C. Garcés-Restrepo and S. H. Johnson III (1997): Impact Assssment of Irrigation Management Transfer in the Alto Rio Lerma Irrigation District, Mexico. Research Report No. 22. IWMI, Colombo.

Mestre, R.J. Eduardo (1997): Integrated Approach to River Basin Management: Lerma-Chapala Case Study – Attributions and Experiences in Water Management in Mexico. Water International 22, pp. 140-152.

Rap, E., P. Wester and L. Nereida-Pérez-Prado (1999): The Articulation of Irrigation Reforms and the Reconstitution of the Hydraulic Bureaucracy in Mexico. Paper presented at the International Researchers' Conference The Long Road to Commitment: A Socio-Political Perspective on the Process of Irrigation Reform. December 11-14, 1999, Hydrabad, Andhra Pradesh, India.

Samad, M. and D. Vermillion (1999): An Assessment of the Impact of Participatory Irrigation Management in Sri Lanka. International Journal of Water Resources Development 15 (1/2), pp. 219-240 [Special Double Issue: Research from the International Water Management Institute].

Stimie, C., E. Richters, H. Thompson and S. Perret, M. Matete, K. Abdullah, L. Kau and E. Mulibana (2001): Hydro-Institutional Mapping in the Steelpoort River Basin, South Africa. Working Paper 17. IWMI, Colombo and ARC-ILI, Pretoria.

Thompson, H., C. Stimie, E. Richters and S. Perret (2001): Policies, Legislation and Organizations Related to Water in South Africa, with Special Reference to the Olifants River Basin. IWMI, Colombo and ARC-ILI, Pretoria.

van Koppen, B. (2000): From Bucket to Basin: Managing River Basins to Alleviate Water Deprivation. IWMI, Colombo.

Vermillion, D.L. and D.J. Merrey (1998): What the 21st Century Will Demand of Water Management Institutions. Journal of Applied Irrigation Science 33 (2), pp. 165-187.

Vermillion, Douglas L., M. Samad, Suprodjo Pusposutardjo, Sigit S. Arif and Saiful Rochdyanto (1999): An Assessment of the Small-Scale Irrigation Management Turnover Program in Indonesia. Research Report 38. IWMI, Colombo.

Wester, P., B. Marañon-Pimientel and C.A. Scott (1999): Institutional Responses to Groundwater Depletion: The Aquifer Management Councils in the State of Guanajuato, Mexico. Paper presented at the International Symposium on Integrated Water Management in Agriculture, 16-18 June 1999, Gómez Palacio, Mexico. IWMI, Colombo.

Wester, P., R. Melville and S. Ramos Osorio (2000): Institutional Arrangements for Water Management in the Lerma-Chapala Basin, Mexico. Book Chapter forthcoming in a Kluwer Academic Publishers Monograph: The Lerma-Chapala Watershed, Evaluation and Management. Draft. IWMI, Colombo.

Agenda 21 §18.65 Earth Summit, Rio 1992

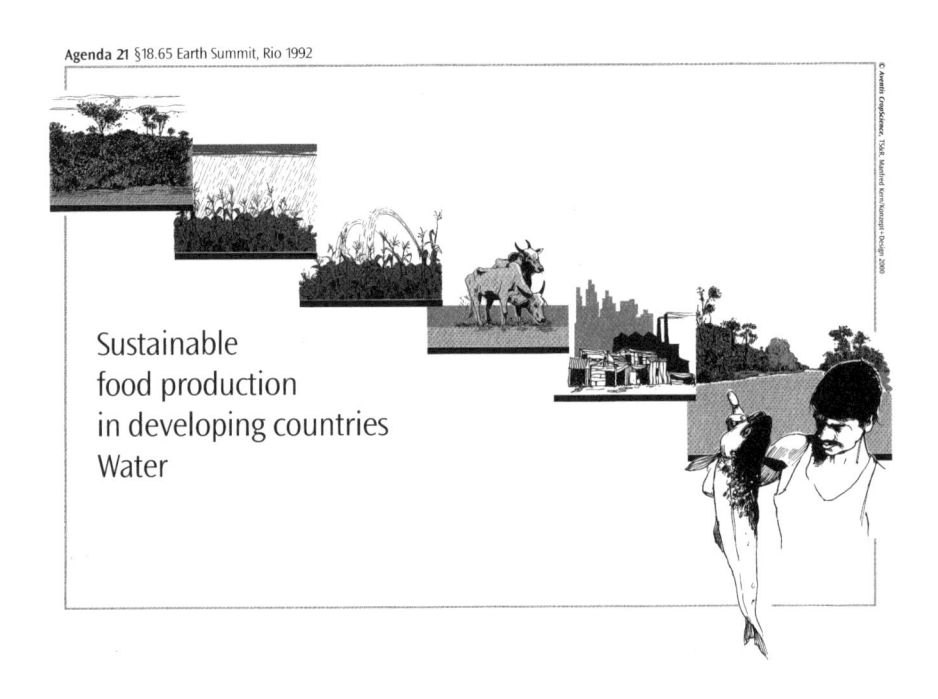

Sustainable
food production
in developing countries
Water

30 More Water for Village-Based Irrigation

Hans P. Aeppli[1]

This paper is not the result of systematic research work. I have consciously written it as a plea for irrigation, because I am convinced that this sector has been unduly neglected by international cooperation over the past two decades. It should therefore be seen as a personal statement, summarizing my years of professional experience in different functions (research, consulting, financing) but always concentrating on rural and irrigation development in Third World countries. My professional career has taught me how difficult and risky it is to simply make rules out of experiences, as all too often these rules cannot be applied to new experiences and perceptions. I will nevertheless attempt to formulate some general conclusions at the end of this paper. All of the elements of a situation must be considered, however, before applying these conclusions to a particular case.

The term village-based irrigation as it is used in this paper is largely synonymous with user-driven systems, only with a somewhat more narrow focus. It refers to irrigation schemes that integrate one or several villages, which in turn build their internal organization largely on their social structures.

The Importance of Irrigation

Without irrigation, our planet could not support its population. Food prices would increase and economic growth would be hampered. Millions of people would starve. In view of the current trends in population growth, anarchic urbanization and exceedingly widespread famine, it is clear that more food must be made available for the growing Third World populations. Today we are well aware of the challenges and difficulties that lie ahead in the attempt to achieve such a goal. The interactions between agriculture, poverty and hunger are extremely complex. Still, Third World countries are hardly in a position to overcome poverty unless they substantially increase local agricultural production. Until now, increases in agricultural production were mainly due to an increase in the areas cultivated with crops. This led to a marked increase in deforestation and/or the cultivation of marginal lands with rather disastrous effects on the environment. Higher yields are, therefore, the only way to increase production without increasing pressure on forests or marginal areas. Experience has clearly shown that it is difficult to

---

[1] Irrigation development is teamwork. I could never have written this paper without integrating the ideas of numerous colleagues and countless farmers. I gratefully acknowledge their contribution, although it is impossible to list them all here. A special thanks, however, is due to my old friend Dietmar Sprenger in KfW and to the many younger colleagues of the renowned consulting firm GFA in Hamburg, all of whom have largely contributed to my knowledge about irrigation.

intensify sustainable agriculture in subtropical and tropical climates without at least some supplementary irrigation.

There is yet another important argument in favor of irrigation that should be emphasized here. In areas with market accessibility, irrigation has proved to be a highly effective instrument in the effort to increase farm incomes substantially and, hence, to strengthen the economy of rural areas. We all know how difficult it is to create jobs and boost incomes in the rural areas of developing countries. It is therefore all the more surprizing that the public is largely unaware of the direct and indirect benefits of irrigation.

Dramatic Water Availability Scenarios: Are they Justified?

The depletion of freshwater resources worldwide over the past few years is cause for growing concern. A tremendous and continued increase in the demand for freshwater over the last century; dramatic floods and extended drought periods; unfavorable changes in river flow patterns; depletion of aquifers and the ever-increasing pollution of freshwater reserves are all clear indications that water can no longer be considered an unlimited and free commodity. Consequently, existing habits and attitudes with regard to the use of water must change fundamentally.

It should be noted that although these general statements are undoubtedly correct, the dramatic scenarios of future trends that are discussed in public and the conclusions drawn do not always reflect reality. First of all, there is a strong tendency, particularly of opinion leaders, to oversimplify these proclamations and to apply them indiscriminately without taking enormous regional differences into account. Secondly, these scenarios leave ample room for misinterpretation as very often simplistic projections are made that take into consideration currently available resources while disregarding existing storage potential.

Crops and Water Consumption

It is understandable that the growing concerns about the depletion of water resources lead to scepticism about the plea to increase irrigation, which is by far the greatest consumer of water. Some 70% of global freshwater withdrawals are directed towards agriculture.

This is an impressive figure, especially since water resources are so scarce. However, this figure should also be considered together with the water requirements of various crops. In the case of cereals, the plant consumes roughly between 1,000 and 5,000 liters of water to produce one kilogram of grain, depending on the climate and yields. This kilogram is roughly the minimum equivalent (when considering losses due to processing and other factors) covering the daily food requirements of one person, so the crop water consumption figure mentioned above may be directly compared to the 20 to 100 liters consumed per day per person. It is a fact that much more water is required to produce enough food for humankind than to satisfy its direct needs for domestic water.

Once crop water requirements can no longer be satisfied by rainfall in a given region, a very high percentage of the exploitable freshwater resources must be dedicated to agriculture[2] – always with the goal of achieving self-sufficiency in food production on a local basis. This may not be very sensible from an economic point of view, as it is probably cheaper to transport one kilogram of grain than several tons of water. However, this argument notwithstanding, each kilogram of food imported must be grown in other regions, adding additional pressure on their resources.[3]

Regional Focus

Doomsday scenarios about available freshwater resources do not reflect reality in most Latin American, in many African and still in quite a number of Asian countries. In this context, I am not referring to the central tropical regions, where rainfall is high and evenly distributed. Their tropical forests should be preserved by all means, and the development of their rather modest agricultural potential should be limited to the needs of local populations. However, adjacent to these tropical core regions are vast areas with basically high agricultural potential. These areas are still under-exploited due to seasonal droughts. In these regions, irrigation would ensure spectacular increases in agricultural production, contrary to the pessimism expressed by some. My further comments refer to these regions.

The Need for More Dams

The climates of the countries discussed in this paper are characterized by marked seasons with periods of high rainfall and extended dry spells. The uncontrolled expansion of agriculture into marginal areas and the subsequent destruction and conversion of forests into overstocked pastures have changed river water regimes. High discharge periods are becoming shorter and show higher peaks, whereas low water periods are getting longer. Even large rivers are often going dry. The water from many rivers is diverted for irrigation, mostly using simple systems and with no storage facilities, thus contributing substantially to seasonal water shortages in the lower parts of the catchment areas.

The creation of a greater capacity for water storage in the upper catchment areas is necessary to smooth the discharge regimes of these rivers and thus

---

[2] There is obviously not only competition between irrigation and domestic use of water; industrial use, hydropower, water for ecological needs etc. must be considered as well. The conclusions of this chapter, however, are not affected by these factors.

[3] I don't want to join the debate on the still only marginally known sustainable production potential of our oceans in this context.

increase water availability during the more critical seasons. Reforestation, erosion control and flood retention basins are important within this context. It is extremely difficult and costly, however, to achieve these measures, especially in the political and economic environment of most developing countries. Moreover, these measures can only be effective if implemented on a long-term basis. In many cases, the construction of dams is the only way to guarantee a sufficient water supply for irrigation without increasing water stress in downstream areas.

Water Distribution and Use

Simply building dams is not enough to improve water supplies for the different groups of users and to promote irrigation: the water must be brought to the right places. In the case of irrigation, there are three levels involved, each one playing its part in order to make optimum use of water resources:

- *The farm level*: Farmers worldwide have not only adapted their production systems to the prevailing ecological, economic and social environments, but they also have proven their ability to react quickly to changes in these environments and to be open to innovations. This observation is particularly valid for irrigated farming. Cropping patterns as well as farm sizes within traditional irrigation schemes reflect the local water availability to a remarkable degree. New cash crops can spread like wildfire if their cultivation results in recognizable advantages. Water-saving, on-farm techniques are accepted quite easily if the necessity of their use is well founded and reflected in an improved operational farm outcome.
- *The irrigation scheme*: A well-operated irrigation scheme must be service-oriented. Its main purpose is to deliver water to farms at times and in quantities agreed upon in scheme regulations.

Traditional schemes fulfill this objective quite well, obviously within their technical limitations. They mostly use diverted water, with no or with only limited storage capacity. Their design is well adapted to the prevailing water regimes, making good use of high water discharges. However, they are less efficient during low water periods. Water availability within the scheme varies according to location and season with tail-enders being at a disadvantage. These systems work well and satisfy the rather limited expectations of their owners, but they face difficulties when confronted with seasonally decreasing river discharges or growing populations.

Modern schemes, many incorporating dams, are implemented with the explicit goal of reducing conveyance and distribution losses. The aim here is not only to make better use of scarce water resources but also, due to economic considerations, to ensure that as much as possible of the water provided is channeled towards agricultural production. Considerable sums have been invested in primary and secondary structures in order to guarantee an efficient and even distribution of water. Unfortunately, all too often the results are discouraging. Although there are indeed modernized schemes that work to the

full satisfaction of their users, in many cases their performances do not meet expectations.
- *The catchment area*: Policy makers should give priority to the distribution of water within and eventually between river basins. It is their responsibility to decide on priorities and to establish an optimal mix of competitive utilization. In many developing countries, however, no effort has been made in this area. Water and land use are hardly regulated, leading to anarchic situations that follow the rule of *"first come, first served,"* thus placing downstream areas at a distinct disadvantage. Although there is clearly an urgent need for generally accepted and respected regulations, neither governments nor international cooperation efforts seem to have recognized the importance of sound water resources management practices.

Shortfalls of Irrigation Projects

There are many traditional and modern irrigation schemes that work to the full satisfaction of their users. Unfortunately, these positive examples are overshadowed by the underperformance of a good number of nationally and internationally financed projects. These failures have resulted in creating a suspicion in the greater public that any investment in dams and irrigation is potentially dangerous. However, closer scrutiny shows that these problems are the result of obvious and avoidable errors in planning, implementation and/or operation.

Wrong Attitudes towards Farmers

Many failed schemes have been planned without even considering the existing farming systems. Starting from hypothetical *"with and without project"* situations, cropping patterns have been assumed which on paper allow for a nice internal rate of return but have nothing to do with reality. The quality of planning procedures has improved somewhat in the meantime, but farming systems are still too rarely well enough understood to enable realistic predictions about the effects of future irrigation projects.

Many planners and even more governments did not see farmers as the heads of small enterprises but rather as uninformed individuals who needed to be told what to do. It was the scheme operator, normally a state or parastatal officer, who decided what to grow, when to sow, etc. Neither planners nor those implementing the scheme were aware of existing land and water rights. This created additional tension between all parties.

Heavy subsidies (for water, fertilizer, seed, credit, etc.) were common. They were felt necessary to overcome apparent "bottlenecks" at the farm level; the bottlenecks were not recognized as the results of incorrect analyses of traditional farming systems. When farmers then did not adhere to the unilaterally declared scheme regulations, this was not understood as an indicator of bad rules and a lack of understanding of farmers' needs. Rather, it was seen as a further argument to justify subsidies. Soon these subsidies proved themselves to be poisonous gifts, with farmers considering them rightful compensation for losses incurred because of bad advice and poor services provided by the scheme. They therefore contributed to maintaining the vicious cycle of inadequate operation of schemes and low performance of farmers, creating an atmosphere of general frustration. Consequently, quite a number of schemes did indeed end in disaster.

Policy-Related Issues

I have already mentioned that water resources management matters are poorly dealt with in many of the countries discussed in this paper, leading to anarchy in water use. Water laws are quite often outdated, and only seldom have outspoken and coherent policies to promote irrigated agriculture been formulated. Many of the failed schemes have not sufficiently taken into account the difficulties and uncertainties related to the legal deficiencies.

Central and local governments do not always welcome participatory approaches in irrigation schemes. They find it difficult to accept the idea of an emancipated farmer, or taking it one step further, they prefer to keep the farmer under close supervision. However, even when they agree on the principles of farmer participation, the government agency's administrative procedures and general inefficiency often make rather incompetent partners for the implementation of irrigation schemes.

Last but not least, donors have had their share of failures, too. Methodological approaches to the planning and implementation of irrigation projects have been largely developed by the donor community. Reluctance to admit errors and bureaucratic internal procedures do not allow for adequate adjustments when things go wrong. Most donor agencies are either state or parastatal bodies, so they prefer to collaborate with state organizations rather than to rely on the private sector. Therefore, inefficient or even corrupt local administrations remain the partners of international cooperation, especially since "political correctness" does not encourage outspoken criticism.

How to Achieve Better Irrigation Schemes

There is no general formula for implementing good projects. Sound professional knowledge, experience, broad vision and common sense are still the most important factors for achieving good results. The following topics should, therefore, be seen as some important principles to be duly considered when planning and implementing projects. How they should be dealt with and what solutions are to be proposed depend upon the respective case and its particular circumstances.

1. *Proposed technologies must be within the reach of farmers*: Farmers adjust to changing environments, and farming systems adapt surprisingly well to new incentives and developments. However, a farming community cannot be expected to make major changes overnight. One must therefore carefully evaluate whether or not the proposed changes in technology can be handled and absorbed by the existing social structures in a given society. For instance, going from rainfed subsistence agriculture to market-oriented irrigated agriculture is an enormous step. In sub-Saharan Africa, for example, most schemes failed when modern irrigation systems which required a high degree of organizational skills were introduced in traditional societies. On the other hand, African farmers in many areas have spontaneously begun to produce vegetables for home consumption and local markets, relying on small-scale structures like wells and small dams. Contrary to sub-Saharan Africa, in Latin America, where there is an old, rich tradition of irrigated farming, even complicated schemes can be perfectly handled by farmer communities. This holds true if the scheme matches the expectations of the farmers and if they are given the full responsibility for its operation without undue intervention from central or local governments.
2. *Village-based schemes as solutions for a weak legal and institutional environment*: The legal and institutional frameworks of developing countries are hardly in a position to solve the numerous problems that arise during the implementation and operation of irrigation schemes. Therefore, schemes must be designed so that they can be handled as much as possible by the farmers on their own. This normally causes little problem with small-scale schemes. The bigger the scheme the more important it is that all users respect the rules and regulations. In societies with a tradition in irrigation, we find a surprising willingness to adhere to such regulations and to exercise social pressure on any member not behaving according to the rules. It is of the utmost importance to build on these social structures when planning and implementing a project. In fact, village-based irrigation schemes have shown enormous potential for dealing with internal problems as well as with problems between neighboring villages. It is indeed not unrealistic to expect village-based irrigation schemes to compensate for many of the shortfalls at both the central and local government levels, at least in the Latin American context.
3. *Small is not always beautiful*: The smaller a scheme, the easier it can be handled by its users. Micro- and mini-schemes[4] are therefore quite popular among donors, all the more so as they are often implemented in poor regions. Still, their disadvantages should not be underestimated. Mainly based on diversion structures without any water storage capacities, their effect on agricultural production and income is limited, not least because their marketing perspectives are rather modest. Although investment costs are low, project overheads and technical assistance add considerably to total costs, particularly on a per hectare basis. Their economic rate of return is not spectacular and does not a priori outmatch bigger schemes. Perhaps the biggest disadvantage is an

---

[4] From under ten to a few hundred hectares.

ecological one: without reservoirs, the schemes contribute significantly to water stress in downstream areas especially when implemented in great numbers.

In countries with a tradition in irrigation, village-based small to medium-size schemes[5] in combination with the construction of dams offer a much higher potential for increasing agricultural production and rural income than micro- and mini-schemes. Total control of water not only allows production all year round but also enables the farmers to make the best use of the markets' peak demand periods. This creates additional income due to good prices. If properly designed, their reservoirs are filled during high discharge periods and allow irrigation during the dry seasons with water resources that would otherwise be lost. Additional water stress in the downstream areas can thus be minimized or even avoided.[6] The construction of reservoirs increases the potential water availability for all concerned. It also helps to reduce the occurrence of competition or conflicts concerning domestic water supply, especially since only marginal quantities of stored water are consumed by the domestic sector compared to what is consumed by irrigation.

Large-scale irrigation schemes have a particularly bad reputation, but again, this is not always justified. Unfortunately, during public discussions, the failed examples get much more attention than those that operate successfully. The potential benefit of a properly designed and managed large-scale scheme is by no means lower than that of a small- to medium-scale system. However, the planning, implementation and operation of large-scale systems require a degree of institutional proficiency that many developing countries are ill equipped to provide. Village-based user organizations also play important roles in these larger schemes, although they have obviously less influence in compensating for the legal shortfalls and institutional weaknesses of central and regional governments than might be the case in smaller schemes. Scepticism about the social and institutional feasibility of big schemes is, therefore, often well founded, but this should not result in their general or even apodictic rejection without carefully analyzing the case in question.

4. *Principles of farmer participation*: Farmers should be involved in the planning, implementation and operation of irrigation schemes in order to avoid possible low performance or even failures. While farmer participation is not an end in itself, it is an important and often imperative tool to efficiently convert irrigation water resources into increased agricultural production and farm income. A good irrigation scheme is always a compromise between the different and sometimes conflicting interests of all parties involved. This compromise is all the more difficult to achieve as one must not only consider the individual capabilities and the political, legal, institutional, and social environments; one also has to respect binding constraints due to the biophysical

---

[5] From some hundred up to several thousand hectares.

[6] I cannot discuss pollution aspects within this paper. Intensification of agriculture, irrigated or not, is not free of pollution. Its effects must be weighed against improved food supply, higher rural incomes and, particularly in the countries discussed, against a reduction of pressure on marginal lands and forests.

environment, technology, finances and markets. Farmers' interests play an important role within this set-up, although clearly not all their desires can be fulfilled. According to my experience, there are a few rules that can help in reaching a constructive and generally accepted arrangement with farmers:

- *No rights without obligations*: Transferring the operation and maintenance of a scheme into the hands of farmers must not be undertaken without clearly stating the associated responsibilities and obligations of the beneficiaries.
- *Clear and legally binding arrangements*: Rights and particularly obligations have to be respected by everybody. In order to allow their endorsement by law, they must be regularized in the form of legally binding documents and contracts before construction starts.
- *Good information is more important than excessive participation*: The efficient flow of information from farmers to planners and vice versa is a vital element of good planning. The quality of a planning team is reflected by a balanced mixture of information-gathering campaigns, internal planning phases and professionally developed public presentations to support the mediation process. Such an approach allows farmers to express their needs and wishes, to understand the pros and cons of the proposed alternatives or solutions and to be fully aware of the consequences of their decisions process.

5. *Professional project preparation and implementation*: The complexities and difficulties of the preparation and implementation of an irrigation project require a proficient team with extensive specialists' knowledge and managerial capacities. This team must also be able to communicate with and moderate the mediation process for all the parties involved. State or parastatal organizations rarely have the capacities to cope with such tasks due to budget restrictions, political influence, bureaucratic procedures and badly conceived incentive systems which curtail the creativity, initiative and flexibility of the most capable professionals. It is, therefore, recommended to outsource such a task to an experienced planning team, especially since it is difficult for a government, as an involved party, to act as an impartial adviser and mediator.

6. *Water tariffs and farmer contributions*: Water tariffs in irrigation are a major topic for discussion, and not all the proposals made are viable solutions for village-based irrigation schemes. As a principle, it is nowadays widely accepted that operation and maintenance must be taken care of by the users themselves. Depending upon the layout and size of the system, many models are possible for organizing operation and maintenance. The same holds true for the contribution of the individual user. In most schemes we find a combination of monetary water fees and contributions in the form of work. There is no need to insist dogmatically on one form or the other, as long as the user association handles and maintains the system properly.

The additional request, that farmers contribute to the investment costs of an irrigation scheme, is fully justified as their income as well as the value of their farmland will increase substantially. Finally, a marked contribution of every farmer to investment costs is an unmistakable proof of his co-ownership. This strengthens both internal adherence to the scheme and the farmers' position against the authorities' undue interference.

The collection of these contributions through water fees, however, has some serious disadvantages: The amounts are much higher than operational and maintenance tariffs, involving risks a user association should not be burdened with. The money belonging to the state, the user association would be misused as a tax collector. Moreover, the application of graded tariffs is very complicated, progressive contributions for bigger farms are therefore hardly possible. To charge the share of investment costs to be paid by every farmer as a mortgage on his farmland[7], which has to be repaid in fixed annuities, seems to be a more adapted solution in village-based schemes. Contributions in kind or work can be deducted from the initial amount, graded contributions may easily be applied and, as the handling of all financial transactions can be contracted to a bank, the user association does not have to assume tasks that are hardly compatible with its basic functions.

7. *Markets are stronger than agricultural extension*: Donors tend to overestimate the beneficial effects of agricultural extension. Greater investments in extension are, therefore, quite often seen as a remedy for badly performing irrigation schemes. Such a view underestimates the effects of markets as well as the capabilities of farmers to adapt their farming systems to changing environments. There is no doubt that farmers must be trained to derive the greatest benefit from a new irrigation scheme. However, training and extension must be limited to key issues for the operation of the scheme at all levels involved – obviously including the individual farm – and must have clearly defined targets. As far as agricultural and marketing topics are concerned, more emphasis should be placed on the private sector. A strong user association is in a much better position to contract competent technical advisers to help its members solve specific problems than the typical government or donor-financed extension service. When evaluating the needs for extension in an irrigation scheme a basic rule must not be forgotten: agricultural extension may help to speed up adaptation processes within a farmer community, but it cannot change or create markets.

Irrigation and Water Resources Management

Water resources management issues are still a low political priority in most of the countries referred to in this paper. This situation not only adds to increasing water stress and environmental damage, it also slows down irrigation development and endangers existing schemes. Some efforts have been made to strengthen political and institutional frameworks, but tangible results are hardly visible. There is a tendency among donors to halt further investments in additional irrigation projects until these countries have formulated a coherent policy and improved their legal and institutional frameworks. However, convincing arguments show that such an

---

[7] As an alternative, the amount to be paid can also be charged as an active dept to the individual water right title, applying the same criteria as with mortgages.

attitude may not be very constructive. Top-down approaches in a field as complicated as water resources management policies run the risks of not being accepted or even provoking resistance, because the developers of these approaches lack knowledge and experience that can only be generated at the operational level. Without knowing the practical needs of water users, without reflecting professional standards of water-related investment projects and without a better understanding of water-related problems within the greater population, there is the risk that the formulation of policies and laws will remain a theoretical exercise with little practical benefit but with considerable political dangers.

Investments in village-based irrigation schemes offer an ideal opportunity to generate irrigation-related knowledge at the levels of the farmer, the professionals and the government. Deplorably, most policy makers still ignore these experiences. Major efforts should be made to encourage a better flow of information from bottom to top and to systematize the lessons learned. However, reducing or slowing down investments in village based irrigation will result in the stifling of rural development. It will also increase the difficulties and stretch the foreseeable time horizons in the formulation and installation of a coherent sector policy.

Outlook

Analyzing errors and proposing ways to avoid them in a sector as diverse and complex as irrigation development are not easy tasks. It is essential to avoid placing exaggerated emphasis on possible problems as well as disseminating undue optimism. Either would lead to the conclusion that investments in irrigation are risky and should only be embarked upon with the utmost care, if at all. Such a position, however, would discount a phenomenon that can be observed in many schemes: the consequences of even serious errors in preparation, design, implementation and organization do not have to be permanent. Improved marketing conditions, new legislation, the delegation of power to users, etc., have all recently led to dramatically better performances in numerous projects. In many of them, it was possible to correct the errors of the past. The self-healing powers of village-based irrigation schemes are often underestimated. As a planner, it is good to know that they can work for you, as long as you are ready to give them room to develop and adjust interventions accordingly. In this context, I would like to venture a plea for greater optimism and confidence when looking at the future of irrigation development in Third World countries.

Agenda 21 Principle 20, Earth Summit, Rio 1992

Principle 20
Women have
a vital role in
environmental
management and
development.
Their full
participation
is therefore
essential
to achieve
sustainable
development.

31 Access, Participatory Parity and Democracy in User Based Irrigation Organizations

Margreet Zwarteveen

The question of how to ensure "enough water for all" is largely political and centers around contentious distribution issues. What is "enough" water, and who should determine this? How do we set priorities in the many beneficial and competing uses of water, and who should make these decisions? Discussions about these issues are fraught with ideological tensions between those emphasizing optimal productive use of water (plus trickle down effects); those emphasizing the need for a more fair distribution of wealth and those emphasizing environmental concerns. These tensions are exacerbated by the fact that water is a finite good.

For irrigation, the resolution of such tensions used to be considered the domain of engineers, managers and planners. This reflected a view of irrigation management and planning as a non-political and "rational" phenomenon. In the last decades, the political nature of water distribution issues has gained greater recognition. Policy focus on the establishment of user based irrigation institutions can be seen as a reflection of this. However, as I argue in this article, the liberal view of politics (and of democracy) that seems to inform much irrigation policy making does warrant some critical reflection.

User based irrigation institutions are widely held to be an important element in ensuring the fair, efficient and effective management and distribution of water. Central to the belief in the success of user based institutions are: (1) the ideal that all users affected by operational rules are included in the group who can modify these rules; and (2) the ideal that user based organizations are democratic and representative of the interests of all members of the water users community.

This belief in and rhetoric of accessibility, democracy and participation by all users sits uneasily with the empirical evidence upon which most user based irrigation institutions, including those that are evaluated as successful, are founded. Importantly, they are constituted by a number of significant exclusions. A key parameter of exclusion, and the one on which I will focus in this article, is gender.

The feminist preoccupation with exclusion because of gender in user based water management organizations stems from concerns about social inequality. Women's lack of meaningful participation in the management of irrigation systems is seen as an indicator of gender inequality. Women's (non-) participation is also seen as an underlying reason for the fact that women benefit less from the fruits of irrigation development. The search for ways, tools and methods to more fully involve all potential beneficiaries in user based management organizations is the obvious avenue for discovering solutions to these problems.

I suggest that this problem analysis linking gender inequality in access to water with women's exclusion from user based management organizations, although convincing enough, might not go far enough. The problem analysis leaves the ideals of full user participation and democracy unaffected as it is based on the belief that, in principle, gender based exclusions can be overcome. However, the available evidence suggests that the linkages between recognition (representation and participation in users organizations), participation parity and democracy are much less straightforward than current irrigation management models (and the theories on which these are based) predict. I suggest that such evidence calls into question the very model of participatory irrigation management.

In this brief article, I shall not try to provide a full-fledged critical discussion of this model from a feminist perspective. Rather, I will focus on two central assumptions about the functioning of the "public sphere" and the democracy on which the current model of participatory management is based:

- The assumption that it is possible for participants in user based irrigation organizations to bracket status differentials and power inequalities and to deliberate "as if" they were social equals: the assumption, in other words, is that social equality is not a necessary condition for successful irrigation management organizations.
- The assumption that the existence of a multiplicity of competing "publics" or "domains" of interaction, deliberation and decision making is necessarily a step away from, rather than toward, greater democracy (and more successful water management) and that a single, comprehensive forum organizing all water users is always preferable to a nexus of multiple domains.

This discussion is heavily inspired by the chapter in Nancy Fraser's book, "*Justice Interruptus,*" entitled "*Rethinking the Public Sphere. A contribution to the Critique of Actually Existing Democracy*" (Fraser, 1997:69-98). It reflects part of a first and rather preliminary attempt to more systematically think through the wider topic of "*gender and participation in natural resource management.*"

Inclusion/Exclusion, Participatory Parity and Gender Inequality

The most easily recognized gender barriers to participation in user based organizations stem from membership rules that more or less directly exclude women. Such membership rules either stipulate that only formal rightholders to irrigated land can become members or require head-of-household status in order to be eligible. Since men tend to occupy these categories more often than women, women are not considered eligible for membership. Would a redefinition of membership rules or a removal of the "entrance barriers" for women lead to greater participation parity?

The evidence suggests that simply making it formally possible for women to participate is not enough to guarantee women's full and equal participation. To give just a few examples: in the Buttala irrigation system in Sri Lanka, although 32% of the legal cultivators (who are entitled to membership in the water users organization) are women, only 21% of the organization's members are female (Kome, 1997); in a small-scale irrigation project in Ecuador, almost as many women as men participate in the users organization. Nevertheless, observations during meetings showed that while on the average regular male members spoke about 28 minutes, female members only spoke 3.5 minutes. The women said that they were reluctant to voice their concerns at meetings, because they were afraid to make mistakes and to be ridiculed. They also thought that they lacked experience and knowledge (Krol, 1994). Female farmers in Nepal said that they felt uncomfortable with the aggressive tone in which matters were discussed at meetings, and they also referred to the gender ideology which discourages outspokenness in women (Neupane and Zwarteveen, 1994). Similar evidence emerges from participatory forest organizations (Sarin, 1995) and agricultural cooperatives (Mayoux, 1995)

Feminist research done in western countries suggests that such examples may be part of a familiar syndrome that many women and some men have experienced in mixed meetings: men tend to interrupt women more often than men; men also tend to speak more than women, taking more and longer turns; and women's interventions are more often ignored or not responded to than men's. In response to the kinds of experiences documented in this research, an important branch of feminist political theory has claimed that deliberation can serve as a mask of domination. Jane Mansbridge has argued that:

The transformation of "I" into "we" brought about through political deliberation can easily mask subtle forms of control. Even the language people use as they reason together usually favors one way of seeing things and discourage others. Subordinate groups sometimes cannot find the right voice or words to express their thoughts, and when they do, they discover they are not heard. [They] are silenced, encouraged to keep their wants inchoate, and heard to say "yes" when what they have said is "no" (Mansbridge, 1990).

Mansbridge's analysis shows that gender inequalities (just like other social inequalities) can "infect" deliberation, even in the absence of formal exclusions. It shows that the achievement of participatory parity also requires an examination of the process of discursive interaction within the domain of the water users association. It is here that the ideal model of a water users association merits critical reflection. This ideal model presupposes the absence of social inequalities, as is reflected in phrases like "*social divisions should not be so serious as to disrupt communication and decision-making between farmers*" (Vermillion and Sagardoy, 1999:72). The ideal water users association is to be an arena in which participants set aside such characteristics as differences in birth, wealth and gender and speak to one another as if they were social and economic peers. The operative phrase here is "as if." In the reality of many irrigation reform programs, the social inequalities among participants are not eliminated but only bracketed. But can they be effectively bracketed?

The evidence for gender inequality suggests they cannot. Women's freedom to publicly interact with men is constrained by social practices and norms that define what kinds of interaction are permissible, with which men, in what contexts, and the modes of conduct. Discursive interactions within the domain of water users organizations are governed by protocols and styles of decorum that are correlated and markers of gender inequality. These function informally to marginalize women and to prevent them from participating as peers. In the above cited examples of Ecuador and Nepal, to be outspoken and opinionated are positive characteristics for men, markers of masculine "distinction" in Pierre Bourdieu's sense, a way of defining and reconfirming masculinity and male superiority. This prompts the question whether or not the ethos of the ideal water user organization is constructed around a "masculine" austere style of public speech and behavior, a style deemed "rational", "virtuous", and "manly." Could it be that masculine gender constructs get built into the very conception of the public water management decision making domain? I am not preaching a return to an universalist and essentialist gender perspective here that uniformly labels certain characteristics as "male" and others as "female", but I do think that discourses and styles of deliberation are often gendered in such ways as to positively evaluate those styles of speech and behavior that are associated with men and masculinity (or with dominant groups). If this is even only partially true, then the water users association itself may be (or become) a way of defining and reconfirming existing gendered norms and practices and of distinguishing a separate male "public" domain from a female "domestic" domain. It implies that, for women, to be able to actively and fully participate in water users organizations would involve challenging prevailing gender practices, a new valorization of female identity and work and a rejection of norms and regulations which tie women to specific roles. It would involve a struggle to occupy spaces previously reserved for men.

The bracketing of gender inequalities (or proceeding *as if* women and men can speak to each other as social and economic peers) may thus work to strengthen and reinforce the norms and practices that support and legitimize women's unequal access to water. I therefore suggest that rather than *bracketing* these social inequalities, it would be more appropriate to "unbracket" them in the sense of explicitly thematizing them. The questions whether and how it is possible, even in principle, for members of water users associations to deliberate as if they were social peers in the domain of the water users association (if these associations are situated in a larger societal context that is pervaded by structural social inequalities) is a question that merits attention. It is a question that goes much beyond the identification and removal of formal entrance barriers.

What is at stake here is the autonomy of the water users association vis-à-vis the surrounding social, political and economic context. One salient feature that distinguishes the liberal political model upon which the ideal water users' organization is based from other political-theoretical orientations is that liberalism assumes the autonomy of the political sphere. Liberal political theory assumes that it is possible to organize a democratic form of political life on the basis of socioeconomic and gender based structures that generate systematic inequalities. The problem of democracy, viewed from this liberal perspective, becomes the problem of how to insulate irrigation management decision making processes from what are considered as non-political or prepolitical processes, those characteristic for instance of the economy (or the market) or the family (or the household). The problem as perceived by many irrigation professionals is how to strengthen the barriers separating water management institutions that should be based on equality from those economic, cultural, or gender based institutions premised on social inequality.

I would argue instead that in order to have a public domain in which water users and irrigators can deliberate as peers, it is necessary for systematic social and gender inequalities to be eliminated. To assume that such inequalities do not exist by bracketing them certainly does not actually eliminate them. I do not mean to say that everyone must have exactly the same access and right to water and irrigated land, but I do think that some degree of equality is an important precondition for achieving some degree of democracy. Addressing inequitable water distribution requires more fundamental political and economic changes than can be achieved through administrative and management reforms.

Equality, Diversity and Multiple Publics

How then do we ensure that water in irrigation systems is distributed equally if such systems are situated in stratified societies or in societies whose basic institutional frameworks generate important gender (and other social) inequalities? In such societies, full participatory parity and democracy in water users organizations are not possible. What institutional arrangements will then best help to achieve the goal of "enough water for all?" I suggest that pushing towards the ideal of one single, overarching water management institution may in fact not be the most suitable way to achieve distributional fairness or equity. This ideal may be more closely approximated by arrangements that allow for competition among a plurality of groups than by one single, comprehensive, water management organization. This follows from the argument in the previous section. There I doubted the possibility of insulating the water users organization from the surrounding political, social and economic context and its effects of social inequality. In societies which are highly gender stratified, the deliberative processes and practices that take place within water users organizations may even tend to operate to the advantage and reinforcement of male privileges in access to water. These effects may be exacerbated if the water users organization in actual fact becomes the only domain in which water management decisions are made. In such a case (which is the ideal model) women would have no alternative arenas for deliberation about their needs, objectives and strategies. They would not have venues in which to undertake communicative processes that were not under the supervision of more powerful male groups, and they would have to resort to ways to get water that are seen as illegal and illegitimate.

This argument lends support to the cases already cited above. Even though women do not participate fully and equally with men in water users organizations, they often do have their own ways and strategies to access water. In the Buttala irrigation system in Sri Lanka, most water problems are not solved through the water users association. Farmers, both male and female, obtain (extra) water by directly manipulating the water flows. They can do so either by widening and lowering their pipe outlet structures so that the gate opening is increased and more water flows into their fields, or they can raise the upstream water levels in the distributary canals by putting obstacles (usually banana stumps and trees) in the canal. Farmers may also block additional upstream division gates. Whether or not farmers try to solve irrigation related conflicts through the water users association depends on the quality of their personal relations with the office-bearers. If those are not good, they may either directly approach the staff of the irrigation department or the local administrators (Kome, 1997).

In the Buttala case, the water users association would be judged unsuccessful on many counts. The evidence from Nepal and Ecuador is more interesting, since these are both examples of relatively successful water management organizations. In Nepal, female farmers in the head-end of the system never attend the general assembly meetings; nor do they attend the village meetings of the water users organization. Women directly approach the irrigation committee presidents to obtain water turns at times appropriate for them. They also often simply steal water. Women strategically use identities of exclusion and vulnerability to successfully argue for priority in water supply, to justify the "stealing" of water, to avoid night irrigation and to win exemption from contributing labor or cash to system construction and maintenance (Zwarteveen and Neupane, 1996). In Ecuador, women use their social networks and contacts to obtain the right amount of water at the right time. Many women say they prefer solving water related problems outside the organization, because it saves them the hassle of writing letters and making presentations in public. At the level of the secondary canals, female farmers' access to water is theoretically not different from that of men. Gender specific problems occur mainly because the timing of water turns may make it difficult to combine irrigation tasks with domestic tasks. If a particular irrigation turn is inconvenient for a woman, she has multiple possibilities to exchange part or all of her turn with other irrigators along the same secondary canal. Maintaining good social relations with "canal neighbors" and proper knowledge about rotation schedules are important for women so that they can assure themselves of a timely and adequate supply of water (Krol, 1994).

These and other examples suggest that for women it may sometimes be more advantageous (in terms of obtaining access to water) to develop alternative ways for making decisions about water. Women make use of different social channels to voice their concerns and needs, and they often also refer to different normative and legal frameworks to justify their claims to water. Water management related decisions and interactions thus often take place in a number of coexisting and partly overlapping "domains of interaction", such as households, female networks, saving and credit groups and family networks. The domains may form parallel discursive arenas where women invent and circulate ideas which permit them to formulate alternative and sometimes oppositional interpretations of their identities, interests and needs. In summary, the above accounts suggest that the view of women as "excluded" from water management decision making may in fact be ideological. It rests on a gender-biased notion of public decision making, one that accepts at face value the claim of irrigation policy makers and researchers that water users associations constitute the public decision making realm.

The recognition of multiple domains of water management interactions suggests that limiting efforts to improve irrigation management to strengthening one public domain (the water users association) risks weakening the effectiveness of other domains; this is to the detriment of those groups who rely more on those domains for voicing their concerns and needs. The identification of fruitful avenues for public action to improve water management and distributional equity requires the explicit recognition of existing water use and distribution practices. It also requires a much less rigid view of sovereignty and "the political" in favor of more Foucauldian notions of control and power.

References

Fraser, N. (1997): Justice Interruptus. Critical Reflections on the "Postsocialist" Condition. Routledge, New York and London.

Kome, A. (1997): Gender and Irrigation Management Transfer in Sri Lanka. (Second Component). MSc Thesis. Irrigation Management Research Unit, Irrigation Department. International Irrigation Management Institute and Wageningen Agricultural University. Colombo and Wageningen.

Krol, M. (1994): Irrigatie is mannenwerk. Genderverhoudingen in een kleinschalig irrigatieprojekt in de Ecuadoriaanse Andes. MSc Thesis. Wageningen Agricultural University, Wageningen.

Mansbridge, J. (1990): Feminism and Democracy. American Prospects. Spring (1990) 1.

Mayoux, L. (1995): Beyond Naivety: Women, Gender Inequality and Participatory Development. Development and Change 26 (1995), pp. 235-258.

Neupane, N. and M. Zwarteveen: Gender and Irrigation Management in the Chhattis Mauja Irrigation System in Nepal. Research Report. International Irrigation Management Institute, Kathmandu and Colombo.

Sarin, M. (1995): Regenerating India's Forests: Reconciling Gender Equity with Joint Forest Management. IDS Bulletin 26 (1995) 1, pp. 83-91.

Vermillion, D.L. and J.A. Sagardoy (1999): Transfer of Irrigation Management Services. Guidelines. FAO Irrigation and Drainage Paper No. 58. FAO, Rome.

Zwarteveen, M. and N. Neupane (1996): Free-riders or Victims: Women's Non-participation in Irrigation Management in Nepal's Chhattis Mauja Irrigation Scheme. Research Report No. 7. International Irrigation Management Institute, Colombo.

32 Water Conservation and Demand Management in Africa

André Dzikus

As the African continent follows patterns of unprecedented urbanization, the demands for water to satisfy the needs of the industrial, commercial and domestic sectors continue to rise and outpace the capacities of governments resulting in gaps which have steadily widened over the years, threatening sustainable development and the environment of cities. Unfortunately, today there are still too many governments, multi and bilateral agencies who are clinging to an outdated paradigm of Water Supply Management as an approach to water resource development. A growing economy, a needy and increasing population, expanded access to water supplies and our increasing environmental obligations presuppose a need for more water, not less. But with finite freshwater resources available, and these decreasing due to pollution, we have to rethink our philosophy of endless growth. Unless we plan and start working at good water management, less is what we are heading for. The new paradigm is water conservation and demand management (WC&DM). A new approach to WC&DM will reduce:

- overuse of water resources,
- overcapitalization,
- resource wastage,
- pollution problems, and
- other problems of varying severity and complexity.

Water conservation can be described as *"the minimization of loss or wastage, the prevention, care and protection of water resources and the efficient and effective use of water"*. Water conservation is both an objective and a strategy for water resource management. Water demand management can be viewed as a strategy by water institutions, such as catchment management agencies to influence water demand and water usage in order to achieve greater economic efficiency, social development, social equity, environmental protection, sustainability of water supply and services, and political acceptability. The task of water resource planners and managers is to reconcile demands and supply. In the past, conventional water resources strategies, such as the creation of storage, have been utilized to meet growing water demands. It is now recognized that resource capture through conventional surface and groundwater development (supply management) is not the only option available, but the implementation of water conservation strategies can successfully achieve the same objective of reconciling demand and supply and providing sustainability. The optimal long-term solution is most often to be found in an Integrated Resource Management (IRM) approach, which combines conventional supply management strategies with water demand strategies promoting effective and efficient use. This section describes each of these management strategies and discusses the benefit of an integrated approach.

Water Demand Management Tools

Water demand management places much more emphasis on the socioeconomic characteristics of water use. For example, traditional approaches have generally limited the use of resource economics to analyses that focus on benefits and costs, cost effectiveness, and other passive types of analyses that assume water needs are requirements that must be met. In contrast, demand management is much more aggressive in its use of micro-economic principles, such as economic efficiency to influence the origin of water demands in the first place and to provide incentives for satisfying given ends in the cheapest possible manner. In other words, the various water uses are seen as demands, in the economic sense, which can be influenced and governed by incentive structures, public education and other means, utilizing principles commonly used in the social sciences.

The essence of water demand management is that of promoting the efficient and equitable use of water. By reducing the need for continuing expansion of conventional water supply systems, Water Demand Management can "buy time" by delaying large capital investments for the development of new conventional water resources.

A broad range of approaches and instruments for Water Demand Management exists, from economical to socio-cultural and technical tools. The following chapter is meant to illustrate the various different approaches to managing the demand for water. As any comprehensive Water Demand Management program needs a supporting framework, some more general management tools, such as institutional and legal reforms and private sector participation, are also included in this chapter. Although these tools do not have a direct influence on water demand, they are essential to any water management program and are therefore a part of Water Demand Management.

Market-Based and Economic Instruments

Economic techniques rely upon a range of monetary incentives (e.g. rebates, tax credits) and disincentives (e.g. real cost, penalties, fines) to convey to users an accurate and clear message about the value of water. The aim is to promote more efficient water use practices that are oriented towards increasing conservation and sustainability. The most important of the market-based instruments are:

- water pricing/tariff setting,
- abstraction charges,
- effluent charges.

Realistic *water pricing* is one of the most fundamental keys to Water Demand Management and is central to many of its options.

Prices perform two essential roles in a market system: rationing and production motivation. Rationing is necessary since scarcity precludes both the satisfaction of all needs and the unlimited production of goods and services. Goods and services must be rationed to consumers, and factors of production must be rationed to producers. The price system allows bidding for scarce goods and services and factors of production, thereby ensuring that goods and services are allocated to the highest valued users and that factors of production are allocated to the uses where they bring the largest return. Prices perform their production motivating role by indicating what consumers are willing to pay. Simply stated, prices send "signals" to both consumers and producers about the economic value of the resource use. Many empirical studies in a variety of country and local settings have analyzed and confirmed that water consumers respond to price changes according to the individual price elasticity of water consumption.

However a pricing system based solely on a flat rate regardless of the volume used or a system based on property value would not have the effects predicted. Pricing based on use requires some means of measuring usage- normally through water meters. The reasoning for this is straightforward. Once the flat-rate is paid, the price for volumes used is essentially zero. Therefore, there is no incentive to conserve on use. Nevertheless in some cases, volumetric pricing for low-income groups, small volume users, is not an economically attractive option. This is not the case, however, for most users and almost never the case for industrial and high volume domestic and commercial users (UNCHS, 1999b).

The *abstraction charge* is another economic instrument which supports the regulating activities of Water Demand Management. Germany, among others, can serve as an example for the use of the water abstraction charge, proving its possibilities. In 1988, Baden-Württemberg became the first German *Land (state)* to introduce the "water penny" (*Wasserpfennig*), and by 1995, 12 of the 16 *Länder* had followed this example. The various *Länder* use different terms (taxes, charges, fees, levies) to denote water resource taxes, and no simple classification of the instruments as to their fiscal, financial and incentive functions is possible as they are often integrated and inseparable. However, the focus now appears to be shifting towards setting incentives to save water (rather than retiring water rights), both directly through rate increases and indirectly through the decrease in the blanket reduction allowed for public water suppliers. Due to this, the water abstraction charge became one instrument to indirectly manage the demand for water (UNCHS, 1999b).

As an expression of the "polluter pays" principle, *effluent charges* can be used as an economic instrument to indirectly influence water consumption and thus manage demand. Although this instrument is mainly used in order to foster water resource protection and to mitigate the environmental impact of pollution, it also has an indirect impact on water consumption. In the early 1970s, for example, the federal Effluent Charges Act was passed in Germany. In principle, all water uses in Germany, including effluent discharges, require a permit (*Erlaubnis*) or a licence (*Bewilligung*) which are thus the primary instruments for water resource protection and management. Established water quality requirements, often reflecting the need to maintain quality for specific uses, can justify denying a permit or licence. In this context, the federal effluent charge is designed to help prevent water pollution rather than to finance the management of pollution. However, with the implementation of the Federal Effluent Charges Act, the quality and amount of effluent discharges became more important to the water consuming industry and to commerce as they were forced to treat effluents before discharge. Consequently, production and processing technologies using less water and resulting in less pollution were sought, and water recycling or re-use systems were installed in many plants. Germany, therefore, may serve as an example to prove that effluent discharges in combination with other water management measures may aid in the management of water demand even though their main focus is on water resource protection (UNCHS, 1999b).

Technical Measures and Water Efficient Practices

Besides the market-based and economic instruments, many concrete technical measures as well as various operational and water efficient practices exist, including:

- leakage detection and repair,
- meter accuracy and application program,
- retro-fitting,
- pressure management.

Leakage is often a major source of hidden water loss and is a result of either lack of maintenance or failure to renew aging systems. Leakage may also be caused by poor management of pressure in distribution systems causing high pressure zones which result in pipe or pipe-joint failure. Some leakage manifests itself above ground in areas which are noticed, but in other cases it is in an isolated location or below ground and may go unnoticed. Other leakage is only recognized after studying water use data and inferring losses due to unaccounted for increases above the normal baseline values. Early detection of visible leakage requires good reporting, including some level of public participation. However, leakage detection should always take place after checking the customer meter. Although there are often great losses in the system after the meter, they are frequently neglected as the water already has been paid for. A sustainable change in this area of leakage detection not only increases the efficiency of the whole water supply system, but it also encourages public support by decreasing the individual water bills of the consumers.

Accurate metering of water consumption is one of the major cornerstones of Water Demand Management. Before any Water Demand Management program is implemented, un-metered connections must be reduced to an absolute minimum, and metering coverage must be maximized in all sectors. Especially in developing countries, some of these un-metered connections may be illegal. These connections can occur at any point along a distribution system. For lengthy pipelines which carry water to cities from distant supplies, water may be taken along the way by rural settlements or for irrigation purposes. In city distribution systems, these connections may result from contractors connecting to the water system illegally to supply new housing developments; also, "illegal" settlements may connect to such supplies. However, to ensure that the Demand Management

Program covers all consumers, these unofficial connections need to be made and metered. The most expensive water loss in a distribution system in terms of direct cost is that associated with inaccurate meter measurement, sometimes referred to as inaccurate metering. Incorrect measurement may result from poor meter quality, poor water quality, aging materials, improper sizing, improper application, and improper installation. Therefore any comprehensive meter program should include sustainable control and monitoring of the meters (UNCHS, 1999b).

Retro-fitting provides one of the most effective short-term options for reducing water demand, particularly in the domestic and institutional sectors. Many government buildings or public institutions either do not pay for their water, or the consumers have no interest in conservation. Good examples are university campuses, ministry buildings, government hospitals, etc. With very little capital investment, usually only a few dollars per fitting, water consumption can be reduced by as much as 20%. Typical examples include:

- low-flush/double flush toilet cisterns (even reducing the flush volume of existing ones),
- spring-activated/low flow/aerated faucets, and
- slow-flow shower heads.

In order to monitor the effectiveness of retro-fitting programs, especially when designed as pilot-demonstrations, efficient metering is necessary. Block metering for an entire institutional apartment block can serve as an example. Incentives could be offered to those who participate in the program, including payments for installation which are added to the water bills in order to spread the cost or grants from local authorities (UNCHS, 1999b).

Pressure management is another tool the water supplier can employ to reduce the demand for water. It is mainly used to control leakage levels, as leakage rates vary with pressure. Leakage is more sensitive to pressure than traditional wisdom suggests. Therefore, the introduction of different pressure zones according to the various local needs is an efficient tool to control leakage rates. Pressure management has plenty of benefits. As maximum pressure and pressure cycling strongly influence burst frequency, comprehensive pressure management can reduce the danger of system failure and ensure adequate service. But a comprehensive pressure management program can also help to control customer demand directly. A reduction of pressure can, for example, reduce the water in "open tap" use. However, pressure management is not only restricted to the introduction of different pressure zones. To meet the varying needs of the customers, the use of roof and ground tanks represent another alternative. By introducing roof and ground tanks instead of a full pressure connection to the main distribution system, the water supplier can meet the different levels of demand accordingly (UNCHS, 1999b).

Public Education, Awareness and Involvement

An enormous variety of non-financial measures are available to promote Water Demand Management (Brooks and Peters, 1988). Information and consulting services can be provided; social pressure can be applied; regulations can limit the time or quantity of use; and institutional and legal reforms can establish the necessary framework to aid in the implementation of Water Demand Management Programs.

Public Awareness Campaign

Water demand-side management essentially seeks to influence the consumer to voluntarily use water more efficiently. A public awareness-raising campaign is one of the major tools of a focused demand-side management program. A campaign consists of two distinct components: knowledge and information transfer and education and awareness raising at all levels. Its major objectives are to mobilize support for ongoing and future activities and to ensure the sustainability of the program.

The print and broadcast media are primary influences on public thinking. Therefore, information materials must be fully utilized to ensure an appropriate and continuous transfer of information and knowledge as the basis for awareness raising. However, the dissemination of information requires effective "packaging" to suit the different target groups and objectives. Various means are available to disseminate the information. The use of the mass media is probably the most cost effective in cities as even the urban poor have access to these means of communication.

Apart from the dissemination of information to households, education is the second major cornerstone of an awareness raising campaign. However, the measures to raise awareness should address both the decision makers and the consumers at all levels. To ensure the sustainability of the campaign, consumer education should start already at a very young age. This could be achieved by including Water Demand Management in the school curriculum.

Water Use Restrictions

Although regulations have a bad name, they are often both appropriate and efficient for managing water demand. Exhortation is also more effective than generally believed, particularly in times of drought. The range of options is wide enough to preclude generalization, but one can say that they should be chosen to support, and if possible to reinforce, the effects of market based measures.

Water use restrictions are mainly for emergencies, such as droughts and acute water shortage or in order to protect water resources from pollution. These restrictions can contain instructions forbidding or regulating the introduction of substances or material into surface water bodies and can provide for other measures to avoid surface water contamination. To secure the groundwater, Germany, for example, has developed a zoning approach which is increasingly recognized as an effective model. On the whole, water use restrictions aim at resource protection rather than at demand side management.

Institutional Framework

In many countries responsibility for water management and supply is split between different ministries and agencies at both the national and local levels. In most countries, there is a ministry for water or at least a nodal ministry with responsibility for water. It may be a ministry of the environment or of public works, and its responsibility will be the development of a county's water policy including pollution control and development of resources. The amount of responsibility delegated to local authorities varies depending on their size, but usually a local authority will have a water and sewage department. For smaller urban centers, this responsibility may be combined with other aspects of infrastructure. Local authorities are responsible for setting tariffs, collecting revenue and providing adequate operations and maintenance. Bulk water supply and large scale development projects are usually the responsibility of the national government ministry. This hierarchical and vertical diversity and complexity of water management can cause many problems for the implementation of a Water Demand Management program including:

- competition between political figures, authorities and interest groups,
- overlapping responsibilities,
- competition for donor support, and
- competition for revenue earned from local taxes.

Therefore, a clear definition of roles and responsibilities at all levels throughout the entire program planning process is needed. Institutional reform should be part of the long-term strategy of establishing and sustaining comprehensive Water Demand Management Systems.

Legal Framework

Legislation for water management is often outdated and relies upon the continued use of unrealistic statutes. It can be restrictive to the entry of the private sector and to the application of Water Demand Management, having building codes and standards which restrict the use of water-saving technologies. In addition, pollution control legislation is neglected. There is, however, often little capacity to enforce legislation, for example to stop illegal connections and to disconnect users who do not pay.

Therefore although not necessarily applicable in the short term, in many cities it will be necessary to make adjustments in the legislation. This may be in areas of:

- privatization regulation,
- water abstraction licensing,
- variable tariff structures,
- regulation about private sources exploitation, and
- regulation of the informal sector of private vendors.

Private Sector Participation

Private sector participation arose from the inability of the public sector to provide efficient and reliable service. In many countries both domestic users and industry have had to resort to alternative informal supply systems. Private sector participation is not only helpful in providing good water supply service but also in establishing and maintaining efficient water management systems. By establishing a comprehensive management system, the private sector aids the implementation of a water demand management program. The key however is to identify the right level of private sector participation and to control and regulate this cooperation. So in essence a state agency is needed which has the responsibility for setting an appropriate regulatory framework, standards and prices, for protecting against unscrupulous private sector partners, and for developing an attractive business environment for potential investors (UNCHS, 1999b).

Development of a Water Demand Management Program

While many decision makers and local agency managers assume that one action alone will suffice for Water Demand Management, a comprehensive WDM program usually makes use of a wide range of different management tools. While not dismissing traditional structural approaches, a comprehensive Water Demand Management strategy also calls for the use of policy, laws, economics and finances, technology, etc. For an optimal combination of the different approaches and measures, a comprehensive program to establish a framework is essential. It should not only coordinate the various different approaches but also structure the main elements of the program. The following schematic representation of a program illustrates its main elements and highlights their linkages.

Although the structure is similar in any program planning, the development of every individual program is a dynamic process. It can hardly be emphasized enough that conditions vary significantly from one country to another as well as within countries. Any program plan needs to be tailored down to the local needs and circumstances. This guide offers a general framework for developing a Water Demand Management program. The application of this guide facilitates an open and flexible program process sensitive to individual circumstances and local needs. As the art of programming should be seen as a continuous learning process, the handbook provides a set of practical checklists at each stage of program development to highlight the key steps and to adjust the general program framework to the individual needs. These checklists are not exhaustive, intended rather as pointers to foster the continuous process of program development.

References

UNCHS (1999a): Managing Water for African Cities. Brochure. UNCHS, Nairobi.
UNCHS (1999b): Urban Water Demand Management: A Handbook for Developing a WDM Programme. UNCHS, Nairobi.

Agenda 21 Principle 9, Earth Summit, Rio 1992

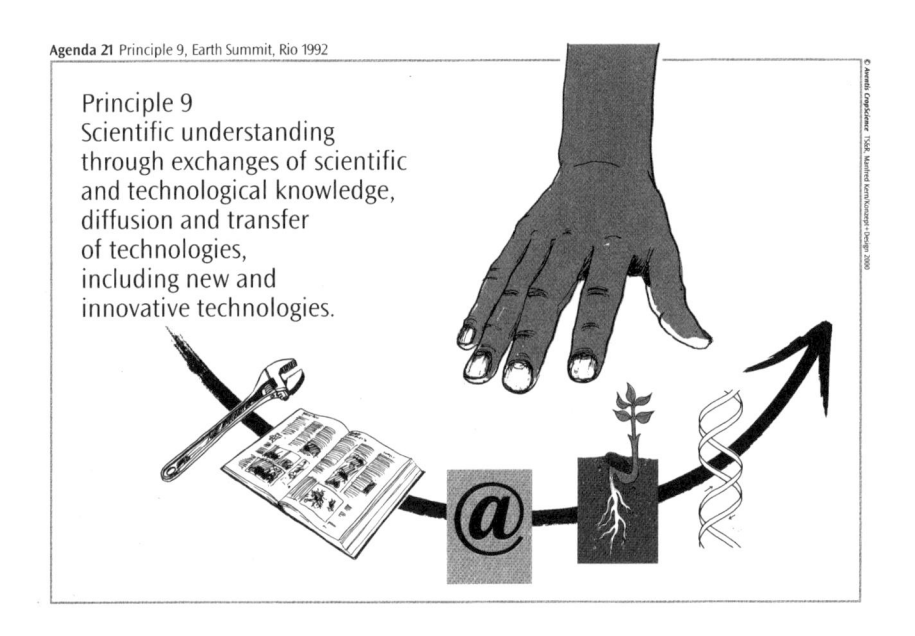

Principle 9
Scientific understanding
through exchanges of scientific
and technological knowledge,
diffusion and transfer
of technologies,
including new and
innovative technologies.

III. BROADENING THE TECHNOLOGICAL BASE

Information and Communication Technologies

Introductory Statement

Subbiah Arunachalam

The widespread availability and convergence of information and communication technologies - computers, digital networks, telecommunication, television, etc. - have led to an unprecedented capacity for dissemination of knowledge and information. The impact of this fourth information revolution is felt in education, research, medicine, government, business and entertainment. But, as *Brun and Mangstl* indicates, the benefits have reached only about 5% of the world's population. A threat presently facing a significant percentage of the global population is of not just being marginalized but completely bypassed by this revolution. The new information and communication technologies (ICTs) have in fact led to a digital divide not only between rich and poor nations but also within nations. Even in the affluent United States of America, as the Rev. Jesse Jackson has pointed out, ICTs have not only widened the digital divide but also deepened the racial ravine. The relative disadvantage suffered by inner city populations (mostly Blacks and Hispanics) is growing. Technology by itself is a great divider. It exacerbates the inequalities in society. Is there then no way for the poor and the downtrodden to benefit from ICTs?

Experts like Monkombu Swaminathan and Bruce Alberts believe that with intelligent intervention, we can make information technology an ally in the movement towards equity. In a cluster of villages near Pondicherry on the eastern coast of southern India, Swaminathan and colleagues have demonstrated how facilitating access to information and innovative use of information technology can make a difference to the lives of the rural poor, especially those whose household incomes are less than a dollar per day. The project is designed to provide knowledge on demand to meet local needs. It uses a mix of wired and wireless technologies, solar power and mainline electricity. The information needs of the community were assessed first, and databases were created in the local language (Tamil) to address those needs. These databases are being updated regularly. As the resources are limited, the provision of telephones and computers to each household is not possible. Therefore, the emphasis was placed on community rather than individual ownership. Thanks to this program, villagers now get information that they can use immediately in a wide range of fields: health care, transport, market, subsidies, employment opportunities, entitlement, crop diseases, etc. In the coastal village of Veerampattinam, fishermen are provided with wave height and other weather information based on satellite pictures of the coast downloaded from a US Navy site. The computers in the village information centers are used to train villagers, especially school children. According to Swaminathan, the villagers take to the new technology like fish to water. All we need to do is give them a chance.

Hudson shows the potential of access to modern technologies to overcome problems in rural areas. It is hoped that the technologies will stimulate the villagers' capacities for innovation (especially by computer and internet access). Telecommunications are also seen as a possible way to overcome major gaps in information access between urban and rural people. The main issue is how to distribute the facilities for granting access to modern, global systems of information in the rural areas.

The single most important idea is that different constituencies – governments, international agencies, civil societies, learned societies, academies, donor organizations, etc. - have a role in bringing the power of ICTs into play in the tasks of development and ensuring the livelihood security of the rural poor. The *Okinawa Charter on Global Information Society* is proof of that public involvement, calling for global participation in utilizing digital opportunities and bridging the digital divide as well as launching the Digital Opportunities Taskforce (dot force), which will report to the next summit on action to bridge the digital divide.

What else can we do with these technologies? As *Ramani* points out, a few years ago when the government of India allowed individuals to set up telephone kiosks, which were then a state monopoly, it led to the creation of a few hundred thousand jobs across India, many of them in rural areas. It also catalyzed a manifold increase in communication from rural areas. If only voice telephony could be allowed over the Internet Protocol, the situation would improve dramatically. According to *Bayes,* in Bangladesh the Grameen Bank has shown that cellular mobile telephones leased to women members of the village-based, microfinance organizations have a positive effect on the empowerment of those women and their households. The services originating from telephones in villages are likely to deliver more benefits to the poor than to the non-poor.

There are many other examples, especially in Latin America and Africa, funded by aid agencies. But considering the magnitude of the task, the undertakings so far have been woefully inadequate. There are some silver linings to the cloud, however. Organizations such as the World Bank, UNDP, and IDRC, to name but a few, are taking considerable interest in this movement. The Secretary General of the United Nations, Mr. Kofi Annan, has emphasized the need to bridge the digital divide in his address to the Millennium Summit. After this recommendation, the United Nations Volunteers initiated a program. The private sector also has incentives to supply the rural areas with ICTs, thereby fostering economic development and overcoming the digital divide. *Heuermann* points out, however, that the private sector's engagement is limited in high risk countries and countries with weak currencies and does not take place in areas with dispersed settlement structures. *Rwayitare* suggests an approach somewhat similar to that of Swaminathan, emphasizing the mobilizing of village communities and the providing of relevant information in sub-Saharan Africa.

33 The Role of Information in Rural Development

Heather E. Hudson

The Importance of Access to Information

Access to information is critical to development; thus information and communication technologies (ICTs) as means of sharing information are not simply a connection between people but also a link in the chain of the development process itself. Information is obviously central to activities that have come to be known as the "information sector@ including education and research, media and publishing, information equipment and software, and information-intensive services such as financial services, consulting, and trade. But information is also critical to other economic activities ranging from manufacturing to agriculture and resource extraction, for management, logistics, marketing, and other functions. Information is also important to the delivery of health care and public services.[1]

For individuals, access to information can have personal, social and economic functions, often accomplished using the same devices. An individual can summon help in an emergency via telephone; she may stay in touch with friends and family members and arrange appointments by telephone or e-mail and may find the Internet a more efficient means of tracking down consumer information on products and services than the mass media. Entrepreneurial sole proprietors, ranging from programmers and consultants to small farmers and craftspeople, can set up global store fronts on the Internet.

In general, the ability to access and share information can contribute to the development process by improving:

- efficiency, or the ratio of output to cost (for example, through use of just-in-time manufacturing and inventory systems, through use of information on weather and soil content to improve agricultural yields);
- effectiveness, or the quality of products and services (such as improving health care through telemedicine);
- reach, or the ability to contact new customers or clients (for example, craftspeople reaching global markets on the Internet; educators reaching students at work or at home)
- equity, or the distribution of development benefits throughout the society (such as to rural and remote areas, to minorities and disabled populations).[2]

---

[1] Hudson, Heather E. (1984): *When Telephones Reach the Village*. Ablex, Norwod, NJ.
[2] Hudson, Heather E. (1997): *Global Connections: International Telecommunications Infrastructure and Policy*. Wiley, New York.

However, many factors may influence whether and to what extent ICTs make an impact. Generally, certain levels of other basic infrastructure as well as organizational activities are required for the indirect benefits of telecommunications to be realized; that is, telecommunications may be seen as a *complement* in development, not a sole contributor. Telecommunications may also serve as a *catalyst* at certain stages of the development process, becoming particularly important when other incentives are introduced such as new curriculum requirements, employment opportunities, or tax incentives.

Information Gaps

In industrialized and other high-income countries, telephone service is almost universally available through fixed lines and increasingly through wireless networks. Ownership of computers is widespread, and use of the Internet is increasing dramatically. In the rural United States, there has been significant progress in access to basic telecommunications; distance no longer accounts for difference in household access to a telephone; income levels are now a better predictor. Yet the gap in access to the Internet persists. Regardless of income level, Americans living in rural areas are lagging behind in Internet access. At the lowest income levels, those in urban areas are more than twice as likely to have Internet access as rural Americans with similar incomes.[3] Those who are connected typically pay more than their urban counterparts for Internet access. Disparities in Internet access are found in the Canadian North, in the Australian Outback and in rural and disadvantaged parts of Europe.

However, this so-called "digital divide" is much more pronounced in developing countries, where access to ICTs remains much more limited. In its Statement on Universal Access to Basic Communication and Information Services, the United Nations noted:

The information and technology gap and related inequities between industrialized and developing nations are widening: a new type of poverty "information poverty" looms. Most developing countries, especially the Least Developed Countries (LDCs) are not sharing in the communications revolution, since they lack:

- *affordable access to core information resources, cutting-edge technology and to sophisticated telecommunications systems and infrastructure;*
- *the capacity to build, operate, manage, and service the technologies involved;*
- *policies that promote equitable public participation in the information society as both producers and consumers of information and knowledge; and*
- *a work force trained to develop, maintain and provide the value-added products and services required by the information economy.*[4]

---

[3] Fact Sheet: Rural Areas Magnify "Digital Divide"
(www.ntia.doc.gov/ntiahome/digitaldivide/factsheets/rural.htm).

[4] United Nations Administrative Committee on Coordination (ACC) (1998): A Statement on Universal Access to Basic Communication and Information Services,@ April 1997. Quoted in ITU: World Telecommunication Development Report, p. 10.

Table 33.1. Access Indicators[5]

Country Classification	Tel Lines/100	PCs/100	Internet Hosts/10,000	Internet Users/10,000
High Income	54.1	22.3	28.1	92.0
Upper Middle	13.4	2.9	8.4	55.9
Lower Middle	9.7	1.3	1.9	19.0
Low Income	2.5	0.2	0.1	0.9

More than 85% of the world's Internet users are in developed countries, which account for only about 22% of the world's population.[6] Of course, Internet access requires both communications links and information technologies, particularly personal computers or networked computer terminals. While there is still much less access to telecommunications in developing countries than in industrialized countries, at present, the gap in access to computers is much greater than the gap in access to telephone lines or telephones. High-income countries had 22 times as many telephone lines per 100 population as low-income countries, but 96 times as many computers (see Table 33.1).

Typically, a high percentage of developing country residents live in rural areas (as much as 80% of the population in the least developed countries), where access to communication networks is much more limited than in urban areas (see Table 33.2).

It should be noted that this table overestimates rural access because the "rest of country" includes everything except the largest city. Also, facilities are not likely to be evenly distributed throughout the country, so that in poorer nations there may be many rural settlements without any communications infrastructure.

Table 33.2. Access to Telecommunications[7]

Country Classification	Teledensity [Tel Lines/100]		
	National	Urban	Rest of Country
High Income	46.0	52.9	43.8
Upper Middle	13.7	25.7	11.5
Lower Middle	9.7	22.1	7.2
Low Income	2.5	6.5	2.3

---

[5] Derived from International Telecommunication Union (1998): World Telecommunication Development Report 1999. ITU, Geneva.

[6] It should be noted that Japan and Australia are included in the Asia/Pacific in this chart; the estimate in the text includes them with industrialized countries of Europe and North America.

[7] Derived from International Telecommunication Union (1998): World Telecommunication Development Report 1999. ITU, Geneva.

Community Access: Telecenters

The term "telecenter" has been used to refer to a variety of means of providing access to information and communication technologies, ranging from cyber cafés to facilities located in public buildings such as libraries and post offices, to stand alone public access centers. In general, a telecenter must meet two essential criteria:

- it must provide access to telecommunications services;
- it must be accessible to the public.

Telecenters originated in Scandinavia in the 1980s with the goal of helping to diversify rural economies by enabling rural residents to become information workers, by taking on projects such as word processing, data entry, telemarketing, and interviewing from their communities for urban clients. With the advent of public access to the Internet, telecenters evolved into a means for community residents to use shared facilities to send e-mail and access the Web. The idea spread to developing countries where, as noted above, access to computers and telecommunications is often very limited and costly, especially in rural areas.

In Africa, telecenters are providing a means for rural residents to gain new skills, stay in touch with family members, and access and share information for community development. A major supporter of African telecenters is the Acacia Program on Communities and the Information Society in Africa, sponsored by Canada's International Development Research Center (IDRC). Acacia is designed *"to empower sub-Saharan African communities with the ability to apply information and communication technologies [for] their own social and economic development."*[8] (The project was named Acacia because the Acacia tree is found throughout sub-Saharan Africa).

Telecenters are only one of Acacia's ICT initiatives, which include such activities as school networks, environmental applications, women's networks, telemedicine projects, and youth projects, as well as policy studies, primarily focused on steps needed to make affordable access to ICTs widespread in Africa. However, telecenters are perhaps the most visible of Acacia's initiatives, as they are designed specifically for community access. Acacia has established telecenters on its own in Uganda, in collaboration with the Universal Service Agency in South Africa, and in collaboration with the ITU and UNESCO in Mali, Mozambique, and Uganda (jointly sponsored telecentres are also planned for Benin and Tanzania).

These projects are referred to as multipurpose community telecenters (MCTs) and are equipped with a variety of facilities including pay telephones, a facsimile machine, a photocopier, and several computers equipped with basic software and connected to a telephone line through which they can access e-mail, and in most cases, the Internet's world wide web. (In some cases, the available bandwidth is too limited for web access. The centers are intended to stimulate and support local

---

[8] See www.idrc.ca/acacia.

capacity for informed decision making, particularly in the areas of health, education, economy, governance and general socioeconomic development; and the production of information to foster local development. In general, the projects are designed to develop sustainable models to meet the information and communication needs of the communities, with the assumption that these models will likely evolve during and following implementation. Women and youth are high priority target groups.

While most of the telecenters are in relatively early phases, they are very popular and have been used for a wide variety of applications. Perhaps the most isolated telecenter in Africa is in the fabled town of Timbuktu on the edge of the Sahara desert. There, a local tour guide sent e-mail from the telecenter to organize a trip on camel back in the desert for tourists. A teacher came to look for a current map of Africa online because the school books had only colonial era maps. A doctor now uses e-mail and the Internet to get advice on treating unusual cases and to get caught up on recent medical innovations. Before the telecenter was available, he said: "*Information is the fuel of medicine. Here we have none. Year by year we are falling behind.*" In Timbuktu, and in telecenters in Uganda, South Africa, and Mozambique, young people are learning computer skills; many volunteer to teach others. Women in South Africa have learned to do desktop publishing to make wedding and funeral announcements. In Nepal, telecenters provide a means of marketing pottery and crafts; in Ecuador, farmers went online to get advice on how to get rid of a pest that was ruining their potato crop

Telecenters are proliferating throughout the developing world. Some are extensions of privately owned "phone shops," others are affiliated with community information facilities such as libraries and schools, while still others are stand alone facilities. In some countries, the telecenters have also become ISPs (Internet Service Providers) to serve residents and small businesses that have been able to buy a computer and get a telephone line.

Telemedicine and Telehealth

Developing countries generally have higher morbidity and mortality rates than industrialized countries; of the 52 million deaths worldwide in 1996, 40 million were in the developing world. Of these, 12 million fatalities were children under the age of 5 years, most of which were preventable.[9] However, conditions are typically much worse in the rural areas. For example,

- a disproportionate number of rural people suffer from chronic illnesses;
- the infant mortality rate is higher than in urban areas;
- the number of deaths from injuries is often dramatically higher.

In addition:

---

[9] International Telecommunication Union (1999): Challenges to the Network: Internet for Development. ITU, Geneva.

- poverty is usually higher in rural areas than in the nation as a whole;
- lack of transportation and few local providers make it difficult to reach health care facilities.[10]

To combat these problems, there are a growing number of initiatives to use telecommunications in rural health care. Telecommunications can be used for several different functions in support of rural health care delivery:

- *emergencies*: to summon immediate medical assistance;
- *consultation*: to give advice to health workers, or directly to isolated patients;
- *training*: of health care workers;
- *education*: of target populations including expectant mothers, mothers of young children, groups susceptible to contagious diseases, etc.;
- *administration*: ordering and delivery of medications and supplies, logistics, patient medical records, billing data, etc.;
- *data collection*: collection of public health information such as epidemiological data on outbreaks of diseases.

Generically, these applications are referred to as "telemedicine", although some researchers and practitioners prefer to use that term to refer to consultative uses, and the term "telehealth" to refer to applications for medical education and administration.

Emergencies

Getting help in emergencies can range from summoning a doctor or paramedic, to getting advice about how to treat an injured person from a distant medical professional, to coordinating disaster relief efforts. Studies in India, Costa Rica, Egypt, and Papua New Guinea all showed that about 5% of rural telephone calls were for emergencies and medical reasons. Public payphones, borrowed telephones, and dedicated radio and satellite networks have been used to save lives. For example, in the South Pacific, the experimental PEACESAT satellite network has been used to summon medical teams during outbreaks of cholera and dengue fever, and to coordinate emergency assistance after typhoons and earthquakes.

Consultation

Telecommunications has been used since the early days of HF radio to support health care delivery. Australia originated the Flying Doctors, a medical service which flew doctors to the aid of ill and injured homesteaders in the Outback, and

---

[10] These relative disparities are also often found in industrialized countries. See Witherspoon, John P., Sally M. Johnstone, and Cathy J. Wasem (1993): *Rural Telehealth: Telemedicine, Distance Education, and Informatics for Rural Health Care*. Boulder, CO: WICHE Publications.

operated an HF network to coordinate their logistics and give medical advice to isolated settlers. The Australian Royal Flying Doctor Service still operates a telecommunications support network using public telephone service and a dedicated HF radio network. In East Africa, the African Medical and Relief Foundation (AMREF) also operates a flying doctor service that combines telecommunications with transportation to provide health services in rural areas of Kenya, Tanzania, and Malawi.

Many developing countries now rely on paraprofessionals for delivery of basic health services, particularly in rural areas. These health workers receive basic training in treatment and prevention of common health problems, but need supervision and assistance in diagnosing and treating uncommon diseases and coping with serious health problems. Telecommunications links between village clinics and regional hospitals or health centers can be used for consultation and supervision. Two-way radio networks are used in many developing countries to support isolated paraprofessional health workers. More reliable networks relying on terrestrial systems or satellites are also used. In Guyana, rural health workers called "medex" use a two-way radio network to communicate with headquarters in Georgetown to check on delivery of drugs and supplies and to receive advice on major health problems. In Alaska, village health aides are in daily communication via satellite with physicians at regional hospitals. More than 100 Alaskan villages are equipped with earth stations that are used for the dedicated medical network, long distance telephone service, and television reception.

Physicians in rural and developing regions may also seek real time advice from distant medical schools. Memorial University of Newfoundland has provided assistance to isolated physicians in Newfoundland and Labrador via audio and computer links. Project Rainforest, a collaborative project of Yale University and the National Aeronautics and Space Administration (NASA), conducted a unique experiment in telemedicine, connecting the jungle of eastern Ecuador with the consulting resources of Yale Medical School. Laptop computers and telephone lines linked a mobile surgery program in Cuenca, high in the Andes, with an isolated hospital ten hours away in Sucua, Ecuador and with surgeons on the Yale campus.[11]

Like the doctor in Timbuktu, medical professionals can also seek expert advice via the Internet. They may consult with other physicians via e-mail or search online libraries for guidance. For example, a physician was able to avoid amputating the leg of a young athlete infected with necrotising fasciitis (flesh-eating bacteria) after finding information on an alternative procedure in the MEDLINE online medical data base.[12] This latter function of information-sharing, now expanding in developing countries via Internet access, may be the most valuable of all medical applications. There have been examples of physicians in China as well as Africa successfully treating undiagnosed patients after seeking advice over the Internet.

---

[11] See http://yalesurgery.med.yale.edu/events/oprainforest/oprainforest.htm.
[12] International Telecommunication Union (1999): *Challenges to the Network: Internet for Development*. ITU, Geneva.

SatelLife of Cambridge, Massachusetts, operates a store-and-forward satellite system, using a low earth orbit (LEO) satellite, HealthSat-2. The satellite's unique polar orbit allows ground stations to transmit and receive data from any point on earth daily. SatelLife's Healthnet network enables medical practitioners to seek advice on treatment of unusual cases from colleagues in other parts of Africa, and to download articles from medical libraries. Burn surgeons in Mozambique, Tanzania and Uganda have used HealthNet to consult with one another on patient treatment and reconstructive surgery techniques. In response to a cholera epidemic in Zambia, the medical librarian at the University obtained literature from her "partner library" at the University of Florida, and then disseminated the information to all HealthNet users in the region.[13]

Training and Continuing Education

Telecommunications networks can also be used for training and continuing education of health workers. The Alaska satellite network is used for continuing education of village health workers.[14] The Peru Rural Communications Services Project sponsored by USAID connected health workers in the eastern jungle of San Martin to Lima via VHF radio and satellite links for consultation and continuing education.[15] In Guyana, the Georgetown training staff run refresher sessions and "grand rounds" over the HF radio network.[16]

In addition to providing information, telecommunications also helps to reduce the feeling of isolation from colleagues. It appears that reducing isolation can help to reduce personnel turnover in rural areas. For example, medex in Guyana reported that chatting over the radio in the evening helped to reduce their sense of isolation and boost morale.[17] Better communications is cited as one of several factors encouraging reversal of the medical braindrain in Navrongo, Ghana.[18] Memorial University's Telemedicine Project also states that one of the main recruitment issues for rural doctors is continuing medical education.[19]

Public Health Education

Numerous projects have used communication technology in support of health education, much of which is targeted to women as expectant mothers, mothers of

---

[13] See www.healthnet.org.
[14] Hudson, Heather E. and Edwin B. Parker (1973): "*Medical Communication in Alaska by Satellite.*" New England Journal of Medicine, December 1973.
[15] Hudson, Heather E. (1990): *Communication Satellites: Their Development and Impact*. Free Press, New York.
[16] Hudson, Heather E. (1984): *When Telephones Reach the Village*. Ablex, Norwood, NJ.
[17] Fryer, Michelle, Stanley Burns and Heather Hudson (1985): "*Two-Way Radio for Rural Health Care Delivery.*" Development Communication Report, Autumn 1985, pp. 5, 16.
[18] SatelLife News, 6th Issue, February 1994.
[19] See www.med.mun.ca/med/telemed/tele/telstory.htm.

small children, and community residents. Successful campaigns have used a variety of media from posters to radio and television, depending on the message and the resources available. Interactive communication is less common but may be used for administrative support and for follow up with health workers and project staff. In general, it has been found that mass media messages plus interpersonal communication are needed to effect changes in behavior as well as attitude, for example in oral rehydration therapy, in which the mother is taught to feed a mixture of clean water, salt and sugar to children with diarrhea.[20] Similar strategies are now being used for AIDS prevention.

Administrative Support

Interactive networks can also be used for administrative functions, such as ordering of medical supplies, scheduling field visits, compiling and updating patient records, and filing insurance claims. For example, physicians in Ethiopia use HealthNet to schedule consultations and referrals, making it unnecessary for ill patients to travel long distances with no guarantee of seeing a physician.[21] Practitioners in Alaska can access medical records to compile lists of children who need vaccinations and women who need PAP smears when they visit native villages. The electronic availability of patient records, which requires very little bandwidth, could have a significant impact on efficiency of patient treatment, and reduce guessing and errors in treating patients with unknown medical histories.

Data Collection

Communication systems can also be used to collect field data on outbreaks of infectious diseases and to issue medical alerts. In Mali, the Radio Administrative de Communication (RAC) network links clinics and hospitals to Bamako to collect epidemiological data. In The Gambia, health workers who once had to travel 700 kilometers per week to collect data for a clinical trial now use HealthNet to send this information via electronic mail. Health care workers in Zaire's Vanga Hospital use HealthNet to send regular dispatches to report on progress in treating Trypanosomiasis to health organizations in the North. Malaria researchers at a remote site in northern Ghana used HealthNet to communicate daily with the London School of Hygiene and Tropical Medicine and the Tropical Disease Research Center in Geneva.[22]

---

[20] Meyer, Anthony, Dennis Foote and William Smith (1985): "*Communications Works Across Cultures: Hard Data on ORT.*" Development Communication Report, No. 51.
[21] See www.healthnet.org.
[22] See www.healthnet.org.

Distance Education

Four Organizational Models

Rural areas typically have lower education levels than the national average; in the developing world, rural literacy rates tend to be much lower than urban rates. Rural teachers also tend to have less training than their urban counterparts, and few have specialized expertise. To address some of these problems, several models have been developed to use telecommunications in rural education. The curriculum-sharing model maximizes the value of scarce expertise by linking schools using telecommunications so that courses available at one school can be taught to students at other schools in the region.

The outside expert model involves identifying course content that is not available in a number of rural schools, developing specialized instructional programming, and delivering the programs to rural schools. These projects are typically regional or national in scope; many use satellites to transmit the courses to the schools and phone lines for interaction with students. The Texas Interactive Instructional Network (TI-IN) began in response to an educational reform initiative by the Texas legislature in 1984, requiring that school districts provide any course requested by 10 or more students. TI-IN's satellite network offered a cost-effective alternative for rural school districts without the funds to hire specialized teachers. Now TI-IN is received at more than 1,000 schools nation wide and is distributed over cable television systems in some areas.

The consortium model has been applied in higher education; several universities may join together to deliver courses to distant students. The National Technological University (NTU) is a consortium of 29 engineering schools that distributes technical courses via satellite to corporate sites. China delivers televised university courses to its workers throughout the country using its own domestic satellite.

A fourth model is the virtual school or university, which enables students to study from home or the workplace via the Internet. This approach has tremendous potential for rural and developing regions. For example, the World Bank is assisting with the establishment of an African Virtual University.

Supplements to Classroom Instruction

Other educational applications of telecommunications supplement classroom instruction. With the explosive growth of the Internet, students from elementary grades to graduate school are frequently using computer networks to exchange information with distant students and to search for information for school projects. In the Canadian North, Inuit children use a web based *"virtual frog dissection kit."* (Frogs are rare in the Arctic.) In Africa, Acacia and the World Bank are supporting school nets which provide computers and connectivity for disadvantaged schools.

Case Study: Alaska and Northern Canada

Telemedicine

The Alaska Telemedicine Project, founded in 1994, is building on Alaska's nearly three decades of experience in using satellite communications to support rural health care delivery.[23] From Kotzebue, north of the Arctic Circle, six doctors serve a dozen villages in an area the size of Indiana. Now, instead of relying only on verbal descriptions from health aides or sending x-rays to Anchorage, they use the Internet. One common application is for diagnosis of otitis media, a common ear infection among children that can cause deafness if not treated in time with antibiotics. Health aides can use an electronic otoscope connected to the computer to transmit images of the ear canal. The equipment can also be used to send digitized x-rays. *"In the past,"* states the information manager of the Maniilaq health center, *"there was a big delay in the process. There would be times when the bone would set before a diagnosis could be made. Now we digitize the film, and it's in Anchorage the same day."* The system may save money as well as time. An evacuation by plane can cost from US $ 10,000 to US $ 25,000. The package of computer, peripheral equipment, and training is estimated to cost US $ 22,000, so that if it saved two evacuations, it would pay for itself.[24] Under a new federal initiative called AFHCAN (Alaska Federal Health Care Access Network), in the next three years, similar facilities will be installed in 235 sites affecting over 212,000 beneficiaries, the majority of whom will be in rural Alaska Native villages.[25]

Education

Alaska has been a pioneer in using telecommunications for education since the installation of satellite facilities for telephone service and television distribution in the 1970s. An early activity involved distribution of educational video programs via satellite. With a limited budget, the state could not afford to produce many video courses. However, based on teacher requests for supplementary material, it obtained rights to a diverse collection of video programs which were transmitted at night via satellite. Teachers could set the timers on the school's VCR to tape the programs and use them in the classroom where appropriate. This approach holds considerable promise for developing regions which have access to satellite facilities. While it does not overcome the problem of developing appropriate content for different cultures and language groups, it offers a very low cost way of greatly expanding the educational resources available to village schools.

---

[23] Taft, Destyn E. *"Telemedicine and Telehealth Services going online for Providence Health System Alaska."* Telemedicine in Alaska News, June 26, 1998.
[24] Personal interview with project director Fred Pearce, Anchorage, August 1999.
[25] Mukluk Telegraph: *Official Publication of the Alaska Native Tribal Health Consortium.* Volume 1, April - May 1999.

Native people in northern Canada are using satellite communications to reach their own people. In northern Canada, the Wawatay Native Communications Society has set up an audio conferencing network enabling Cree and Ojibway people in the villages to talk to teachers and other students, so that they can complete high school via correspondence, rather than having to leave their communities to attend regional high schools. Canadian Inuit have established the Inuit Broadcasting Corporation which transmits TV programs via satellite across the North, including programs using puppets to teach Inuit children in their own language about health, hygiene, and northern living; and programs for teens and adults on Inuit culture and skills for survival in a changing world.

Alaska has more school districts with Internet access than any other state. Several factors have contributed to the enthusiasm for connecting to the Internet. First, in isolated areas, it provides an immensely valuable resource in overcoming the barriers of distance that have made it difficult for rural people to obtain and share information. Second, the investment in communications infrastructure has made it possible for almost all of the most isolated communities to connect to the Internet. A third critical factor is the availability of funds to subsidize Internet access known as the E-Rate Program. The Telecommunications Act of 1996 mandated policies designed to foster access to "advanced services" for schools, libraries, and rural health care facilities through a Universal Service Fund (USF). The E-rate has enabled small school districts such as the Yupik Eskimo village of Kuspuk in the Kuskokwim Delta and the Eagle Public Library, 66 miles up a gravel road just below the Arctic Circle, to get online.

Some Alaskan schools also participate in the Star Schools Program, which uses telecommunications to reach underserved populations, and to improve instruction in such fields as mathematics, science and foreign languages. Through its STEP Star Program run from Spokane, Washington, students in rural Alaska can take courses delivered by satellite in Spanish, Japanese, Chinese and astronomy.[26] Televised instruction is accompanied by interaction with instructors over telephone lines, supplemented by e-mail and web based materials.

Libraries

Alaska has also initiated a project called SLED, Statewide Library Electronic Doorway, which offers access to electronic data bases through public libraries. These services have proved to be very popular; during one week in 1999 there were more than 33,000 accesses to these databases by Alaskans. The state's director of libraries, archives and museums found how small the world could be when a patron at a library asked for information about Russian monasteries, a question that would have taken weeks to answer before the libraries came online. She notes: *"With the Internet, the citizen in rural Alaska can be as tied in to what's*

---

[26] See www.esdtcom.wednet.edu.

happening in the world as a person in Anchorage or New York or Seattle or any place else."[27]

Governance

Since the 1970's, Alaska has been using telecommunications to facilitate participation in state government. The State operates a legislative teleconferencing network (LTN) that makes it possible for rural residents to testify in state hearings and participate in other government planning and consultations. The U.S. Forest Service is also using the Internet to increase public participation in developing a new management plan for the Chugach National Forest, a huge preserve of more than 5 million acres. Alaskans will be able to access the proposed plan including maps online, comment on the plan and read other comments on the website.

Alaskan Business on the Net

Like entrepreneurs elsewhere, Alaskan businesses are turning to the Internet to publicize their products and services. The Alaskan Storefronts site provides links to Alaskan products and services.[28] Some 66 Alaskan businesses list their websites in the current *Alaskan Services Guide*, ranging from wild berry products to hunting lodges, art galleries and handicrafts. The Oomingmak Musk Ox Producers' Cooperative markets knitwear made from Qiviut, the soft underwool from the Arctic musk ox. The cooperative is owned by approximately 250 Native Alaskan women from remote coastal villages who knit each item by hand. In addition to a shop in downtown Anchorage, Oomingmak now has a website so that people can order its knitwear from all over the world.[29]

Conclusion

The above overview illustrates how ICTs can contribute to rural development. Paradoxically, rural villages may strengthen their own cultural identities and economic viability by becoming global villages, so that their villagers can learn about the outside world without having to leave the village. They can also find new markets for their products and opportunities for their children.

---

[27] Karen R. Crane, quoted in Kelley, Tina. "*Internet Showing Its Value in Remote Alaskan Villages.*" New York Times, August 5, 1999.
[28] See www.alaskan.com/mall/mallist.htm
[29] Kelley, Tina. "*Internet Showing Its Value in Remote Alaskan Villages.*" New York Times, August 5, 1999. For the website, see www.qiviut.com.

34 Okinawa Charter on Global Information Society

One of the four documents issued at the G8 Summit Meeting at Kyushu-Okinawa on 21 - 23 July, 2000 was the Okinawa Charter on Global Information Society. The following is an excerpt[1]:

1. Information and Communications Technology (IT) is one of the most potent forces in shaping the twenty-first century. Its revolutionary impact affects the way people live, learn and work and the way government interacts with civil society. IT is fast becoming a vital engine of growth for the world economy. It is also enabling many enterprising individuals, firms and communities, in all parts of the globe, to address economic and social challenges with greater efficiency and imagination. Enormous opportunities are there to be seized and shared by us all.
2. The essence of the IT-driven economic and social transformation is its power to help individuals and societies to use knowledge and ideas. Our vision of an information society is one that better enables people to fulfill their potential and realise their aspirations. ...
    In pursuing these objectives, we renew our commitment to the principle of inclusion: everyone, everywhere should be enabled to participate in and no one should be excluded from the benefits of the global information society. ...
5. Above all, this Charter represents a call to all, in both the public and private sectors to bridge the international information and knowledge divide. ...
6. The potential benefits of IT in spurring competition, promoting enhanced productivity, and creating and sustaining economic growth and jobs hold significant promise. Our task is not only to stimulate and facilitate the transition to an information society, but also to reap its full economic, social and cultural benefits.
    The private sector plays a leading role in the development of information and communications networks in the information society. But it is up to governments to create a predictable, transparent and non-discriminatory policy and regulatory environment necessary for the information society. ...
7. Bridging the Digital divide in and among countries has assumed a critical importance on our respective national agendas. We reaffirm our commitment to the efforts underway to formulate and implement a coherent strategy to address this issue. We also welcome the increasing recognition on the part of industry and civil society of the need to bridge the divide. Mobilising their expertise and resources is an indispensable element of our response to this challenge. We will continue to pursue an effective partnership between government and civil societies responsive to the rapid pace of technological and market developments...

---

[1] The full text can be read at: http://www.g8kyushu-okinawa.go.jp/e/documents/it1.html.

9. A key component of our strategy must be the continued drive toward universal and affordable access.
10. The policies for the advancement of the Information Society must be underpinned by the development of human resources capable of responding to the demands of the information age. We are committed to provide all our citizens with an opportunity to nurture IT literacy and skills through education, lifelong learning and training. We will continue to work toward this ambitious goal by getting schools, classrooms and libraries online and teachers skilled in IT and multimedia resources.
11. Promoting Global Participation: IT represents a tremendous opportunity for emerging and developing economies. Countries that succeed in harnessing its potential can look forward to leapfrogging conventional obstacles of infrastructural development, to meeting more effectively their vital development goals, such as poverty reduction, health, sanitation, and education, and to benefiting from the rapid growth of global e-commerce. Some developing countries have already made significant progress in these areas.
12. The challenge of bridging the international information and knowledge divide cannot, however, be underestimated. We recognise the priority being given to this by many developing countries. Indeed, those developing countries which fail to keep up with the accelerating pace of IT innovation may not have the opportunity to participate fully in the information society and economy. This is particularly so where the existing gaps in terms of basic economic and social infrastructures, such as electricity, telecommunications and education, deter the diffusion of IT.
13. In responding to this challenge, we recognise that the diverse conditions and needs of the developing countries should be taken into account. There is no "one-size-fits-all" solution. It is critically important for developing countries to take ownership through the adoption of coherent national strategies to: build an IT-friendly, pro-competitive policy and regulatory environment; exploit IT in pursuit of development goals and social cohesion; develop human resources endowed with IT skills; and encourage community initiatives and indigenous entrepreneurship.
14. The Way Forward: Efforts to bridge the international divide, as in our societies, crucially depend on effective collaboration among all stakeholders. ... IT, in short, is global in dimension, and thus requires a global response.
15. We welcome efforts already underway to bridge the international digital divide through bilateral development aid and by international organisations and private groups. We also welcome contributions from the private sector ...
16. There is a need for greater international dialogue and collaboration to improve the effectiveness of IT-related programmes and projects with developing countries, and to bring together the "best practices" and mobilise the resources available from all stakeholders to help close the digital divide. The G8 will seek to promote the creation of a stronger partnership among developed and developing countries, civil society including private firms and NGOs, foundations and academic institutions, and international organisations. We will

also work to see that developing countries can, in partnership with other stakeholders, be provided with financial, technical and policy input in order to create a better environment for, and use of, IT.
17. We agree to establish a Digital Opportunity Taskforce (dot force) with a view to integrating our efforts into a broader international approach. ... This high-level Taskforce, in close consultation with other partners and in a manner responsive to the needs of developing countries, will:
    - actively facilitate discussions with developing countries, international organisations and other stakeholders to promote international co-operation with a view to fostering policy, regulatory and network readiness; improving connectivity, increasing access and lowering cost; building human capacity; and encouraging participation in global e-commerce networks;
    - encourage the G8's own efforts to co-operate on IT-related pilot programmes and projects;
    - promote closer policy dialogue among partners and work to raise global public awareness of the challenges and opportunities;
    - examine inputs from the private sector and other interested groups such as the Global Digital divide Initiative's contributions;
18. In pursuit of these objectives, the dot force will look for ways to take concrete steps on the priorities identified below:
    - fostering policy, regulatory and network readiness;
    - improving connectivity, increasing access and lowering cost;
    - building human capacity;
    - encouraging participation in global e-commerce networks.

35 Jobs, Livelihood and Information Technology

Srinivasan Ramani

On July 7, 2000, the Economic and Social Council of the United Nations adopted a resolution recognizing the role of information technology as a tool for development.[1] The meeting was devoted to information and communication technologies, making it the first such occasion in history. It recommended that the UN should undertake a project to increase global connectivity.

Connectivity Is not Enough

In this context, we have to develop a vision for optimizing the value of connectivity. The developing world expects direct benefits from such connectivity, and it further expects that investments in ICTs should be cost effective. Creating jobs would be the best immediate contribution that these technologies could offer. However, in what way can the extension of connectivity directly contribute to an increase in jobs in the rural areas? Any successful approach to this issue will require a mix of idealism and pragmatism. Scarcity of resources, including human resources, places rural populations at a disadvantage. Any recommendations we make to benefit such populations must be grounded in current economic reality. A whole variety of jobs cannot magically be created in rural areas using ICTs. Nor can we visualize any method of creating employment for all segments of the rural population. We have to focus on a few selected types of jobs and endeavor to do what we can. The following are a few points relevant to the utilization of ICTs:

- Voice communication is the easiest technology for rural populations to use. India created a few hundred thousand jobs by merely permitting the creation of telephony kiosks.[2] Small street corner shops to retail access to long distance

---

[1] http://www.un.org/esa/coordination/ecosoc;
http://www.un.org/News/Press/docs/2000/20000707.ecosoc5899.doc.html; *"Deeply concerned that the potential of information and communication technology for advancing development, particularly in developing countries, had not been fully captured, the Economic and Social Council this afternoon called on all members of the international community to work cooperatively to bridge the "digital divide" and to foster "digital opportunity." It took that action by adopting without a vote a Ministerial Declaration on Development and International Cooperation in the Twenty-first Century: the Role of Information Technology in the Context of a Knowledge-based Global Economy."*

[2] Named STD Booths, as they provide for Subscriber Trunk Dialing; that is, long distance calls without operator assistance. Persons managing the kiosks keep the place open, offer assistance to users and handle the payments received. They are not in any immediate danger of being replaced by pay phones accepting credit cards; the use of credit cards has yet to reach significant proportions in the rural areas of most developing countries. The

dialing sprang up, effectively replacing a system in which only the state could retail such services. Communication from rural areas increased manifold.
- Internet connectivity will not directly serve most people in rural areas because of a variety of factors: their script may not be usable on the software available; the content available in their language could be negligible; even if available, such content may be irrelevant to their needs.
- This situation can be dramatically changed if voice communication over the Internet is allowed. The use of this technology, providing low cost voice telephony, is illegal in many developing countries! This is like declaring the use of wheeled vehicles illegal.
- It is now becoming increasingly possible to offer some form of automatic translation of the content of web pages into different languages. This surely has its benefits, but merely translating web pages created for urban users is not necessarily of any use to the rural populations.
- A few small sections of the rural populations in the developing world will benefit from the Internet despite the problems mentioned above. We should recognize the differences among rural populations, not regarding them as a uniform lot. Educational institutions in rural areas, along with banks, hospitals, and branches of companies, could all benefit significantly from access to the Internet.
- People in the rural areas of developing countries obviously behave quite differently from those in the rural areas of advanced countries. Any detailed examination of the ICT related issues would need to differentiate between the two.

Human Resource Issues

Persuading urban ICT professionals to move to rural areas is not an easy task. We need a new strategy for meeting the human resource needs for deploying ICTs in rural areas. The primary effort must go into giving ICT training at selected levels to citizens living in rural areas. Considering that approximately three billion people live in the rural areas of the world, we would need a large number of ICT-trained persons to work in these areas over the next five years. Making certain assumptions,[3] we can estimate the demand for ICT-trained persons in rural areas. Our estimates are two per thousand of the rural population. *We will need to train over six million persons in rural areas to reach this target.* Half of them will serve as teachers. The others would act as pioneers to lead rural populations in

---

few hundred thousand jobs mentioned are those of the employees of the STD booths. Many of them are self-employed.

[3] Assumptions: School-going population: 20% of the rural population. Number of teachers trained in using ICT to serve a rural population of 10000 = 10. Number of graduate level persons trained in using ICT = 2 in 10000. Number of persons with school level education trained in using ICT = 8 per 10000. Total number of those who need training in using ICT = 20 per ten thousand, or 2 per thousand.

exploiting ICTs better in the future. We should further assume that future (beyond five years) requirements would need larger training capacities.

Under these assumptions, we will need to train 1,200,000 persons per year for at least five years. This is necessary to reach the proposed target of six million ICT-trained persons. This is a great opportunity to create 1,200,000 jobs per year! These jobs would contribute significantly to increasing the productivity of people working in rural areas and would improve the quality of life.

Computers, software, networking facilities, etc., would cost an estimated US $ 2,000 per person employed, and training would cost between US $ 500 and US $ 1,000 per person. This would involve an overall investment of about US $ 3,000 per job created. There are very few sectors of the economy able to create profitable jobs with such low levels of investment.

Other Initiatives Required

We may create jobs that will pay for themselves by offering educational services or by meeting felt needs of the economy for ICT-trained employees. This alone will not benefit the rural populations beyond a certain point. The real ICT revolution will arrive in rural areas only once we ensure that the information is provided in the local script and language. It should be shared over networks, or in the form of CD-ROMs. Such content has to be selected to be relevant to the rural life of each region.

There has been a laudable effort to define development-catalyzing information for distribution using CD-ROMs.[4] Such efforts need to be encouraged, and their results carefully evaluated.

A whole variety of services are available to urban but not rural populations. Governments must make commitments not to discriminate between urban and rural citizens when it comes to facilities provided by E-government. Getting information from the government and carrying out transactions with government departments should be as easy in rural as in urban areas. Urban and rural schools should be treated equally in regard to ICT related investments, even if the demand for such training will be greater in urban areas.

Many authors have outlined scenarios in which urban citizens move to rural areas because of E-commerce. They can continue to earn their livelihoods, and their quality of life will improve. This looks like an idle dream in much of the developing world, mainly because the infrastructure is so poor in rural areas. Roads are very bad. Electrical power fails often if it is available at all, and drinking water is usually not of satisfactory quality. Sanitation is poor, leading to a

---

[4] See the sites www.humanitylibraries.net and www.humaninfo.org
  "*We believe that every PC in the developing world should get a complete development library on CD-ROM with 1000 or more publications (of which part could be delivered in their local language). The exposure of tens of millions of people to such better practical information and knowledge would create unprecedented participatory effects.*" Dr Michel Loots, MD, Humanity Libraries Project Director, e-mail for additional info : mloots@humaninfo.org.

variety of avoidable diseases. Good schooling and health care are rare. A practical response to this situation would be an effort to improve rural infrastructure, at least in a few selected places, enabling them to function as incubators of local ICT activities. Movement of some people into rural areas in search of a better quality of life will, in the long run, be a major force for development. Used to infrastructure at urban levels, they will fight to improve rural life.

Development-catalyzing efforts should also pay attention to the needs of underprivileged sections of the population. Particularly prominent is the case of rural women, who have hardly received any of the benefits of the ICT revolution. Special efforts must be made to ensure an equitable sharing of these benefits, especially with such an important section of the population.

36 Worldwide Access to ICT: The Digital Divide

Christophe Brun and Anton Mangstl

The world is undergoing a revolution in information and communication technologies (ICT) that has momentous implications for the current and future social and economic situations of all countries of the world. In March 2000, an estimated 276 million persons worldwide were using the Internet with a growth rate of roughly 150,000 persons per day. 220 million devices were accessing the worldwide web, and almost 200,000 devices were being added daily. Web pages totaled 1.5 billion with almost 2 million pages being added each day. E-commerce, or business conducted over the Internet, totaled US $ 45 billion as recently as 1998. An estimate in January 2000 projected it could explode to over US $ 7 trillion as early as 2004. These are astonishing figures, unprecedented by any measure, but they reflect the activity of less than 5% of the world's population. The gross disparity in the spread of the Internet and of the economic and social benefits derived from it is a matter of profound concern. There are more hosts in New York than in continental Africa; there are more hosts in Finland than in Latin America and the Caribbean. Notwithstanding the remarkable progress in the application of ICT in India, many of its villages still lack a working telephone.

The formidable and urgent challenge facing national governments and the development community is, therefore, to bridge the "digital divide" and connect the remainder of the world's population whose livelihoods can be enhanced through ICT. With each passing day, the task becomes much more difficult. To give just one example, exploding e-commerce ties individuals, firms and countries closer and closer together, while those who do not try to catch the "Internet Express" run the risk of being further and further marginalized. Developing countries have great potential to compete successfully in the new global market. However, unless they promptly and actively embrace the ICT revolution they will face new barriers and the risk of not just being marginalized but completely bypassed.

The challenge of a knowledge-based economy is not a scarcity of knowledge but inadequacies in diffusing and using it. Unlike capital resources, knowledge cannot easily be redistributed as a result of political decisions. It needs to be nurtured – by individuals, communities, and countries. ICT programs are undoubtedly beneficial to this development, but there is no single formula for ensuring their success. Undertaking the following measures is important for ICT initiatives to succeed:

- focused and narrowly defined, realistic objectives for ICT projects;
- establishing a legal and regulatory framework, including intellectual property rights, information technology, and telecommunications acts;
- tax and customs incentives and loans at concessional rates to speed up the growth of the ICT-based services sector;

- early support for ICT initiatives can be gained through the use of entry points such as education, health, public administration and e-commerce. Outreach campaigns, including road shows and competitions, have proved to be effective means of raising awareness and winning support;
- development of a local content as a result of national technological initiatives to develop local language character sets for use in computer interface for the countries where a significant part of the population neither speaks nor reads English;
- a determined effort to use ICT to help to integrate isolated rural populations into the national economy;
- the provision of public access points such as cybercafés, community centers and telecenters has proven very successful and should be a key component of any strategy to extend connectivity;
- the issue of affordable access costs should be addressed by the public sector authorities, taking into full account the benefit that ICT brings in improving performance of the public administration; and
- a strategic psychological approach whereby, first of all, each recipient of ICT hardware, software and services would be required to contribute up to a half of the costs involved, thus creating a sense of ownership, and, subsequently, building on the spreading sense of "envy" in the neighboring communities without comparable equipment.

The potential for ICT to contribute to human development, including the elimination of gender disparities, is currently compromised by an uneven diffusion of these technologies and the differential effects that their growth produces across social structures. Urgent reform and actions are required at both the national and international levels to ensure a fair and equal participation in the information society, such as:

- active encouragement and programs for young people, male and female, to access the new economy and use ICT in schools and in other educational endeavors;
- democratization of ICT policy processes that facilitate the active participation and full integration of advocates of human development concerns. In particular, ICT sector reform and governance processes should involve the full participation of a wide range of civil society organizations;
- strengthening the capacity of civil society organizations, including women's organizations, so that they may participate more effectively in the transformations made possible by the ICT sector;
- active encouragement of partnership efforts to allocate and direct R&D budgets to design and develop ICT services and applications that serve social and development objectives, including applications for non-literate communities, content development, human-computer interfaces that are not based on text and natural language processing systems;
- identification and eradication of factors that restrict equal participation of men and women in the ICT sector, in particular discriminatory and unequal access to education and training, social pressures that limit women's and girls' access to

science and technology activities in general and that limit their access to training and necessary ICT equipment in particular.

Impact of ICT On the Rural World

The digital divide is even wider in the rural world, which is traditionally more isolated and poorer than urban centers. The ever-increasing information gap is particularly sensitive in rural areas, as the agricultural sector moves from extensive and artisanal practices towards information-intensive, highly sophisticated, interconnected industries.

The progress by the ICT industry has made it even more compelling in the rural context. Since many developing nations are not overly encumbered with a substantial legacy of telecommunications infrastructures, they have a clear opportunity to leapfrog directly to new systems based on wireless solutions. The third generation of mobile technology and the Wireless Application Protocol (Internet over the mobile phone) are particularly promising for enabling rural communities to access the Internet. Some new standards, such as VoiceXML will give non-literate users the ability to access on-line information services through voice-lead processes.

It is, however, important to remember that technology is simply a means of accessing information. Many development initiatives have focused too heavily on the provision of connectivity and computers. Clearly, low computer literacy and the high costs of digital equipment are enormous barriers, but significant effort must be put into generating relevant content. Recent experience has shown that when rural communities are provided with access to reliable information sources using new technologies, most user groups rapidly become accustomed to the medium. The agent of change is the provision of previously unavailable content, in particular where that material relates to financial security in the form of market information (inputs and outputs), networking between peers (small farmers' interest groups), and information on technical aspects of primary production. In addition, solving conflicts related to land and water utilization is easier when the stakeholders have good and reliable information.

However, improving access to information and knowledge to rural communities, particularly in developing countries, requires an inter-governmental mechanism for making decisions, establishing standards, and reaching agreements regarding access to information and the sharing of knowledge on an international basis. This clearly falls within the normative role of the FAO, and our organization has responded accordingly: at the beginning of last month, the FAO convened the first Consultation on Agricultural Information Management (COAIM: http:///www.fao.org/coaim/). At the consultation, over 160 representatives from 88 member countries, as well as observers from non-governmental organizations, institutes and other UN agencies, discussed issues and policy matters related to access to information, capacity building in information management, and the importance of establishing and adopting common standards and guidelines.

Based on a decade of expertise in FAO WAICENT's information and knowledge management program, COAIM has a solid basis to become the intergovernmental mechanism for coordinating efforts among the UN agencies and member countries.

37 The Phone and the Future: Village Pay Phones in Bangladesh

Abdul Bayes

*One should hardly have to tell academicians that
information is a valuable resource: knowledge is power.
And yet it occupies a slum dwelling in the town of
economics. Mostly it is ignored.*
(Stigler, 1961)

In Bangladesh, as in many other countries of the developing world, the role of telecommunications in economic development does not seem to be duly appreciated. The telecommunications sector has received scant attention from policy makers, and the country has witnessed only a slow expansion of its network over the years. The country's present infrastructure is considered to be inadequate in scope, technology and quality of services. A report produced jointly by the World Bank and the Bangladesh Center for Advanced Studies (BCAS) presented the limitations of the telecommunications services in Bangladesh (World Bank and BCAS, 1998):

- The telephone density of 0.26 lines per 100 people is one of the world's lowest (India: 1.0, Nepal: 0.5, Pakistan: 2.1, Sri Lanka: 1.0, Thailand: 2.5).
- The waiting time for a connection is more than 10 years.
- The installation charge of US $ 450 for a new line is one of the highest in the world (e.g., Pakistan US $ 90, India US $ 60).
- The charge for calling the United Kingdom, US $ 1.50/minute, is about six times higher than the charge for calling Bangladesh from the UK.
- On average, only 2 of 10 calls are successfully completed.
- The complaint rate averages 50 complaints per 100 lines per year, clearly indicating the poor quality of services.

By and large, the present underdevelopment and poverty of Bangladesh - a country with a per capita income of about US $ 300 and with half of its population living below the poverty line - is related to the underdevelopment of the basic infrastructure. The vast majority of rural areas remain largely inaccessible and are consequently unable to take advantage of opportunities conducive to growth and development, for example the diffusion of modern technology, extension services, and rapid and effective measures for dealing with disasters such as floods. It should be mentioned here that a number of studies have established a causal link between deficient infrastructure and the slow rate of development in rural areas of Bangladesh (e.g., Ahmed and Hossain, 1990; World Bank, 1993).

However, recently Grameen Bank (GB) of Bangladesh stepped in to provide some solutions to the current impasse. GB introduced cellular mobile phones in some rural areas - Village Pay Phones (VPPs) as they are called - to be operated under its microcredit programs and to be its *modus operandi*.

The objectives of this paper are:

- to ascertain how and to what extent VPPs can promote the socioeconomic uplift of villagers, especially of the poor, by affecting socioeconomic parameters which tend to constrain households;
- to evaluate the economic effects of cellular mobile phones at the household and village levels: at the household level, by estimating the net increase in income derived from selling phone services and determining how this increase affects household poverty and food consumption; at the village levels, by estimating the consumer surplus, marketing margins, and changes in productivity levels (based on case studies);
- to assess specific social impacts. Some of these include changes in the social equilibrium, empowerment of the disadvantaged, kinship networks and the law and order situation in the villages; and finally,
- to make some suggestions regarding the design and formulation of public policies which are intended to ensure broader and better access to telecommunications services among the poor.

Research Approach and Data Sources

To determine how VPPs affect rural development and poverty reduction, both owners (sellers of services) and users (buyers of services) were surveyed. The sample of phone owners consisted of 50 persons in 50 different villages at a range of 40 to 50 km from metropolitan Dhaka. This sample constituted about 60% of all VPP owners at the time of the survey. The owners were selected at random from a list provided by the GB Head Office, Dhaka. The sample size for users of VPP services, on the other hand, was stipulated to be 400. The sample was drawn at random from lists provided by the phone owners. In this case, however, the sample comprised only about 27% of all VPP users in the villages (excluding owner/users). This study uses both primary and secondary data. Structured questionnaires were administered at both the household and village levels. Focus group discussions were held, and GB branch managers and local people were requested to provide additional input regarding their impressions of the potential and actual effects of VPPs.

This paper is organized as follows: Basic information on rural phone users are provided in Section 2, while the effects of phones are discussed in Section 3. Finally, policy conclusions are submitted in Section 4.

Basic Information on Phone Use

During the week preceding the survey, the phone users in the sample are reported to have made 1,060 phone calls (Table 37.1). The combined groups of the poor made 268 calls and thus accounted for one-fourth of all calls. On the other hand, the non-poor made 792 calls, which constituted about three-fourths of the total. Within the owner/user group, the share of calls made by the poor and the non-poor appear to be more or less evenly distributed (45% vs. 55%), while among the villagers, most phone calls were found to have been made by non-poor households (roughly 78% vs. 22%). It is noteworthy that the intensity of use by the poor is 50% greater than that of the non-poor.

The gender distribution of phone callers is shown in Table 37.2. Among the owners of phones, 60% of users were women and 40% men. Among the villagers, however, 30% of users were women and 70% men. For the sample as a whole, 65% of all calls are reported to have been made by men and 35% by women. Thus, it appears that the presence of VPPs provides rural women with the opportunity to use this modern communications technology.

Keeping in mind the problems of overlapping calls, we have categorized the calls into several groups: economically-related, health (emergency and advice), social/personal (family- and office-related), remittances and other. The poor and non-poor groups accounted for more or less the same proportion of economic/finance-related calls (46% and 47%, respectively). Within the composite poor group, however, the extremely poor seem to use phones chiefly for economic purposes, making about 54% of their calls with these purposes in mind. The poor group also makes relatively more phone calls for health-related purposes than the non-poor group (18% and 10%, respectively).

Table 37.1. Phone Use According to Economic Status

Users	Economic Status		
	Poor	Non-poor	Total
Owners	76	93	169
	(45.0)	(55.0)	(100.0)
Villagers	192	699	891
	(21.5)	(78.5)	(100.0)
Entire sample	268	792	1,060
	(25.3)	(74.7)	(100.0)

Source: JU / ZEF Field Survey, 1998.
Note: Figures in brackets show percentages.

Table 37.2. Phone Use According to Gender

	Owners		Villagers		All	
Gender	Number of calls	[%]	Number	[%]	Number	[%]
Male	68	40	623	70	691	65
Female	101	60	268	30	369	35
Total	169	100	891	100	1,060	100

Source: Information collected through additional telephone interviews, 1998.

The non-poor group, on the other hand, makes relatively more phone calls for family/personal considerations (35% vs. 32%), remittances (4% vs. 1%) and also for business-related purposes (25% vs. 12%). That the non-poor devote more of their calls to business-related purposes is not surprising, given the fact that most of the business activities in rural areas are carried out by persons falling into the non-poor group. However, even the extremely poor group indicates that about 21% of their calls are made for business-related purposes. This shows that even the poorest segment of the village, which is involved in the petty production of eggs, vegetables, puffed rice, poultry rearing, etc., make phone calls in order to keep informed about the business environment. By and large, the lion's share of the phone calls made by the poor group deals with economic and health considerations.

Effects of Village Pay Phones

Economic Effects

As was explained above, GB leased phones to its most successful members in the target villages. The objective was to supplement household income rather than to replace it (at least in the short run). As is the case with other business activities, the net profit gained by telephone services comprises two elements: the total receipts/revenues from sales and the total payments made on account of the sales. (In other words, P = R-C, where P, R and C are profit, revenue and cost, respectively). To calculate the net profit earned from selling phone services, the following formula was used:

$NP_i = (TM_{1i} * \alpha_1 + TM_{ni} * \alpha_2 + TM_{ii} * \alpha_3 + T_{wi}) - (I_{wi} + OC_i + O_i)$

where,

NP_i = net profit (Tk/week) on phone services sold by household I,

TM_1 = total minutes of local calls/week,

α_1 = returns on local calls (=Tk. 2.4/minute),

TM_{ni} = total minutes of NWD (calls/week),

α_2 = returns on NWD calls (=Tk. 5/minute),

TM_{ii} = total minutes of ISD calls/week,

α_3 = returns on ISD calls (=Tk. 7/minute),

T_{wi} = total tips received (Tk/week),

I_{wi} = installment paid for the phone (Tk/week),

OC_i = opportunity costs of the operator's time (Tk/week),

O_i = other costs, e.g. line rent (Tk/week).

Table 37.3. Net Profits from Selling Phone Service

Net profit [Tk./week]	Number of owners	% of total
≤ 100	8	16.0
101-300	17	34.0
301-500	20	40.0
501-1,000	5	10.0
Total	50	100.0

Source : JU / ZEF Field Survey, 1998.
Note : US $ 1 = Tk. 49

Statistics on net profits are presented in Table 37.3. As the table shows, the VPP owners earn an average net profit of Tk. 277/week. The profit level ranges from as high as Tk. 683/week to as low as (-) Tk. 35/week. However, as was stated earlier, the profits accruing from phone services constitute from about one-fifth to one-fourth of the total income of the sample households. The distribution of net profits can be elaborated further. Half of the sample owners reap a net profit of more than Tk. 300/week, and another one-tenth earn more than Tk. 500/week. However, the net profits of 16% of the sample phone owners fell below Tk. 100.

An estimate of the consumer surplus (CS) yielded by VPPs is presented in Table 37.4. CS is defined as the difference between the price consumers actually pay and the price they would be willing to pay in a transaction. In other words, CS is the difference between the potential and the actual prices paid by the consumer (for information). For the purposes of the present study, CS is defined as the price users would have paid for the alternatives forgone minus the price they actually paid for the alternative they accepted (i.e., phone service). Thus, the surplus estimated could also be considered the consumer's real income benefits.

Table 37.4. Estimates of Consumer Surplus Provided by VPPs (in Taka)

Economic status	Hours required by alternative methods	Transport costs entailed in alternative methods	Opportunity costs of time required for alternative methods	Total costs of alternative methods	Total costs of Village Pay Phones	Consumer surplus (5-6)
All poor	3.67	60.89	34.32	95.21	17.35	77.86
Extremely poor	3.08	54.97	26.41	81.38	20.08	61.30
Moderately poor	4.15	65.82	40.89	106.71	15.07	91.64
Non-poor	2.54	45.80	21.71	67.51	16.73	50.78
Entire sample	2.70	48.02	23.57	71.58	16.82	54.77

Source: Calculated from Field Survey Data 1998.
Note: US $ 1 = Tk. 49

The calculation of the CS is based on the following assumptions:

1. Time and road transport costs are included in the potential or possible alternative. For example, we asked the respondents how much they would have to pay in time and transport costs if their only communication alternative was road transport.
2. The opportunity costs of the time spent were calculated on the basis of prevailing rural wage rates. A uniform wage rate was used for both the poor and non-poor groups as the non-poor group may have the choice of employing wage labor to acquire information at the poor group's wage rate, rather than at their own higher wage.
3. The actual cost to phone users amounted to the price they paid at the time of receiving the services.

From the data in Table 37.4, it can be seen that the availability of VPPs provides users of all income strata with a fair amount of CS. For example, while the cost of communication in the previous mode used to be Tk. 71.58, the present mode costs only Tk. 16.82, yielding a surplus of Tk. 54.77. The most important observation shown in Table 37.4 is that the CS of the poor, Tk. 77.86, is 50% higher than that of the non-poor. This amount, if converted at present rural prices, could purchase 12 kg of coarse rice. Furthermore, the moderately poor gain a CS of Tk 91.64, followed by the extremely poor (Tk. 61.30). The non-poor (Tk. 50.78) extract the lowest amount of CS. That the poor would reap the maximum CS following the advent of VPPs is quite obvious, since the poor usually do not have much in the way of alternatives to communicate with the outside world. They have neither relatives to help them make a phone call nor relatives to provide a ride to the desired destination. For the poor, the advent of VPPs opened up a lower-cost alternative for exchanging information.

One of the important effects of VPPs is that they help to avert sharp swings in demand, supply and the prices of commodities. The proponents of VPPs hypothesize that the dissemination of market information to remote villages would help to raise farm output prices and lower farm input prices through the mechanism of information diffusion. Admittedly, differences that could be discerned (either moderating prices or smoothing supply) might be attributable not only to telephones but also to a variety of other factors, such as roads, mass media, etc. The "perceptions" of the local people about the utility of the services were considered relevant to this research. Similarly, attempts were made to determine what views were held in control villages. A summary of the economic and non-economic benefits, as perceived by the villagers, is presented in Table 37.5. The average prices of agricultural commodities (especially of paddy and eggs) were higher in target villages (with phones) than in control villages (without phones).

Table 37.5. Assessment of Selected Benefits

Variable	Target village (N=50, averages)	Control village (N=10, averages)
Prices:		
Paddy [% of final consumer prices]	70-75%	65-70%
Eggs	Tk. 13/hali	Tk. 12/hali
Exchange rate	Tk. 12.50/Ryal	Tk. 11.50/Ryal
Cost of information/knowledge	Tk. 17	Tk. 72
Chicken/ducks	Higher	Lower
Chick feeds	Lower	Higher
Supply of inputs:		
Diesel	Stable	Fluctuating at times
Fertilizer	Regular	Occasional problems
Others:		
Poultry mortality rate	Lower	Higher
Law and order situation	Improved	Same
Communication during disasters	Quick, effective	Slow, less effective
Communication with relatives home and abroad	Any where, any time, any day	Any where, but fixed time, fixed day
Transmission of new ideas	Improved	Same
Mobility of people	Higher	Lower
Spoilage of perishable products	Less	More
Access to health care services	Faster/effective	Slower/less effective

Source: case studies and discussions with local people during JU / ZEF Field Survey, 1998.

For example, farmers in the target villages received 70 - 75% of the paddy prices paid by the final consumers, discernibly more than the 65 - 70% of the prices received by control villagers. The argument that market efficiency is improved is highlighted by one example in particular: the price of eggs in target villages was reported to be Tk. 13/hali (*hali* means four) during the period of the survey, compared to Tk. 12/hali in control villages. Likewise, vegetable growers in a target village informed us that VPPs helped them by providing easy and instant access to the prevailing market demand and supply situation. Thus, they were able to make more appropriate production decisions. Also, the supply of agricultural inputs such as diesel and fertilizer is reported to be smoother and more stable in target villages than in control villages. Furthermore, VPPs are reported to have increased the productivity, capacity utilization and profitability of these tiny livestock and poultry enterprises by (a) facilitating the regular and easy delivery of inputs at lower cost, (b) disseminating market prices for produce and thus helping to ensure fair payments, and (c) reducing the scope of the risks associated with such businesses.

Table 37.6 indicates the household food situations (i.e. whether they eat better now than before). One-fourth of the respondents reported that their food situation had improved by 4 months (i.e., these households reported that they have 4 more months than before in which they can "eat well"), and one-fifth of the cases responded that their food situation had improved by more than three months. In all, 86% of the phone-owning households reported that their food situation had improved over the years. In other words, they can now eat well during more months of the year than before.

One major benefit of VPPs was demonstrated when Bangladesh was hit by the worst flood in modern history in July to September 1998. Two-thirds of the country remained submerged for more than two months. People and transport got stuck en route. When people got stuck in or near target villages, the mobile phones proved to be invaluable for sending messages to worried relatives, informing employers or calling relief agencies.

In a few of the sample villages, it was related that the law and order situation had improved as a result of the availability of VPPs. In the event of a burglary or theft, people in target villages were able to inform law-enforcement agencies immediately by phone. In addition, law enforcement agencies in the vicinity reported that swift communication with villages contributed to decreasing the number of crimes committed there.

In most villages, a fairly large number of people are reported to be working abroad. Many of them belong to poor families who sold land and other assets to pay for the travel expenses. The inhabitants of target villages informed us that communication with the outside world had become very fast and regular following the arrival of VPPs in villages.

During discussions with focus groups and interactions with the local elite and social workers, it was revealed that the greatest benefit brought about by VPPs was the capability to call doctors and clinics immediately. In a country where the infant mortality rate is 91 per 1000 live births, and the maternal mortality rate is also very high, VPPs' contribution to saving lives cannot be exaggerated.

Table 37.6. Extent of Change in the Food Consumption Situation of Phone-Owning Families

Extent of change in months	Number of respondents	% of Total	Cumulative %
0	7	14.0	14.0
1	4	8.0	22.0
3	12	24.0	78.0
4	3	6.0	89.0
5	2	4.0	88.0
6	3	6.0	94.0
7	2	4.0	98.0
10	1	2.0	100.0
Total	50	100.0	100.0

Source: JU / ZEF Field Survey, 1998.

Sociocultural Effects

Various studies in the context of microcredit programs in Bangladesh have argued that there have been substantial positive developments in the empowerment of women (Amin, Becker and Bayes, 1998). The findings regarding empowerment are presented in Table 37.7. It can be observed that sample household decisions were made by both the husband and the wife. Whenever household-level decisions were made about, for example, family affairs, the utilization of GB credit, or the spending of phone income, both members of the couple participate. For example, 72% of the respondents replied that the decisions on family affairs (e.g., schooling for children, the marriage of daughters or sons, etc.) were made by both partners on the basis of mutual consultation. It was 60% when the utilization of GB credit was at issue. When the matter involved was the spending of earnings from telephone services, however, 58% of the owners made joint decisions, while 36% decided entirely on their own. In other words, women in the sample appeared to have greater latitude in deciding how to spend phone income than was the case when the other two issues were involved. For example, 36% of the respondents decided entirely on their own where and how to spend the profit money, compared to 16% and 30% respectively for the other two cases.

Regarding empowerment, another indicator also could be used: the degree of mobility of women in and around the village. Incoming calls for villagers make it necessary to bring the phone to them. If none of the family members are available, the women themselves have to rush about with the mobile phone. This is a job that they even have to perform during the night. Again, since many of the phones are placed in their shops in a nearby "*hat*" or "*bazaar*" (small market), owners were going to the phone whenever necessary. Thus, mobile phones appear to have enhanced women's mobility both within and outside the village.

Table 37.7. Who Makes Decisions?

Decisions	Decision Makers			
	Self	Husband	Both	Total
About family affairs	8	6	36	50
	(16.0)	(12.0)	(72.0)	(100.0)
About utilization of GB credit	15	5	30	50
	(30.0)	(10.0)	(60.0)	(100.0)
About income from phone	18	3	29	50
	(36.0)	(6.0)	(58.0)	(100.0)

Source: JU / ZEF Field Survey, 1998.
Note: Figures in brackets show percentages.

Two more aspects of the mobility issue should also be mentioned. First, women in the village come to the owner's houses to make phone calls, and second, sample women reported that the mobile phones expanded the scope of their mobility even beyond the local markets. For example, whenever they decided to visit relatives living far away from their villages, they could contact their families by phone and inform them of their time of arrival. This capability reduced family tensions and conflicts.

Interestingly, some of the people who came to the phone owners' premises to make calls were the relatively wealthy villagers who had previously helped them with food, clothing and shelter. Reportedly, patron-client relationships had developed between some of them. For example, about one-fourth of the respondents reported that the people who depend on them for phone calls include some persons who once lent them money to alleviate their economic hardships. Also, 12% reported that some phone users had once hired the phone owners as maidservants. And 34% reported that they were helped in various ways by many of the users (Table 37.8).

The women in possession of cellular mobile phones are proud of their present businesses. According to them, it has not only paved the way to earning additional income, but ownership has also conferred upon them a certain amount of fame. Everyone in their village and in the adjacent villages now knows them and identifies their "*bari*" (cluster of houses) by the name of the technology they own, e.g., Phone Bari (house of the phone), or by the name of the owner, e.g., "*Nurjahan's bari*". This was not the case when these same women previously carried out traditional activities such as poultry rearing, rice husking or grocery selling. Their new economic activities are not only generating more income, which raises their standard of living, but are also earning them more prestige in the eyes of the villagers, who respect phone owners. Some of them are even invited to social functions (even to marriage ceremonies of the village elite), which previously would have been unthinkable in the social context of rural Bangladesh.

Women in the sample group who own phones find that, in addition to providing a regular weekly income, phone ownership also gives them access to previously unknown information. For example, when business people come to them to make phone calls to the various markets, the women, who stand next to the callers, learn the names of the markets and the commodities in which they are dealing. When villagers receive calls from outside the country, the phone owners learn the names of the places from which the calls originate. Sometimes owners hear about the currency or the socioeconomic conditions in that particular country. Thus, each day, owners have the opportunity to learn new things, expanding the frontiers of their knowledge. Ownership of a mobile phone has given these women greater confidence.

Table 37.8. Partial Social Change

Previous Status of Callers	Number	% of Total
1. Used to lend money	12	24.0
2. Used to help maintain family	15	30.0
3. Used to hire owner as household maid	6	12.0
4. Used to help in various ways	17	34.0
Total	50	100.0

Source: JU / ZEF Field Survey, 1998.

The strengthening of kinship networks constitutes another major effect of VPPs. Their advent has enabled villagers to stay in touch with relatives living outside of their villages. In Bangladeshi society, finding out news about the immediate and extended family members in far away places is a perennial concern. In the past, it was not possible to maintain the level of interaction with their relatives that VPPs now allow, and it is both more efficient and less expensive.

Conclusions and Policy Implications

The availability of village pay phones results in substantial sociocultural benefits to rural areas. By owning phones, relatively poor households are able to raise their social status and thus bring about change in the social structure. Mobile phones enhance women's mobility both within and outside of their villages. In addition, increased income from phones also raises their exchange entitlements in the market and thus helps to reduce their poverty. Villages with phones also report their usefulness in coping with natural calamities and improving law-enforcement. Given these multidimensional benefits of village phones, the study arrives at the following general conclusions and policy implications.

The study argues that the telephone should be regarded as both a consumer good and as a production good, especially in poor rural areas. Here, telephones are not the preserve of the relatively wealthy, as some have thought, but rather represent a market into which lower income groups can gain access and have substantial stakes. These findings, therefore, call for a rapid reorientation in the thinking of policy makers. The development of telecommunications services in rural areas gives every indication of powerfully supporting the objectives of rural development and poverty reduction. Thus, additional research on the role of communications technology is needed to further explore these links.

References

Ahmed, R. and M. Hossain (1984): Development Impact of Rural Infrastructure in Bangladesh, IFRI/BIDS, Dhaka.

Alfian, D., G. Chu and M. Rauf (1984): Social and Economic Impact of Rural Telephones in Indonesia: A Pilot Study Research Report. East West Center and Jakarta: National Institute of Cultural Studies, Honolulu.

Amin, R.S. Becker and A. Bayes (1998): NGO-Promoted Micro credit Programs and Women's Empowerment in Rural Bangladesh: Quantitative and Qualitative Evidence, The Journal of Developing Areas, vol 32, No. 2.

Antle, J.M. (1983): Infrastructure and Aggregate Agricultural productivity: International Evidence. Economic development and Cultural Change vol 31.

Bayes, A., J. von Braun and R. Akhter (1999): Village Pay Phones and Poverty Reduction: Insights from a Grameen Bangk Initiative in Bangladesh. ZEF Discussion Paper on Development Policy, No. 8, Bonn.

Howe, J. and P. Richards (1984): Rural Roads and Poverty Alleviation. A study prepared for the International Labor Office within the framework of the World Employment Program. Intermediate Publications, London.

Huda, N. and S. Uddin (Undated): Village Pay Phone: Information Revolution for Rural Bangladesh. Grameen Telecom, Dhaka.

Saunders, R.J., J. Warford and B. Wellenus (1983): Telecoms and Economic development. Johns Hopkins University Press, Baltimore, MD, USA.

Stigler, G. J. (1961): The Economics of Information. The Journal of Political Economy, vol LXIX.

World Bank (1994): World Development Report. Oxford University Press, Washington, D.C.

World Bank (1999): Knowledge for Development. Oxford University Press. Washington, D.C.

World Bank and BCAS (1998): Bangladesh 2020-A Long-run Perspective Study. University Press Ltd., Dhaka.

38 Local Participation in the Information Society

Arnulf Heuermann

Until now, the world's ICTs have been concentrated in high-income countries. For instance, it is estimated that only 10% of the revenues from telecommunication services comes from the 128 low- and lower-middle-income countries where 76% of the world population lives. Furthermore, unlike in industrialized countries, ICTs are heavily concentrated among the urban population in developing countries. The share of ICTs infrastructure in rural areas in developing countries is low (e.g., 19% of the telephone lines), although 72% of the population in low-income countries live in rural areas.

Despite these facts, ICTs offer a chance for village populations to be a part of the information society Empirical research suggests that ICTs are a necessary, though insufficient, infrastructure for economic development in rural areas. For instance, by avoiding the costs of physical transport and migration, the economic and environmental effects of ICTs are substantial. Furthermore, there are positive effects on business productivity and consumer surpluses on a microeconomic level.

The crucial question for the future is, *"How can ICTs be brought to villages in developing countries?"* The major regulatory instrument to connect villages to modern ICTs is the "Universal Service Concept," which is *"a set of defined services that must be made available at a defined quality to all users independently of geographical location, and, in the light of specific national conditions, at an affordable price."*[1] The idea is, to serve rural areas for social reasons if the market does not provide the services, enabling at least a telephone service nationwide that is non-discriminatory (in terms of geography as well as race, religion, sex) and affordable.

Many of the services can be provided by the private sector. Large parts of the ICT business are very profitable, including rural public phones or telecenters. The official waiting lists of at least 40 million telephone applicants in low-income countries (willing to pay the actual prices) suggest that these countries face public sector inefficiency rather than lack of purchasing power.

Liberalization and privatization can attract investments in the core network by mobilizing resources from the private sector to bring ICT services to rural areas. They are, however, not always sufficient. ICT services will not always be developed by the private sector without development aid. Private financing does not work in high risk countries, in countries with weak currencies, or in rural areas with very dispersed settlement structures. Hence, although the ICTs sector has high market potential globally, development aid is required to foster the spread of these services through institutional development, e.g., for regulatory authorities, regulatory procedures and professional privatization.

---

[1] New EU directive COM (2000) 392.

39 Telecommunications in Sub-Saharan Africa

Miko Rwayitare

Fixed telecommunication lines have been available to only about 1% of Africans. Unsatisfied demand for fixed lines led African subscribers to welcome cellular phones as a viable alternative. Only in 1987 was the very first cellular call in Africa made in the DRCongo. About 40 African countries now have cellular operators, and although this leaves several countries unserved, it is a major achievement for Africa. The penetration levels are still so low that even the high costs of cellular phones, when compared to the GDP per capita in African countries, are not considered a barrier, because cellular phones have become the substitutes for the unavailable fixed phones. Telecel's subscribers in sub-Saharan Africa have grown from 5,700 in 1993 to 350,000 now. The forecast is to have 3 million subscribers by December 2002.

Africa has an ironic advantage over other parts of the world. Since it has had no real telecommunications systems until now, it has the chance to leapfrog into starting with the latest ICT technology. Unfortunately, so far less than 8 African countries have agreed to liberalize their telecommunications markets in accordance with the World Trade Organization (WTO) and the General Agreement on Tariffs and Trade (GATT). Most governments seem to be continuing their protectionist policies which have constrained the expansion of telecommunications services to the rural areas in the past. Even in those countries that claim to have liberalized, most did not truly open up the telecommunications market to competition. They simply invited a foreign strategic operator to come in and operate the public telecommunications company with continued protection of the existing monopoly and, in most cases, without imposing obligations to invest and expand into rural villages.

In addition to implementing and adopting overall economic development strategies, African governments should introduce investor-friendly policies, fully liberalize and allow competition in telecommunications, thus promoting the introduction of modern ICT technologies. It is only with full liberalization and competition that rural areas can eventually have coverage at cheaper prices. Once the basic transportation and telecommunications infrastructure has been set up, then a platform exists from which to aggressively promote education and human resource development in areas such as protecting the environment, enhancing trade in agricultural and mineral production, tourism and other income-generating ventures. This would have a positive impact on rural life and benefit education, skills development, trade, and, therefore, economic, social and political development.

Agenda 21 §14.25 Earth Summit, Rio 1992

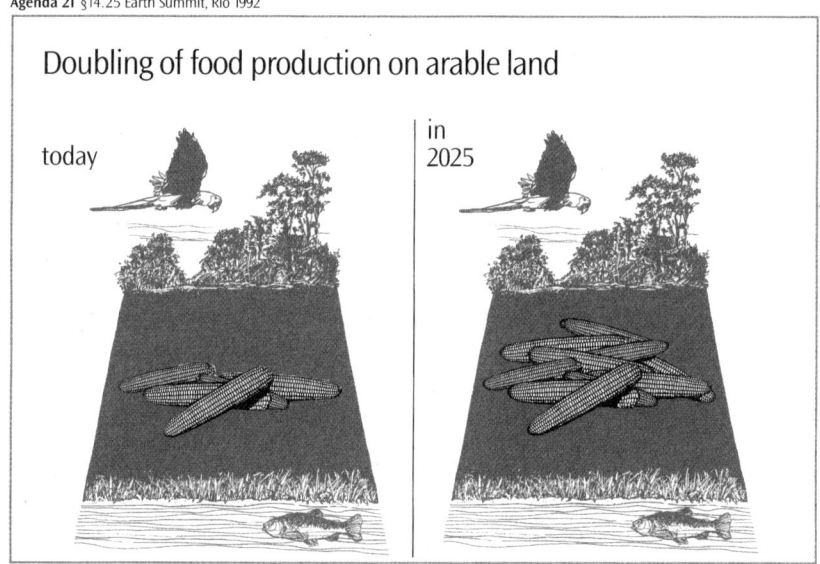

Doubling of food production on arable land

today

in
2025

The Future of Agriculture

Introductory Statement

Per Pinstrup-Andersen

Although the world will change from a global village to a global city in the next years and urban agriculture may increase, most of the food will still be produced in rural areas. Agriculture will remain the primary economic activity and income source for rural areas worldwide. So it is only logical to have a section of this book focusing on the future of the rural areas and agriculture.

In which direction will agriculture move? Globally, the main goal of agriculture is to provide three things: the increase of food production, the reduction of poverty and the sustaining of natural resources. The multifunctional and ecological functions are becoming increasingly important in many countries, especially the industrial ones.

In the last decade, world food production has been increasing faster than population growth. However, will this trend continue, and does it mean that the objective of world food security will be accomplished in the near future? The future challenge for farmers all over the world is manifold. The population growth forces agriculture to produce more food sufficient for about 80 million additional people each year. Furthermore, the desired economic development leads to an increased food demand, in quantity as well as in quality. The farmers will have to increase food production from better use of the existing or even diminishing agricultural areas, because area expansion is no longer a feasible option in most of the world. In tackling this challenge, agricultural production technologies are the focus of the discussions. Besides economic growth, natural resources management, rural infrastructure and related policies, the appropriate technologies will be the determinate factors in global trends in food production and agricultural development over the next twenty years.

As *Kern* predicts, agriculture in the industrialized countries will be determined and shaped by the new agricultural technologies. Precision farming, biotechnology and genomics/functional genomics will be the main elements of the Integrated Crop Management System and Integrated Plant Nutrition System, providing long-term sustained productivity of safe and sufficient food in an environmentally sound manner. According to *Kern*, countries with a food deficit and having to import food, and with their economic and technological limitations, must focus on a more sustainable, "low external input" agriculture. A combination of locally available inputs with selectively applied external inputs will increase the effectiveness of their agriculture. Agricultural biotechnology is seen as one of the most important key technologies which should be incorporated into the local agricultural production systems, enabling the increase of food production while sustaining the environment in all countries.

To solve significant problems of the next millennium, the linkages and interactions between the improvement of crop productivity, food safety and human health and the sustainable use of all natural resources have been stressed at the 3rd International Crop Science Congress 2000, where the participants adopted the *Declaration of Hamburg*, of which an excerpt can be read in this context.

The number, especially of politicians, criticizing the direction of modern agriculture is increasing. They question whether or not modern agricultural technologies will enable agricultural production to meet its objectives. They suspect that the increased "industrialization" of agriculture has reached a dead end. Even if the critics of modern agriculture are correct, which alternative approaches in agriculture can lead the way?

Lockeritz believes that alternative approaches to agriculture will become increasingly important in industrialized countries. The use of these approaches will, however, remain limited, and modern agriculture will continue to lead agricultural production and will become even more industrialized.

Howard-Borjas and Kees Jansen focuses in her contribution on the socioeconomic dimensions of organic agriculture. She considers the implications for farm households and the rural labor force of increased labor requirements, changes in the quality of labor, and shifts in livelihood that can result from the introduction of sustainable technologies. She shows that organic agriculture is highly labor intensive, thus helping to generate rural employment. However, the additional labor requirements are mainly for unskilled workers, and farmers in general cannot afford to employ anyone beyond their unpaid family laborers.

Concerning organic agriculture, crucial questions are not yet answered: Will rural people in developing countries be able to protect their natural resources through organic agriculture, and can they afford it? Are labor-saving technologies essential, and with what inputs can organic agriculture satisfy the increasing food demand?

It must be emphasized that the outcome of the debate about modern biotechnology for agriculture in Europe and in the United States will have tremendous implications for agriculture as a whole and thus for poor people in developing countries. It is vital that all potential technologies are developed. However, it is not only important to make the technological options available but also to give a choice to the people who will benefit or lose from their adoption. While making the important decisions about which technologies should be chosen and promoted, it should always be remembered that agricultural policies in industrialized countries will influence the future of agriculture in developing countries. For instance, subsidized imported food from industrialized countries is posing a threat to the agriculture and livelihood of rural people in developing countries. The import of these subsidized foods puts tremendous pressure on the local farmers.

The crucial question for the future of agriculture is how to return agriculture to a high place on the political agenda to address the existing challenges, mainly of appropriate technologies and policies. A good starting point in this direction is the *Dresden Declaration*, which formulates a vision of agricultural research for development.

40 What Can We Expect from Agrotechnologies in the Future?

Manfred Kern

Future developments and breakthroughs are hard to predict. They are even harder to produce to order. Nevertheless it is reasonable enough to think of a shopping list of what we want and what we need, and of what we can and, ultimately, should do. When trying to answer these questions the following statement from Saint-Exupéry in "*The Wisdom of the Sands*" (1948) should be kept in mind: "*As for the future, your task is not to foresee, but to enable it.*"

The key question at the beginning of the third millennium is: *How can we improve and increase the carrying capacity of the earth to feed the present and future world population?* Agriculture is the way in which people have increased the carrying capacity of the earth for humans. Technology has been an integral part of agriculture, and technological changes have molded the shape of agriculture. It will remain a key factor in the future. For more than 10,000 years, imagination and the use of agrotechnologies, (from the wooden plow, waterwheel, ridge drill, improved seeds, new horticultural crops, horse-drawn vehicles, dung, fertilizers, hybrid seeds, plant protection agents, tractors and combine harvesters to genetically modified plants and satellite-controlled tillage and harvest) have contributed to the availability of food for 6 billion people.

Our first priority must be to create the technical basis for producing sufficient calories and energy-rich food to meet human needs throughout the world. We must bear in mind that over the next thirty to fifty years, world food requirements will more than double. This will make it necessary to double – or even treble – agricultural production and supplies. At the same time, we will have to compensate for reduced farmland areas, water shortages and the transition from plant-based to meat-based diets. This not only requires a different sort of "*Green Revolution*", it also requires a "*blue revolution*" in terms of water conservation and sustainable use. In the next thirty years we will have to produce more food worldwide than over the whole of the last 10,000 years. And we will have to do all this in a sustainable and environmentally compatible way, requiring sustainable agricultural development on both high and low potential lands. Sustainable development means continuous innovations, improvements and utilization of environmentally friendly technologies with the aim of reducing environmental damage and consumption of resources. Improving sustainable agriculture means dematerialization and a rearrangement of resources. In other words: "*Do more with less!*"

Increasing Diversity of Agricultural Production Systems

Considered against the background of the exponentially increasing need for food caused by population growth, economic growth and changes in consumption patterns, the agriculture of the future will need to be increasingly multifunctional, heterogeneous, complex, knowledge-driven, technology-driven and adapted to available resources. To accomplish this, it will be necessary to encourage more fluid transitions between the various cultivation systems such as: rationalized agriculture, value-added agriculture, high-output agriculture, low external input farming, precision farming, prescription farming, site-specific farming, subsistence farming, organic farming, integrated farming, progressive farming, pioneering farming, and knowledge intensive agriculture. Synergies and symbioses will have to be studied and put into practice. More sophisticated technology will bring about a significant increase in the diversity of agricultural production systems.

The optimized and integrated use of all available technologies should be implemented at the level of the farmers. Extreme positions, such as organic farming focusing only on the environment and cost-intensive farming focusing on production, have their niches, but they cannot serve effectively from a global point of view (Kern, 1998).

When considering the high diversity of an agriculture adapted to available resources, it must be remembered that resources such as soil, climate, water, technical know-how, knowledge, etc., are distributed unequally throughout the world. Also, progress in agricultural technology will be implemented very rapidly in some regions, but hardly, if at all, in regions severely limited in their resources. Ultimately, there are two worlds when it comes to agricultural production: the haves and the have-nots.

"The Haves"

The "haves" – North America, Europe, Brazil, Argentina and Australia – possess sufficient cropland to meet most of their food needs and efficient agricultural production systems to allow sufficient food to be produced for export.

Agricultural functions perform well in preferred agricultural areas, with fertile soil, enough water, generally favorable climates, and farmers who are well educated, trained and informed and with access to all agrotechnologies. They are going the way of research- and knowledge-driven precision farming. In North America and Europe, less than 2% of the populations are working in agriculture, and the people are paying less than 15% of their monthly incomes for their "vital" foods. Out of a total of 1.9 million farmers in the USA, for instance, less than 300,000 account for more than 80% of the output. The other 1.6 million produce less than 20% (McCalla, 2000).

In some modern industrial nations, agriculture has changed from its traditional elementary function of food production to production of renewable raw materials, preservation of agricultural landscapes and nature conservancy. It ultimately has no other function than to stabilize the infrastructures of rural areas.

Technological changes are key forces in these countries and shape agriculture. The technological driving forces which will push forward sustainable and essential high-output agriculture are precision farming, biotechnology and genomics/functional genomics.

Precision farming is a strategy that employs detailed and site-specific information for managing production inputs. Precision farming increases productivity by using fewer inputs, i.e., a more sustainable agriculture. This includes precise application of herbicides, fungicides, pesticides and fertilizer as well as finder adjustment of seeding rates.

"Precision farming" is a current buzzword in agricultural circles. The term implies the careful tailoring of soil and crop management to fit the different conditions found in each field. It has brought attention to the use of the Geographic Information System (GIS) and the Global Positioning System (GPS) as well as the Near Infrared Vegetation Index (NVI) System. Some people incorrectly use the term "GPS" to imply precision farming only. We have literally taken *"agriculture into the space age"* (Johannsen, 2000). Farmers have access to services in which satellites collect data, transmit locational information, and provide data from a variety of sources. Farmers can analyze the satellite information themselves or rely on companies to perform this service for them for a fee.

In the near future, GPS will makes use of satellites that can identify the location of farm equipment within a few square centimeters of an actual site in the field. The value of knowing a precise location within inches is that:

1. locations of soil samples and the laboratory results can be compared to a soil map;
2. fertilizers and pesticides can be prescribed to fit soil properties (clay and organic matter content) and soil conditions (relief and drainage);
3. tillage adjustments can be made to allow for various conditions across the field; and
4. yield data can be monitored and recorded by simply moving across the field.

The application of chemicals and fertilizers in proper proportions are of environmental and economic concern to farmers. Using GPS together with a digital drainage map, the farmer can apply pesticides in a safer manner and in a more sustainable way. Weed information may also be detected remotely from aerial photographs. The NVI Spectrum allows weed infested areas to be distinguished from stubble and subsequently mapped; selective sprays can be produced (Kiernan and Bryant, 1999). Seeding hybrid seeds perform best when spaced to allow the plants to benefit from maximum sunlight and moisture. This is best accomplished by varying the seeding rate according to soil conditions such as texture, organic matter and available soil moisture. Since soils vary across an individual farm field, the ability to change seeding rates while moving across the field allows the farmer to maximize this seeding rate depending on the soil conditions. A computerized soil map of a specific field on a computer fitted on the tractor along with a GPS can tell farmers where they are in the field. This allows them the opportunity to adjust the seeding rate as they go across their fields. Furthermore, the ability to vary the depth of tillage depending on the soil is very important to proper seedbed preparation, control of weeds, fuel consumption and, therefore, cost to the farmer.

The real value for the farmer is the ability to adjust seeding rates, plan more accurate crop protection programs, perform more timely tillage and know the yield variation within a field. These benefits will enhance the overall cost effectiveness of crop production. The wide use of satellite-controlled sowing, fertilization and harvesting (2010) as well as of robots for cultivation, harvesting, sorting and post-harvest processing (2014 - 2018) is forecasted by Frauenhofer-Institute (Delphi, 1998). Digital technology clearly will open up wide new economic vistas in rural areas.

In 2000, worldwide more than 70 *transgenic plants* were registered (cotton, chicory, potato, pumpkin, corn, soybean, oilseed rape, papaya, tobacco, tomato, clove). Until now, more than 25,000 field trials have been carried out. Today genetic modifications of more than 100 plant species are growing in laboratories, greenhouses or in the fields throughout the world.

The first wave of biotechnological crops is already in the field. These crops provide agronomic traits that simply make it easier for farmers to grow their crops and to make a profit. In addition, the biotechnological crops contribute to more sustainable agriculture as farmers use fewer agricultural chemicals, conserve topsoil and water, and reduce fuel use.

Furthermore, agrobiotechnology makes an important contribution to integrated crop production (ICM) and leads to a further dematerialization of agriculture, precision farming and, last but not least, sustainable agriculture. Dematerialization stems from a better understanding of agricultural systems which leads to optimal use of scarce and expensive inputs, greater efficiency and less waste (Kern, 1999a).

In North America, South America, Australia, India, and China, commercial realization of biotechnological crops is under way. In 2000, the global area under genetically optimized crops was 44 million hectares, and it will reach 85 million hectares (15 million hectares in Europe/Asia/others, 25 million hectares in Latin America and 45 million hectares in the USA/Canada) by the year 2003 (6% of the total global arable land). In Europe, the issue is hampered by unresolved political deliberations. Genetically optimized crops in North America (1996 - 2003) will be grown on more than 60% of the total crop area. It is forecast that in the year 2003, around 60% of the crop area will be genetically modified (GM) corn, soybean and cotton, 20% GM potato, 30% GM tomato, 40% GM sugar beet, and nearly 75% GM oilseed rape.

The second wave of agrobiotechnology will reach the fields in 2002 - 2005, and here, the qualitative (output) traits will be modified. These traits will change the nature of the crop itself. "Commodity crops" (wheat, corn, canola, soybean, and rice) will be of better quality as well as have a higher market value. It is only a matter of time before new crop biotechnological products come on the market.

The third wave, i.e., the production of genetically optimized plants, nutriceuticals, probiotics or, ultimately, even medicines, can be expected after 2005. Around 2010, there will be genetically modified plants which produce special chemicals, for example potatoes which will be used for the production of paper, biological adhesives or detergents.

New plant breeding technologies will make it possible in about 10 to 20 years to use gene switches to initiate the production of special substances, e.g., pharmaceutically active ingredients and improved resistance management. It will be possible to eliminate or control pests with crop plants possessing built-in control mechanisms which can simply switch certain genes on or off as soon as particular pests appear. If this new "*gene and chemical switch*" technology is properly developed, it will undoubtedly provide a major value creation for chemistry in agriculture.

Genomics is concerned with the mapping and sequencing of genes (structural genomics) and the assessment of gene function on the basis of the information thus obtained (functional genomics).

The importance of genomics to the future of crop production is well described by McLaren from Inverizon International Inc., USA (McLaren, 2000). Genomics, especially functional genomics, will prove to be one of the major keys to opening up the promising benefits of crop biotechnology. The use of new tools in crop breeding will allow more precise, more easily measurable and more beneficial traits to be added to crops around the world. This will enhance the quality of life for billions of people.

The scope of possibility is impressive: one thousand grams of DNA (dissolved in one cubic meter) would have a larger storage capacity than all the computers which have ever been built, or about 100 billion times the storage capacity of the human brain. Thirty grams of DNA work one hundred times faster than the fastest computers we know today (Kaku, 1998).

The genetic code of life is a gigantic biological manuscript that we have hardly begun to examine or decode. An important goal will be to determine the function of a gene, in silico, in vitro and in vivo. Although this will be a task for the future, the following research fields will be established within the next few years:

- genetic mapping for identifying and localizing traits on the chromosomes;
- cytological mapping for visualizing chromosomes and bands corresponding to traits and for localizing genes on the DNA; and
- physical mapping describing the complete nucleotide sequences.

Genomic research will enable us to identify and validate new targets for insecticides, herbicides, fungicides and nematicides, and we will be able to determine the mode of action of active ingredients. Crop protection products will be essential to maintain the levels of production and the security of the food supply in the coming century (Yudelman et al., 1998; Kern, 1999b; Brooks and Roberts, 1999).

In the future, sustainable pest control will depend greatly on developing new strategies and tactics. Functional genomics will offer an effective box of tools. It will also put us in a position to identify new genes and gene functions to enable us to discover and validate new traits for seed improvements (Kern, 2000). Over 15 organisms have been fully sequenced, and the genomes of more or less all relevant plants will be analyzed within the next 5 to 10 years. Functional genomics will enable us to understand how the genome relates to the phenotype. Marker-assisted breeding will be an essential instrument for future seed improvements. It will shorten the breeding program from the current 10 - 12 years to only 2 - 4 years. The ability to introduce enhanced traits more rapidly to elite germplasm will be a key to future crop improvement (McLaren, 2000).

Besides the key technological driving forces discussed above, the value creation chain in plant production will realize a significant improvement. The sweeping changes in the market make it necessary to offer new preparations in the form of plant protection solution packages. Companies in the future will form scientific platforms which they will share with their customers. The changes in distribution structures resulting from e-business will reinforce customer binding. New products will become innovations if they are able to perform within the framework of an overall value creation chain (Stübler, 2000). On the other hand, the agribusiness consumer will raise much more specific questions about the quality and identity of products, e.g., food, and will want to know how they have been produced. The companies producing agrochemicals and plants will take the demands of the processing industries and the consumers into account by means of intensive relationship management concerning their new products.

The integration of the seed business into modern plant production and agribusiness industries, and also the drastic changes brought about by e-commerce or e-business, will lead to marked changes in value creation in agricultural business processes. It will no longer be new products alone which are created, but overall package solutions. Specialized supply chains will spring up. Rural communities that hitch themselves to the new forms of agriculture will benefit from the jobs brought by processing activities, as well as the prospect of higher incomes for large local producers (Drabenstott, 1999).

The "Have-Nots"

The "have-nots" make up most of the developing world. These countries contain over three billion people, and, for many reasons, they cannot produce or import enough food for their populations. In many cases, their food production capacity is deteriorating in the face of rapid population growth, misdirected agricultural policies, wars and widespread land degradation.

These countries or areas are poor in resources, highly heterogeneous and risk-prone. In the humid or subhumid areas and in the semi-arid tropics and subtropics, the farming systems are complex, the soil fragile and the weather highly variable. Also, these are the regions most likely to be negatively impacted by global warming. Most poverty is found in arid or semi-arid zones and in steep hillslope areas that are ecologically unstable. Agriculture in these areas suffers from little rainfall and from limited irrigation potential. Thus, agricultural productivity is very low. The arable land is of low fertility, and production is mostly for their own consumption. Most of these regions have large deficits in basic needs, such as nutrition, education, health and water. They have no or only partial access to agrotechnologies.

In such subsistence farming areas, more than 80% of the people are engaged in agriculture, most of them women and children. Nearly 90 - 100% of their "monthly income" has to be spent on food. In most developing and least-developed countries, the size of small family farms has been halved over the past four decades as plots are divided into smaller and smaller pieces for each new generation of male heirs. In 57 developing countries surveyed by the FAO in the early 1990s, over half of all farms were found to be less than one hectare in size, not enough to feed a family with four to six children (Hinrichsen, 2000). To make matters even worse, erosion and desertification now threaten 40% of Africa's non-desert land and 30% of Asia's production area.

Reflecting the limitations in these countries, agricultural technology will have to emphasize more sustainability with "low external inputs." For the short- and midterms, the most effective technologies will involve a combination of locally available inputs with selectively applied external inputs. Besides green manure, dung and compost, chemical fertilizers will be intensively used to increase crop production. Farmers who use the right amounts and types of chemical fertilizers suffer far less crop loss from competition for light and nutrients or from infestation. When no fertilizers are applied, diseases easily move into the fields.

The balanced use of organic and chemical fertilizers, herbicides and pesticides described by Ruben and Lee (2000) will help farmers consistently to raise land and labor productivity and maintain sustainable resource management practices (Yudelman et al., 1998). Nevertheless, the high labor requirements of subsistence or low external input agriculture may reduce returns to labor, and family labor constraints may hinder its adoption. Combining internal and external inputs will be essential for a sustainable intensification of food production.

Seed is the basic agricultural input for a sustainable agriculture. *"Seed is in fact the hub around which all other strategies to improve productivity revolve,"* as Sehgal (2000) states. More than 50% of the rate of yield increase is due to genetic improvement. Improved farming techniques, agrochemicals and machinery are only as effective as the germplasm they support. This means that farmers everywhere require a secure source of good quality seed for successful harvests. Access to improved seeds adapted to local demands is important for sustainable intensification. Significant contributions will be made by the public and private sectors. The seeds that farmers require must be bred for the areas in which they are to be grown and extensively evaluated there before being released on a large scale. Furthermore, an improved hybrid or variety is useless unless the seeds reach the farmers in sufficient quantities and are of high quality and purity. Farmers in the developing world need to establish local industries involved in the breeding, production and distribution of seeds in order to meet their particular needs.

Many crops that are important to the developing world are not interesting to multinational seed companies, even to those operating in developing countries. For this reason it will be important to support the development of an indigenous public and private sector seed industry. Smallholder agriculture will benefit significantly from such seed improvements. For example, hybrid rice (average yield of 6.8 tons per hectare) which has revolutionized rice production in China will be available in almost all Asian countries. Seeds tolerant to abiotic stress will be developed and sold or distributed to farmers living in low potential areas – a significant contribution to sustainable agriculture and livelihoods.

Unfortunately the speed of development, implementation and utilization of genetically modified crops in the developing world is unsatisfactory, although the enormous potential of biotechnology is well known.[1] For example, FAO (1997): *Biotechnology, particularly the genetic engineering of plants or animals to meet specific needs, holds much potential for meeting the challenge of increasing productivity and conserving natural resources.*

AGENDA 21 (1992), Chapter 16: Biotechnology, expressly propagates gene technology: *Biotechnology, an emerging knowledge-intensive field, is a set of enabling techniques for bringing about specific man-made changes in deoxyribonucleic acid (DNA), or genetic material in plants, animals and microbial systems, leading to useful products and technologies. By itself, biotechnology cannot resolve all the fundamental problems of environment and development, so expectations need to be tempered by realism. Nevertheless, it promises to make a significant contribution in enabling the development of, for example, better health care, enhanced food security through sustainable agricultural practices, improved supplies of portable potable water, more efficient industrial development processes for transforming raw materials, support for sustainable methods of afforestation and reforestation, and detoxification of hazardous waste.*

The quantitative contribution of biotechnology to securing global food supplies in the various regions of the world - which is not the same thing as eliminating hunger in the world - will proceed along the following lines: In 2025, 28% of the food production in the developed world (USA, Europe, Australia, Canada, CIS) will be made up of genetically modified materials. Asia will achieve 20% of its food production with genetically modified organisms (GMO) food, Latin America 17%, and Africa only 6%.

There are two reasons for the very low share in Africa. The first is that there are hardly any lucrative areas for biotechnology there. Especially while this technology is still at the introductory stage, private companies will tend to concentrate on profitable markets, mostly in the industrial nations. The second reason is that in the western world the risks of GMOs are the subject of very heated and often contradictionary public discussions. They are sometimes condemned entirely. The conclusions drawn from this can be described very briefly as follows: as long as the risks have not been one hundred percent clarified, there will be no transfer of this technology to the Third World. This will be yet another example of the way we so often make use of new technologies but leave

---

[1] Sasson (1993) presents an important overview of the status of biotechnology research and development in the developing world through a regional and national survey. While describing the situation in the early 1990s, it also includes forecasts for the end of the decade and contributes in this way to the understanding of the economic and sociocultural implications of biotechnologies. A further overview is given by de Kathen (1999).

the Third World 20 to 30 years behind. Many years of experience in scientific research have shown that starting gene technology research now to help solve a typical African problem will mean at least 10 to 15 years before a solution comes on the market. A delay in introducing biotechnology will have the effect of keeping its contributions in Africa down to only about 6%.

The crucial questions are: What are we going to do about it? How is this key technology to be put into the hands of small farmers in, for example, the Himalayas? What value enhancements in plants are necessary and beneficial in the long-term? In January 2000, the potential benefits of genetically modified rice, called "golden rice" because of the color caused by the modification, was announced. Vitamin A was added to the grain - which could cure the vitamin A deficiency of 124 million children worldwide. The deficiency causes blindness. "Golden rice" was developed by the Swiss Federal Institute of Technology by inserting three genes into the rice; these make the plant produce beta carotene or provitamin A. It will be made freely available to farmers in developing countries.

Other new scientific breakthroughs were published recently. Genetically modified rice which could boost yields by up to 35% was developed jointly by Washington State University and agricultural researchers in Japan (Brown, 2000). Genetic material from maize was inserted into the rice, boosting the rate of photosynthesis so that the plant was able to produce more sugar and higher grain yields. Such rice plants could play a major role in future agriculture, especially in developing countries. Insect-resistant cabbage, for example, is cultivated in the same way as non-insect-resistant cabbage. Insect-resistant cabbage is not devoured by cabbage moths, and thus the harvest is protected. Here, then, is a possible way to increase and safeguard yields.

Joint research by Aventis CropScience and the Biotechnology Institute of the University of Cambridge in England recently achieved more rapid growth by an increase in the rate of cell division. After an Arabidopsis gene responsible for cell division had been integrated into a tobacco plant, it became possible to obtain a doubling of the growth rate. This achievement opens up the future possibility to accelerate vegetation periods and secure several harvests per year, thus bringing about an overall increase in productivity. This result would be a simplified and more cost-effective contribution from active ingredient production in plants.

An intensive implementation of biotechnology in developing countries will be an important contribution for safeguarding world food security and feeding humankind. Despite the low quantitative contribution of biotechnology to securing global food supplies in Africa, studies in Kenya and other African countries (Wambugu, 1999; Qaim, 1999) have shown that the technology and products were easily adopted by farmers. The "*quality/technology is in the seed,*" and farmers do not need to alter their traditional farming practices to obtain tremendous benefits.

The Key Element of Future Agriculture

All over the world a key element of future agriculture will be the concept and the realization of Integrated Crop Management (ICM). This will meet the requirements of sustainable development and agriculture by managing crops profitably without damaging the environment or depleting natural resources for future generations. This dynamic system uses the latest research, technology and experience in ways suitable to local conditions in order to optimize food production, enhance energy conservation and minimize pollution worldwide. ICM will enable Integrated Resource Management (IRM) as well as Integrated Plant Nutrition Systems (IPNS), which bring together all controllable agricultural production practices to provide long-term sustained productivity of safe and wholesome food in an environmentally sound manner.

In this context, it is interesting that today it is generally reckoned that 1.7 billion people "owe" their food to fertilizer inputs, another 1.7 billion to intensive irrigation, and 2.4 billion to plant protection and plant production together with the specialized knowledge and expertise of farmers throughout the world. This picture will change. On the whole, the use of fertilizers will increase dramatically, especially in developing countries. However, this will not be the case in Western Europe where the fertilization level is about 100 kg per hectare, which is already the optimum limit. Irrigation systems will be extended, and there are major reservoir projects around the globe. The amount of plant protection will decline. It is assumed that by 2025, 1.6 billion people (18% - 20%) will owe their food to the use of genetically modified organisms. Insect resistance, modified fatty acid spectra in plants, tolerance to biotic and non-biotic factors, and renewable raw materials in the form of starch modifications or bio-plastics and bio-detergents are just some of the steps along this road.

Conclusions

The history of technological innovation and development in agriculture suggests that changes are not progressing in a linear way but in a positive, non-linear acceleration. Fortunately, the increasing speed at which agrotechnologies are being implemented corresponds to the exponential growth of world food demand and safeguards world food security, today and in the future.

The technological components can only be effective when considered in conjunction with the ecological, economic, political, social, ethical and demographic factors. By means of an intelligent integration of all agrotechnologies and of suitable social and political conditions, it will be technically possible to make sufficient food available and to reduce hunger significantly at regional levels. To do this, as Qaim and Virchow (1999) rightly insist, it will be necessary to have a constructive dialogue or *cooperation* between farmers, scientists, the economy, NGOs, politicians and consumers - with all sides being prepared to make concessions. Governments will have to support public research facilities but at the same time encourage the private sector, i.e., industry. Seed breeders, biotechnology companies and many other companies associated with agriculture are becoming partners. Only if we all work together will we be able to find solutions that will enable sufficient food to be produced in a sustainable way.[2] The private sector will have to develop or present new solutions to problems in markets surviving on short or medium term profits. For this purpose, market structures must already exist or be easy to establish, and investments must have sufficient protection. Private research, especially from the industrial countries, will find itself increasingly under an obligation to develop new technologies and to help adapt these to local conditions in the developing countries. In all cases, adaptation of the private sector's technology to the developing countries requires:

- a proper regulatory environment to secure global deregulation standards;
- enforceable intellectual property protection;
  national/international technology transfer infrastructures including analytical capabilities to monitor GMOs;
- development investments; and
- local seed industries.

Achieving food security is a political and not just a technical matter. In a democratic process, people must be consulted about matters which affect their right to food. We will double the size of the world economy in just 12 years, doubling it again in 25 years. By the year 2025, more than 30% of all foodstuffs worldwide will be exported or imported. While the new agrotechnologies increase in importance or even assume top priority, *trade* will also come to assume a key position in making good quality food available to people in sufficient quantities. WTO negotiations, GATT agreements and intellectual property rights will decisively influence crops, jobs and livelihoods.

The integration of policies and all technologies, including biotechnology and genetic engineering, will make it possible to meet the increased needs for food without extending agriculture into areas of high biodiversity. It will also promote agricultural systems which conserve diversity within the system itself. In the future, we can expect a more sustainable agriculture, which will be the foundation of better crops, better nutrition, advanced jobs, and better livelihoods in an intact environment.

References

Brooks, G.T. and T.R. Roberts (eds) (1999): Pesticides, Chemistry and Bioscience. The Food- Environment Challenge. The Royal Society of Chemistry, Cambridge.

---

[2] A valuable and informative UNESCO documentation of international cooperations under the title "*Biotechnologies in developing countries: present and future Vol 2: International cooperation*" has been provided by Sasson (1998).

Brown, P. (2000): Unveiled: The GM Rice that could Feed World. Internet: http://www.mg.co.za/mg/news/2000mar2/31mar-genetic.html.
de Kathen, A. (1999): Transgenic Crops in Developing Countries. A Report on Field Releases, Biosafety Regulations and Environmental Impact Assessments. Umweltbundesamt Texte 58, Berlin.
Drabenstott, M. (1999): Meeting a New Century of Challenges in Rural America. Internet: http://minneapolisfed.org/pubs/region/99-12/rural.html.
Frauenhofer-Institut für Systemtechnik und Innovationsforschung (1998): Delphi '98. Studie zur globalen Entwicklung von Wissenschaft und Technik.
Hinrichsen, D. (2000): Feeding a Future World. Internet: http://www.oneworld.org/patp/pap_7_1/hinrichsen.html.
Johannsen, C.J. (2000): An Overview on Precision Farming. Internet: http://www.giannottitech.com/prefarmoverview.htm.
Kaku, M. (1998): Zukunftsvisionen. München.
Kern, M. (1998): Sustaining High Output Agriculture. The World Bank, American Society of Agronomy, Crop Science Society of America, Soil Science Society of America, Sustainability in Agriculture Systems in Transition, October 20-22, 1998, Baltimore, Maryland, USA.
Kern, M. (1999a): An die Grenzen des Notwendigen. Biotechnologie als integraler Bestandteil nachhaltiger Landwirtschaft im 21. Jahrhundert. In: Haushalt & Bildung Heft 4/1999, pp. 14-18.
Kern, M. (1999b): Global Perspectives on Pesticides, Biotechnology and Seeds. National Academy of Science and Technology (NAST), 21st Annual Scientific Meeting, Food, Population and Environment, July 7-8, 1999, Manila, The Philippines.
Kern, M. (2000): Entomology for the Third Millennium. Scientific and Technological Challenges. Closing plenary lecture. XXI International Congress of Entomology, August 20th to 26th, 2000, Foz do Iguassu, Brazil.
Kiernan, R. and M. Bryant (1999): Precision Patch Spraying of Weeds. Internet: http://www.rinex.com.au/news.html.
McCalla, A. (2000): Agriculture in the 21st Century. CIMMYT Economics Program. Fourth Distinguished Economist Lecture. CIMMYT, Mexico.
McLaren, J.S. (2000): The Importance of Genomics to the Future Crop Production. Pest Management Science, 56, pp. 573-579.
Qaim, M. (1999): The Economic Effects of Genetically Modified Orphan Commodities: Projections for Sweetpotato in Kenya, ISAAA Briefs 13. Published in Collaboration with Center for Development Research (ZEF), Bonn.
Qaim, M. and D. Virchow (1999): Macht Grüne Gentechnik die Welt satt? Herausforderungen für Forschung, Politik und Gesellschaft. Gutachten Friedrich Ebert Stiftung, Bonn.
Ruben, R. and D.R. Lee (2000): Combining Internal and External Inputs for Sustainable Intensification. IFPRI 2020 Brief 65. IFPRI, Washington, D.C.
Saint-Exupéry, A. (1948): The Wisdom of the Sands.
Sasson, A. (1998): Biotechnologies in Developing Countries: Present and Future (vol 1/2), UNESCO Publishing, United Nations, Imprimerie des PUF, Vendome, France.
Sehgal, S. (2000): Agricultural Biotechnology and the Seed Industry: Some Implications for Food Production and Security. In: Qaim, M., A. Krattinger and J. von Braun (eds): Agricultural Biotechnology in Developing Countries: Towards Optimizing the Benefits for the Poor. Kluwer Academic Publishers, Boston.

Stübler, H. (2000): Forschung und Entwicklung. Neuorientierung und Aufbruch am Beispiel einer Life Science Industrie. Vortrag anläßlich der Handelsblatt-Jahrestagung, 4. Mai 2000, Frankfurt.

UN (1992): Agenda 21: Programme of Action for Sustainable development. Rio Declaration on Environment and Development, Rio de Janeiro.

Wambugu, F. (1996): Case Study: North-South-Transfer of Biotechnology to East Africa. In: International Workshop on Biotechnology for Crop Production. Its Potential for Developing Countries. Berlin, 9.-13. December.

Yudelman, M., A. Ratta and D. Nygaard (1998): Pest Management. The Expected Trends and Impact on Agriculture and Natural Resources to 2020. IFPRI 2020 Vision Food, Agriculture and the Environment. Discussion Paper Series. IFPRI, Washington, D.C.

Agenda 21 §14.74 Earth Summit, Rio 1992

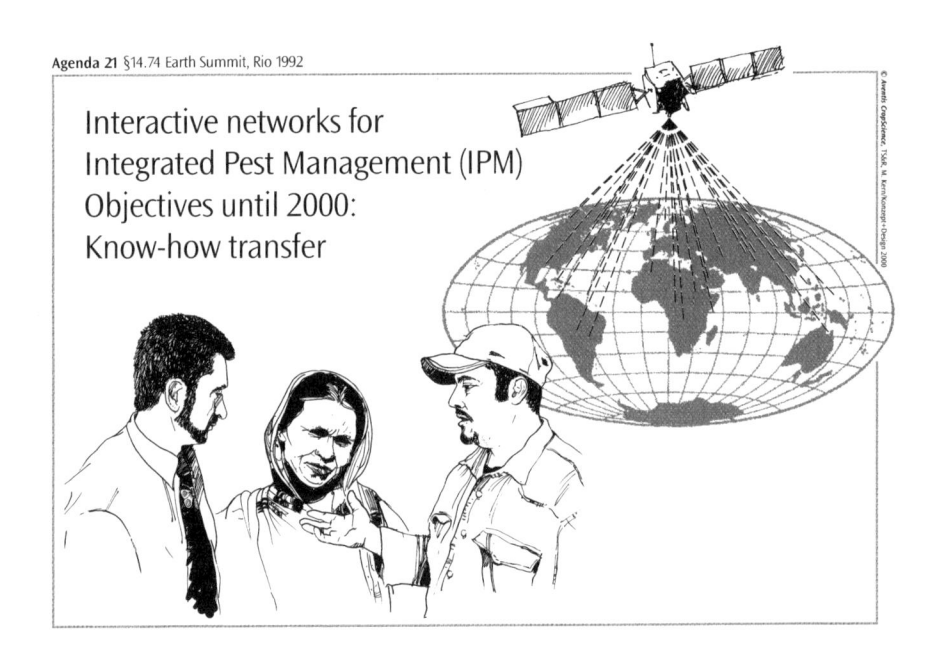

Interactive networks for
Integrated Pest Management (IPM)
Objectives until 2000:
Know-how transfer

41 Declaration of Hamburg: Meeting Future Human Needs

The participants from the 3rd International Crop Science Congress 2000, held in Hamburg from August 17 - 22, 2000, adopted the following declaration, paying tribute to a more holistic, interdisciplinary approach, exploring the linkages and interactions between the improvement of crop productivity, food safety and human health and the wise use of genetic and natural resources to solve the problems of the next millennium. The following is an excerpt[1]:

... At the present, agriculture is able to feed the majority of humans. But scientists are becoming very concerned that this will not be the case when we have 8 billion people on the planet in about 20 years time. Failure to feed 8 billion people in a sustainable way will lead to enormous environmental damage, social dislocation and reduced economic growth that will affect the whole world. Scientists, gathered at the 3rd International Crop Science Congress, alert the society and policy makers to take action on the following concerns:

- The lack of awareness on the gravity of food security and poverty issues for the next 20 years on the global level.
- The urgency of protecting genetic resources and biodiversity.
- The scarcity and degradation of natural resources, such as land and water.

Strengthening agricultural research and education at national and international level is a prerequisite to fulfill future human needs. We believe there are grounds for cautious hope in our ability to feed ourselves via improved education, modern technological developments and most importantly a shared appreciation of the problem. Failure of the world's agricultural scientists to communicate this message would be an abandonment of one of their most important professional and ethical obligations.

The third International Crop Science Congress highlighted the fact that sustainable development of plant production and resource conservation is essential for achieving and maintaining food security. This requires a better and more comprehensive insight into ecologically sound crop production processes, especially in fragile environments and resource-poor countries.

The participants at the 3rd International Crop Science Congress decided to foster the necessary dialogue on the potential contributions of new technologies, such as molecular biology and gene technology, to production agriculture and resource use management. Science-based knowledge is essential for a well-informed dialogue and for an effective policy and regulation of new technologies.

---

[1] The full text can be read at: http://www.cch.de/CROPSCIENCE.

Concerns and Prospects:

1. *Crop sciences* provide the key-knowledge base for increasing the production and quality of human food, animal feed and biomass for industrial use and the provision of energy. Crop sciences play a vital role in improving the quality of life for all human beings by producing food and renewable resources in a sustainable and safe way that meets the needs and the standards of the Global Village.
2. *Food security* continues to be a growing concern globally, due to widespread poverty and the need for more and better quality food for a fast growing population, particularly in developing countries. The world's food supply depends in many regions on the *availability of land and water*, which is becoming increasingly scarce as demands for it increase. Water saving strategies and a better adaptation of crops to limited water and nutrient availability in semi-arid regions are key issues for research.
3. To overcome the large losses due to *pests and weeds* and to reduce the use of pesticides and herbicides, advanced technologies for integrated crop management and ecologically based sustainable cropping systems should become available.
4. When developing new crops and cropping systems knowledge of *genetic resources and plant-biotechnology* should be developed and applied with a focus on sustainable and efficient plant production. An integrated and multidisciplinary approach to crop science is needed for innovations and improvements in agricultural production and resource conservation at the field, farm/household, regional and global level.
5. To meet the concerns of society, participation of scientists in the public debate on the potential benefits and risks of the application of *modern technologies* should be encouraged by all public and private research organizations.
6. For the sake of human health, a high priority should be given to improvement of the quality and monitoring of the *safety of human food and animal feed* throughout the production, storage, processing and delivery chain of plant produce.
7. To bridge the *gap in knowledge and modern technologies* between developed and developing countries, more financial resources should be made available for education, training, access to public and private research as well as for technology transfer.
8. To strengthen the role of farmers in the *stewardship* of the rural landscape, natural resources and environment, more emphasis should be given in research and education to forms of land use other than agricultural production.
9. To foster a greater understanding of *environmental impacts*, and to promote efficient *resource use*, knowledge and research tools in crop science should be combined with advances in information and communication technologies, such as GIS and DSS.

42 The Future of Agriculture in the Industrialized World

William Lockeretz

Predicting the future of agriculture is notoriously treacherous, and I do so only hesitatingly. For example, more than a few years ago who would have predicted the current situation for genetically modified crops, one of the most notable technological changes in agriculture in recent decades? On one hand, this technology was adapted remarkably rapidly in several countries. In the US, after barely five years, some 30 to 40% of major crops such as soya and maize are planted with genetically modified varieties. On the other hand, this change has evoked massive protests from consumers and environmentalists around the world, protests that in agricultural matters have hardly been matched in intensity and impact. It also is dangerous to offer generalizations regarding "the industrialized world". This oversimplified label encompasses a wide range of agricultural systems and cultural values.

These caveats notwithstanding, I would like to offer two broad-brush predictions concerning where agriculture is likely to go in most industrialized nations. The first seems fairly safe; the second is slightly more venturesome.

First, I expect agriculture as a whole to become increasingly industrialized. But at the same time I expect continued growth of the much smaller but still significant sector that is loosely called "alternative agriculture." This sector rejects treating agriculture as just another industry. Instead, it seeks to preserve what is most worth preserving from traditional agriculture and combining that with the best of current scientific knowledge and technology to create an agriculture that meets the diversified demands of modern society better than one that blindly follows the industrial model.

Paradoxically, therefore, I see two opposing tendencies, with agriculture splitting even more sharply into two diverging streams. The mainstream, which already has moved substantially towards industrialization, will remain the mainstream and will move even further toward industrialization. Meanwhile, the alternative stream will both gain strength and find itself separated even more from the mainstream. Of course, it is also possible that the two streams will partially converge, with alternative agriculture influencing the course of the mainstream - or simply getting absorbed by it. However, as I discuss at the end, I do not expect this to happen.

Increasing Industrialization of Agriculture

The characteristics of industrial-style agriculture have been well described by many observers and need only be summarized here:

- fewer but larger and more specialized farms;
- substitution of machinery and other capital investments for labor;
- greater use of inputs from industry, especially fertilizers and pesticides, and with it, greater dependence on non-renewable resources, especially fossil fuels,
- greater distinction between farm management and labor, with the latter often employed in temporary, low-skilled and low-paying jobs;
- greater separation of livestock production from feed production, often with livestock raised in confined, factory-style facilities using only brought-in feeds, often with heavy reliance on antibiotics and growth hormones;
- greater concentration of production (both internationally and within individual countries) in regions with lowest costs of production, to the possible detriment of more marginal regions;
- longer marketing and distribution chains from producers to consumers, both geographically and in the number of intervening handlers and processors;
- more vertical integration of farms with processors, either through contracting or through outright ownership of the former by the latter; and
- more emphasis on interchangeable commodities rather than regionally characteristic speciality foods. (However, there also has been a recent interest in developing highly specialized varieties intended for particular processing applications or as sources of industrial raw materials).

Rapid movement in all these directions has characterized the industrialized world for at least a half century and in some countries for more than a century. Some would say that these changes are inexorable. However, they also have been fostered by public policies that have given capital favored tax treatment over labor, for example, and that have subsidized a farm in proportion to how much it produces rather than in accordance with the farm family's income needs. National agricultural research establishments certainly have not been neutral in the kinds of changes they foster. They have been an important force for industrializing agriculture. More recently, industrialization has received additional impetus from the liberalization of world trade, which favors production at the lowest possible cost over production that gives more weight to food quality and safety, environmental protection, and equitable treatment of workers.

Continued Growth of "Alternative" Approaches

But regardless of how much the industrialization of agriculture has been the result of free market forces and how much the result of public policies, I see little indication that, for most of agriculture, the future holds anything substantially different.

Although greater industrialization of agriculture may be inexorable, it is not necessarily welcomed, certainly not by everyone. Thus, especially in the past quarter of a century–although its origins go back much further–a conscious attempt has been made to offer an alternative vision of agriculture. It could be characterized by what it does not want–all the characteristics listed above. But it is much better characterized positively, by what it does want:

- production of safe, wholesome, high quality foods, especially those that embody "authenticity";
- integration of many diverse enterprises into a balanced farming system striving for closed nutrient cycles;
- reliance on the farm's own feeds, plant nutrient sources, and other resources in preference to those brought in from outside;
- less intensive production methods that may give lower yields but that also have lower production costs, sometimes lower by enough to leave the farmer with comparable net income;
- preservation and enhancement of soil quality through livestock manures, green manures, and appropriate crop rotations;
- enhancement of natural means of pest control, such as predators and parasites;
- humane treatment of animals;
- protection of wildlife, biodiversity, water quality, attractive landscapes and other environmental values;
- farms that are big enough to support the farm family but not much bigger;
- emphasis on the craft of farming, with sophisticated management by committed, knowledgeable, ecologically aware farmers;
- closer contact between consumers and producers;
- maintenance of farmers as the heart of the food production system, rather than subordinating them to processors, marketers, and input suppliers;
- decent and equitable treatment of farm laborers; and
- farms that contribute to the social and economic vitality of rural communities.

Alternative systems take varied forms that embody these characteristics to a greater or lesser degree. Particularly noteworthy is organic farming, especially when it is carried out according to systematic, enforceable standards, such as those of the International Federation of Organic Agriculture Movements. Once insignificant and ignored, its increasing acceptance in the past decade or two has been truly remarkable. Consumer demand for organic products is growing rapidly throughout the industrialized world, in contrast to the generally stagnating demand for food in general. Also, organic products are now offered in a much greater number and range of outlets. Legally binding standards have been developed not only at the national level but globally as well. Considerable public money is being spent on subsidies and organic research. The result has been that in some countries a big debate within the organic community is whether the growth has been too rapid, rather than how it might be accelerated.

I expect that organic and other alternative approaches will continue to gain support because they serve three distinct but complementary sets of interests: those of consumers, of farmers, and of society as a whole. I say this while also recognizing the arguments that agriculture will continue to become more industrialized, because that way it will best serve other important interests, such as consumers' interest in cheap food, farmers' in making money, and society's in having a reliable and adequate food supply. Competing forces are at work simultaneously, which is why I see agriculture moving in two directions, not one.

Interests of Consumers, Farmers, and Society

Consumers' choices are complex and multifaceted. If consumers were concerned only about getting food at the lowest possible price, alternative agriculture would have a hard time taking hold, because industrial-style farming produces food cheaply. In industrialized countries, the cost of food as a share of income has been declining for a long time. It is now typically around 15%. But other things matter to consumers, too. For many the price of food is low enough that they need not search for the cheapest version. They can and will pay extra for something they perceive as superior, whether because it is fresher, safer, or more interesting and varied (e.g., regional speciality cheeses or "heirloom" fruits and vegetables). Therefore, even if the products of alternative agriculture are more expensive, although they are not always, this is not an absolute barrier to their acceptance, as shown by the growing demand for organic foods. Organic foods also get a boost every so often from one or another food safety scare. Recent well-publicized examples are BSE[1], regarded by many as the quintessential example of industrial agriculture gone awry, and last year's contamination of Belgian meat and eggs with dioxin from tainted commercial feeds.

Like consumers, *farmers* are motivated by more than just money. If that were not so, alternative agriculture would have a problem gaining acceptance, because it sometimes is less profitable. (Sometimes, though, its lower yields are more than offset by lower production costs, higher product prices, or government subsidies, such as those under the European Union's agri-environmental program.) But the assumption that farmers strive only to maximize their profits–however much this is beloved by agricultural economists–is easily refuted. Very simply, if farmers were motivated only by profits, most would stop being farmers. They almost always could earn much more by putting their skills and their capital to work elsewhere in the economy. Certainly, farmers cannot go on losing money forever. But provided they make enough to stay in business, many other motivations come into play. They farm because they love it, because they take pride in their abilities and in the importance of what they produce, and because a farm is a good place to raise children. But farming offers all these attractions only if it is the right kind of farming: the kind they really do love, the kind that lets them use their abilities, the kind that produces products they really can be proud of, the kind that keeps the farm clean and healthful and does not threaten their children's well-being with toxic chemicals. While some farmers take pride in their ability to manage an ever-larger operation, others find that the continued need to expand puts them on a perpetual treadmill that makes farming too stressful to be enjoyed. While some derive pride and satisfaction mainly from being able to get ever-higher yields,

---

[1] *Bovine Spongiforme Enzephalopathie*, the "mad cow disease".

others seek a different kind of satisfaction, such as comes from hearing their customers tell them how much they appreciate their products. Thus, as with consumers, alternative agriculture is not for everybody, but it is for somebody. As industrial-style agriculture comes to dominate even more, I expect that more and more farmers will say, "*This isn't why I farm*", and they will seek alternatives. Even more important, many countries are finding that the most likely future farmers, the children of current farmers, are turning away from it. In its current form, farming is not as attractive as other careers. However, many would stay in farming if it could be made more appealing, something that alternative agriculture promises. Likewise, alternative agriculture holds considerable appeal to potential entrants from non-farm backgrounds. Many of these people otherwise would not have considered farming at all. They constitute an important countercurrent to the prevalent exodus from farming.

Of course, every member of any *society* is also a food consumer. However, society's interest in how food is produced goes beyond the consumer-related aspects of food discussed earlier, such as price. That is why, through both government policies and the activities of non-governmental organizations, the public in the industrialized world increasingly supports alternative agriculture, as has already been noted regarding organic farming.

Public Support of Alternative Agriculture

The reasons for public support of alternative agriculture are many, and they include both the national economic interest and environmental goals. Once, the industrialization of agriculture might legitimately have been said to be in the national economic interest because it "freed" farmers to work in industry (although the particular farmers forced to leave probably did not think of themselves as being "freed"). But, at most, that was true when growing economies had plenty of jobs for ex-farmers, which has not generally been true more recently. That alternative agriculture employs more people is now recognized as a social benefit, not a drawback. So, too, is the fact that it often is particularly well suited to agriculturally marginal areas. These increasingly are being abandoned by industrial-style agriculture, leaving them depopulated and economically depressed because of the lack of other employment.

Alternative agriculture also offers potential economic benefits in the international arena, especially when its products have a distinct identity, such as certified organic. Trade liberalization increasingly will mean that countries producing undifferentiated bulk commodities will have to compete on price alone. This places international competitiveness in direct conflict with farmers' incomes. Not everyone can win with this "beggar thy neighbour" strategy. For some, an attractive alternative will be to offer higher quality products than the competition and to cultivate a "green" image internationally, such as Sweden's "*On the way to the world's cleanest agriculture*". (Of course, such sloganeering works best if your agriculture truly is green.) Moreover, quite apart from food safety risks to your own citizens, your economic well-being is not well served if other countries shut out your food exports because of safety fears, as with the European Union's ban on British beef because of BSE and on Belgian animal products because of dioxin. Nor does it help when your overseas customers massively reject the products of your newest industrial-style production methods, as the US has painfully learned regarding genetically modified foods in European markets. (In contrast, GM foods are strictly forbidden in organic farming.)

The environment is another important source of support for alternative agriculture. For example, the environmental benefits, along with the potential for enhancing exports, are a major reason behind Denmark's aggressive action plan to promote organic farming. This includes a serious evaluation of a *total* conversion of the country's agriculture.

Among the environmental problems that alternative agriculture can help solve, a particularly widespread one is water pollution by pesticides and by nitrogen from inorganic fertilizers and factory-style livestock facilities. This problem is not just a matter of intangible environmental values. It has a clear monetary value too, because it often entails heavy costs for public drinking water supplies. Eliminating the problem at the source is clearly the preferred strategy, economically and in every other way, as water districts in agricultural areas throughout Europe have discovered.

Besides contaminating water supplies, pesticides have long been recognized as damaging to wildlife, including many endangered species. Consequently, significant restrictions on pesticide use have been imposed in industrialized countries. However, industrial-style agriculture also threatens wildlife in less obvious ways than direct poisoning. When large areas are planted to a single crop, it deprives wildlife of important food sources and habitat. Alternative agriculture, in contrast, features diversified rotations, often including pasture and meadow that harbor much more varied fauna. Also, hedgerows and other non-cropland landscape elements are more likely to be found on a farm with many small fields. In addition, a diversified agricultural landscape will be much more attractive to people, whether they are there for "agrotourism" or are just passing through.

The Future Trend for the Different Approaches

All these sources of support, working together, make it highly likely that organic and other forms of alternative agriculture will thrive throughout the industrialized world. Yet this does not mean that today's alternatives will become tomorrow's mainstream. As discussed earlier, I expect that the main change in agriculture as a whole will be toward an even more industrial model. I am under no illusion that alternative agriculture will ever be anything but an interesting, significant and growing side stream, but only a side stream, while the mainstream remains the mainstream. (If Denmark indeed decides to adopt the goal of 100% organic, that would certainly be a dramatic counterexample, although one that probably will not be followed by many other countries).

But what about the possibility that the mainstream, while remaining that, will all least shift its overall direction toward alternative agriculture by selectively adopting some of its features? There have been some developments along these lines, such as more stringent restrictions on the use of pesticides and fertilizers. These developments are likely to continue. However, they involve only specific production technologies. In contrast, in areas such as the number and size of farms, the subordination of farmers to input suppliers and food processors and the separation of farmers from consumers, I expect the mainstream to continue to pursue the industrial approach, moving even further from the alternative model.

There also has been an ostensible "greening" of the food system downstream from the farm itself. But this usually does not involve a shift in the direction of the mainstream as a whole. Rather, it typically involves large national or multinational food companies buying up organic and other alternative enterprises or developing their own "whole food" or "natural" or organic product lines. In the meantime, the bulk of their business continues unchanged.

This kind of mainstreaming of alternative agriculture may please some of its supporters, at least for a while. But it will hardly meet the full range of goals that can be achieved only with a full commitment to alternative approaches. While some organic consumers will be satisfied with the new organic products offered by the food industry, others will not be able to ignore that the products have come from thousands of kilometers away or were so thoroughly processed and refined that they lost any possible advantage from having been made from organic raw materials. Nor will they be satisfied with buying their organic foods in a gigantic supermarket rather than from, say, a local grower at a farmers' market.

Similarly, some farmers will not be satisfied with just doing the minimum required to obtain organic certification–as they no doubt would have to do when they grow under contract to a processing company concerned only about minimizing its costs. Rather, they also will want to farm in a way that reflects not just the letter but also the true spirit of organic farming–such as making a special effort to preserve biodiversity. And some policy makers will realize that although certain national economic goals may be served when mainstream processors and distributors handle more organic foods, this will do nothing about depopulation and depression of marginal areas if the governing principle is "*buy it wherever you can get it the cheapest.*" Nor will the goals of wildlife protection, humane treatment of animals, or equitable treatment of labor be served without a strong commitment to these goals. So far, there is no evidence that the mainstream food system has that commitment or that the new-found interest in "green" food is much more than a corporate marketing ploy (although I grant there are individual corporate exceptions).

Therefore, although I would love to be proved wrong, I do not expect that the mainstream food system will shift significantly in the alternative direction. Rather, in the face of more "greening" by multinational giants, I expect that the original motivations of consumers, farmers, and citizens will reassert themselves, and some members of all these groups will continue to demand a real alternative to how most of our food is produced and marketed. This interest clearly is not a fad, as it once was derisively characterized. It is here to stay. Its growth has been impressive already, and I expect it to offer an evermore attractive and significant vision for the most basic of human enterprizes.

43 Ensuring the Future of Sustainable Agriculture

Patricia Howard-Borjas and Kees Jansen

Sustainable Agriculture and the Green Revolution: Parallels between Paradigms

A few decades ago, Green Revolution technologies were seen acritically as providing the "solution" to a perceived crisis in agriculture, especially related to the need to increase yields, agricultural incomes and total food production. The Green Revolution has since been critiqued by many for its unanticipated social and environmental costs. Its promoters have been accused of failing to consider trade-offs entailed in terms of equity and the environment. This failure can be attributed, at least in part, to the fact that early research tended to focus on the "obstacles" to the adoption of Green Revolution technology and on the best means to promote it rather than on the potential socioeconomic and environmental consequences of its adoption.

Two decades of research on the socioeconomic and environmental processes entailed in the Green Revolution should result in a more critical perspective on the application of any new technical prescriptions in agriculture, no matter how popular. But today, policy makers, activists and researchers are promoting the "Sustainable Agriculture" (hereafter SA) paradigm with little critical forethought. This is occurring even though, like the Green Revolution, SA entails "new and different" packages of technologies, techniques, knowledge and inputs that are also meant to overcome a crisis in agriculture, albeit defined in different terms. In fact, an assumption underlying most of the contemporary literature on SA is that its technical prescriptions *can* be applied acritically since they take as their starting point everything that the Green Revolution was not. "*Virtually everything that is perceived as being "good" or benign is included under the umbrella of "sustainable agriculture*" (Conway and Barbier, 1990:10). Examining positions within the SA paradigm that assume that conversion to SA brings only benefits to those directly involved, this article argues that the new technologies also involve trade-offs between equity, efficiency, productivity and sustainability.

The focus here is on one dimension of SA that has been neglected in the literature and in research – the increased labor requirements and intensification of labor that can result from the introduction of SA technologies. This article explores the potential for SA to generate new processes of social differentiation in agriculture, as compared to contemporary claims that it is accessible and beneficial to all. It brings a gender perspective to bear on the SA sector in order to highlight what may be potential obstacles to its diffusion. It will also highlight the social and economic trade-offs that will have to be carefully negotiated if SA is to live up to its human and environmental potentials. There are numerous lacunae in the literature in this regard; therefore, there are serious issues that must be researched in the short-term.

In the SA literature reviewed, positive claims are made about improvements in the environment and in food quality with the employment of a wide range of SA techniques. These claims are also made about social and economic benefits of conversion to SA, which are often presented as major goals of SA movements. For example, the International Federation of Organic Agriculture Movements (IFOAM) enumerates its principle aims as: *"To consider the wider social and ecological impact of the organic production and processing system and to allow everyone involved in organic production and processing a quality of life which meets their basic needs and allows an adequate return and satisfaction from their work, including a safe working environment to progress toward an entire production, processing and distribution chain which is both socially just and ecological responsible"* (IFOAM, 1998).

SA technologies differ from conventional ones in that they are more dependent on natural processes and local agro-ecological conditions as well as on human labor, knowledge and skills. In this sense, they in some ways recall a return to pre-industrial techniques that were subordinate to the local environments and the knowledge and skills of the producer. Many would also claim that they were subordinate to values other than technical rationality (aesthetic, religious, ideological) (Ellul, 1964:64-74). Nevertheless, the socioeconomic conditions under which SA techniques are employed bear little resemblance to those prevailing in pre-industrial societies. The driving forces behind technical rationalization have as much influence on SA as they do on "conventional" (modern) agriculture. In fact, these driving forces are principally determined by the conventional agro-industrial system. Labor processes differ substantially in SA when compared to conventional or traditional agriculture, which has numerous repercussions for the quantity, timing, skills and quality of labor. However, labor relations (e.g., with wage workers and among farm family members) and the cost of labor may differ little from those prevailing in conventional agriculture. While resource use also may differ substantially from conventional or traditional agriculture, differences in resource access (including land, labor and capital) and market competition still influence the socioeconomic differentiation process in SA. The significance of the prevailing socioeconomic conditions are often overlooked by proponents of SA who tend to emphasize technique at the expense of social relations, leading to a theoretical and practical naiveté: *Sustainable Agriculture can represent economically and environmentally viable options for all types of farmers, regardless of their farm location, and their skills, knowledge and personal motivation. Sustainable farming can be compatible with small or large farms, and with many different types of technology (*Pretty, 1998:89*)*.

The claims about the efficacy of SA technologies in terms of their biophysical outcomes are increasingly founded on well-grounded experimental and field research. However, claims about the social benefits of SA are founded on thin empirical evidence which, further, is both partial and constructed to emphasize what proponents desire to demonstrate - e.g., that SA is healthier and more satisfying for producers, economically feasible, and generates greater overall social benefits such as closer ties between producers and consumers, as well as greater rural employment and social justice.

Labor in Sustainable Agriculture

Most efforts to develop or recover environmentally sustainable forms of agricultural production involve the adoption of techniques and technologies that are more labor and knowledge intensive than those in existing farming systems, whether "traditional" or "conventional". However, virtually none of the literature on SA or its variants (e.g., organic agriculture) stresses the increase in labor requirements as a fundamental limiting or conditioning factor in its wide-scale adoption or viability.[1] Labor is also not considered as a factor that has far-reaching consequences for farm/household livelihoods or processes of social differentiation. However, these latter are critical to the success of such conversion efforts and thus to the future of agriculture, the environment and human welfare.

The rapid growth in SA and organic farming has been welcomed by policy makers, environmentalists and consumers as representing progress toward sustainable rural development (Altieri and Hecht, 1990; Fischler, 1999; Lampkin, 1994a). In organic agriculture, chemical inputs are replaced by new knowledge, management practices and labor. On average in the EU, 10% - 20% more labor is used on organic farms than on similar conventional farms (Offermann and Nieberg, 1999). The data, however, are very limited and may underestimate total labor requirements on organic farms since they focus on the crop or field level rather than on whole farms or farm households (Howard-Borjas and Jansen, 1999; Jansen, 2000), in spite of the fact that these entire systems tend to change. It is assumed that the sector promotes rural employment (Lampkin, 1994c; Marino et al., 1997; Pretty, 1998) and better working conditions (Birnthaler and Hagen, 1989; Rapp, 1998b), but research is very scarce. It is known that, in the EU, labor requirements vary greatly between types of organic farms: grazing livestock systems tend to use less labor than conventional counterparts on a per hectare basis due to reduction of stock numbers (although labor use may be the same or greater on a per head basis), whereas crop and organic vegetable production can require approximately 40% more labor than conventional farms (Offermann and Nieberg, 1999). Experts are now beginning to realize that the higher labor demand may be an important reason for variation in growth of the sector across regions and types of farming systems. They cite the recent rapid growth of the relatively low-labor demanding livestock sector versus the slower expansion of more labor-intensive crop and vegetable production across much of the EU.[2] In any case, substantial changes in the organization and quality of labor are inevitable and not always positive. In part, this is due to the introduction of more labor-intensive production techniques and crops, as well as on-farm processing and the need to find new market outlets (Lampkin, 1994b; Rapp, 1998a).

---

[1] Exceptions can be found in highly technical literature on specific topics such as weed control, fertilizers, soils conservation and swidden agriculture, where labor bottlenecks are sometimes pointed out as significant constraints.

[2] Personal communication with Dr. S. Padel, Welsh Institute of Rural Studies, University of Aberystwyth, UK.

SA involves a wide range of techniques oriented toward environmentally stabilizing farming systems. Therefore, it is difficult to make generalizations of any kind without specifying the types of techniques or farming systems involved. However, generally there is a consensus that the conversion to SA from conventional farming systems involves the substitution of chemical inputs (e.g., synthetically compounded fertilizers, pesticides, growth regulators, and livestock feed additives) with management techniques and labor processes that improve biophysical interactions in these systems with respect to the quality of soils, the management of pests and diseases, soils and water conservation, and energy and feed management (Pretty, 1998; Altieri and Hecht, 1990). From the literature it can be inferred that particularly plant nutrient recycling, hand weeding, and husbandry methods to control pests imply increases in labor requirements (Bouwman, 1996; FAO, 1984; Lampkin, 1994b; Marino et al., 1997; Nieberg and Pals, 1998; Padel and Zerger, 1994; Zanoli and Micheloni, 1998). The importance of preventative measures increases. Optimum timing is very important and more labor may be required to manage it and carry out more small tasks. In organic pest control, farmers spend more time observing crops, and the prevention of pests and diseases through crop rotation is crucial. A larger variety of crops in rotation may increase labor intensity. Also, parcels may be smaller, increasing labor for cultivation. Mixed cropping can require more labor since mechanization is difficult. High labor requirements for hand weeding are often mentioned as an important characteristic of organic crop production, and farmers are experimenting with techniques to reduce labor for weeding and to enhance the quality of this labor. Synthetically compounded fertilizers are not used so more labor is required to recycle plant nutrients, compost and apply manure. The need for nitrogen-fixing crops and green manure increases which may mean labor-intensive cultivation of low-value crops. Other factors increasing labor requirements include specialization, management of greater genetic diversity, on-farm processing and direct marketing, as well as more training and participation in organizations. Soils conservation measures, such as terracing, hedgerows and bunding, are known to increase labor requirements not only for construction, but also for maintenance.

The higher demand for labor in SA generally contrasts with historical trends toward a reduction in labor costs and in the absolute amount of labor employed in agriculture as well as an associated decline in the rural population. One argument put forth for supporting SA in low- and high-wage countries has been that it requires more labor and leads to higher rural employment. However, the increased labor demand in SA may result in a shift in the supply of labor towards greater use of unpaid family labor and low-paid seasonal or casual labor rather than contributing to stable, high quality rural employment. This could accentuate the trend toward a highly segmented rural labor market, especially increasing that sector characterized by low wages, poor working conditions and high labor turnover, which is dependent in large part on migrants, ethnic minorities, women, children and refugees. On most organic farms in the EU, the increased labor demand is mainly covered by seasonal workers from non-member states in Eastern Europe and North Africa (Offermann and Nieberg, 1999). Further, there is growing evidence that uncertainty about potential increases in labor requirements constitutes a real constraint to the growth of the organic sector. Increased labor demands or labor quality problems may create a disincentive or an inability to convert on the part of resource poor farm households. Changes in the labor process are not anticipated by converting farmers which can affect the viability of their farms over the short and long-term. Also, higher labor requirements particularly in high wage regions could have negative effects on the competitiveness of organic farming. It is especially in these areas that agriculture relies heavily on external inputs. Evidence suggests that there are trade-offs in the conversion process with respect to the quality of labor: healthier conditions may prevail, but farm work may be more physically demanding, with greater intensity and drudgery. However, as yet little is known about the quality of labor and working conditions in SA since research is almost entirely lacking.

Furthermore, with the conversion to SA, not only is the entire farming system likely to be altered but also the entire farm-household (livelihood) system can be expected to change. This occurs as farm families assume new tasks in agriculture and, particularly in the growing organic sector, new collateral activities (e.g., processing and direct marketing). Much literature maintains that, in high wage countries, lifestyle satisfaction increases with conversion to SA. However, research in this area has focused almost exclusively on male heads of farms and, even in these cases, has not looked specifically at work satisfaction or its relationship to the division of labor on the farm. There is evidence to suggest that the roles and relationships of different household members may be realigned, with general consequences for the distribution of costs and benefits within farm households. Finally, higher labor demand influences which farm households convert to SA and the relative success of different types of farm/household systems after conversion. Thus, it influences social differentiation both between conventional and SA households as well as among SA households.

The authors conducted a literature review focusing mainly on Western Europe but including other high wage countries and a collection of case studies concentrating on low wage regions. In the most recent literature on the topic, the interest in labor in SA is principally or exclusively as a factor of production, essentially equivalent to all other factors (e.g., land, external inputs) and then, "*...not many studies present a detailed analysis of this subject*" (Padel and Zerger, 1994:108). Even with these limitations, these studies begin to highlight certain important issues. Two collections of studies, one on Western Europe (Lampkin and Padel, 1994) and one on developing countries (UNDP, 1992), serve as illustrations.

Western Europe

Research in Germany indicates that increased labor inputs in organic farming, where labor costs are nearly twice as high as on conventional farms, is due to the increased production of root crops and vegetables plus direct marketing and processing. Also, there are wider variations in labor costs among organic farms, where these costs are much higher on small than on large organic farms (Padel and Zerger, 1994:110).

In Denmark, *the organic sample [of farms] uses about twice as much labor per hectare as the conventional counterparts. Some of this difference is explained by the considerably larger share of labor-intensive crops - such as vegetables - in organic farming and the greater importance of labor-intensive dairy production. If these structural differences were eliminated, labor requirements of organic farming would probably exceed conventional averages by about one-third "only"* (Dubgaard, 1994:124).

In Switzerland, 20% more labor is used on the organic farms studied. There, much of the additional labor is related to direct marketing activities and processing (Muhlebach and Muhlebach, 1994:134). Additional labor needed for crops runs 49 - 72% higher (Ibid.:139). The supply of this additional labor is problematic and, together with lower yields in comparison with conventional systems, is largely responsible for the higher costs of production on organic compared to conventional farms:

In Britain, wage costs were on average 40% higher on a per hectare basis and twice as high per farm [excluding horticulture] reflecting differences in farm size. In Switzerland and Germany, where farm sizes were more comparable, wage costs were two to four times as high as conventional, although total labor use was only around 20% higher. In all three cases, family labor use was similar; the higher wage costs reflecting the need to meet additional labor requirements through hired rather than family labor. Difference in family labor use per hectare in Britain were again primarily due to differences in farm size. Only in Denmark, where wage costs were 2.5 times as high as conventional, was family labor use also higher at 40% more than comparable conventional farms (Padel and Lampkin, 1994:212-213).

Although higher labor costs appear to be intrinsic to arable organic farming, higher total returns to labor are also evident, due principally to the "premium" prices paid for organic produce. After conversion from conventional to sustainable or organic agriculture, there are frequently decreases in yields and also overall decreases in production costs.

Low-Wage Countries

UNDP compiled a series of case studies examining experiences with organic and SA farming in various non-Western countries. In Indonesia, on a six ha organic farm, labor costs ran two times higher for cabbage and four times higher for carrots than on conventional farms. *"The higher labor investment can be a disadvantage for farm families with a shortage of labor, especially during labor peaks. Cash payment of hired labor can then be a problem"* (UNDP, 1992:39). A plantation in Tamil Nadu, India, has 312 ha in organic tea production. Prior to conversion, the total labor required for one field was approximately 84 days, whereas the total after conversion was 414 days. The increase was due to labor for trenching (196 days), weeding (135 days) and shade (24 days). *"The conversion towards organic cultivation is a mid-term investment. It requires more labor that needs to be financed. A financially successful conversion therefore depends on getting a premium price in the market,*" which is feasible in this case since the tea is exported to the European organic market (Ibid.:52-53).

In the Azua Valley of the Dominican Republic, a commercial firm contracts farmers with 3 - 4 ha each to produce organic bananas, although production is not wholly organic. Total labor input is 30% higher, and labor costs are 50% higher on semi-organic versus conventional farms. It is 66% higher for both indicators on semi-organic compared to traditional farms. In this case, "*A lower yield and higher labor input ...are compensated by much lower costs for external inputs and a premium price*" (Ibid.:119).

In Chiapas, Mexico, the Finca Irlanda produces organic coffee, which was compared with nearby traditional coffee production and large conventional estate coffee production. Labor inputs for organic production were 40% higher than in traditional systems and 47% higher than in conventional systems. "*For wider acceptance, it is important that new farmers can convert to organic relying on their own family labor. Small farm owners cannot afford the cash investment of external labor*" (Ibid.:134). The net return was 10% lower than in conventional systems and 80% higher than in traditional systems, but the price for organic coffee (exported) was double that of conventional coffee (Ibid.:135).

In Nicaragua, Finca Esperanzita was established as a center for experimentation and propagation of SA. Of the 300 farmers who participated in its courses, 240 "*did not adopt the system proposed, mainly because they are not full-time farmers (for example, they also work for timber companies), so they are not motivated to improve their farming income*" (Ibid.:141). "*Many farmers know no other life than migrating to another area when the soil becomes exhausted. Applying a sustainable system, which includes soil conservation, demands the investment of labor*" (Ibid.:142). However, in this case the UNDP failed to recognize the fact that most poor farmers in low-wage countries do not depend solely on agriculture for their livelihoods. Thus, technologies that are incompatible with off-farm work may not be viable for them.

The UNDP's summary of 21 case studies included several references to labor. The conclusions cannot be said to be well developed or consistent: on the one hand it is argued that, with respect to SA, "*Investment levels are lower and of a different nature. Generally speaking they involve more labor and less capital. This is an advantage for motivated starters*" (Ibid.:172-173). The UNDP did recognize that, in certain cases, higher labor requirements present a constraint to adoption where other alternative uses of labor are present. Further, they stated: "*It is not clear from the available reports how the higher labor requirement affects gender distribution*" (Ibid.:174). They also recognized the need for more data: "*In family farm situations, the emphasis is on monetary outlays, and family labor is often not yet explicitly treated as a cost. The case studies would have gained in comparability and clarity if all labor had been valued in some way*" (Ibid.:173).

Failing to consider the complex relations entailed in the adoption of new techniques and technologies such as those presented by SA means risking non-adoption, partial adoption giving rise to unanticipated consequences in terms of expected benefits and yields, and negative impacts from adoption, particularly in low-wage countries. Negative impacts on men and women in farm households can give rise to broader processes of social differentiation. There may be an increase in the number and poverty of households headed by women, and a poor farm sector may emerge that is increasingly marginalized or relegated to virtual landlessness because it is unable to adopt either conventional or SA technologies.

These precautions are not only applicable to organic farming. They are equally applicable to the introduction of other SA techniques and technologies. The example of Malawi serves to demonstrate both class and gender-based implications. Malawi is affected by land competition and degradation. Female-headed households constitute 42% of the "*core poor*" and 34% of the "*other poor*" households which "*shoulder a disproportionate burden of the impacts of poverty and land degradation*" since they depend on smaller plots of land found on steeper slopes and less fertile soils (Burgess, 1991:29). The "*smaller the land holdings the higher is the percentage of female-headed households*", in part due to male out migration. Women cannot "*finance agricultural inputs, rotate annual crops, or undertake soil conservation,*" so that soil fertility and crop yields are declining over time, which "*further exacerbates their poverty, increases their dependence upon the land and vulnerability to its degradation.*" Extension advice does not reach poor women since they are not members of the organizations providing these services. Projects in Malawi have introduced soil conservation measures to farmers. "*The achievement of targets for adoption of soil conservation is much lower by women in comparison with men*" (Ibid:31), although women's farms have the greatest land degradation problems.

Belsky (1991), researching rice farming systems in Indonesia, noted the effects of a government program subsidizing bench terrace construction to promote soil conservation. She demonstrated that household resource access and labor allocation patterns are critical to the success of these measures even with government subsidies for labor costs (50 days/ha). At least 100 - 150 days are required for terrace construction (Ibid.:218-219), and maintenance requires 12 - 36 days per year to repair walls plus several hours per week to prune grasses. Poorer farmers built fewer terraces than richer ones, in part because they feared that yields would be reduced (in fact, both yields and cropping areas declined). To compensate, the government promoted a change from cassava to higher value peanut crops. Many farmers who constructed terraces failed to maintain them due to labor bottlenecks when maintenance activities conflicted with other activities (Ibid.:237-238). Gender relations are particularly important in explaining why farmers fail to maintain, and often abandon, terraces. "*The burden of terrace maintenance frequently falls to women. In general, women performed a large percentage of farming tasks on upland farms located close to the house, which...also were the ones most likely to be terraced*" (Ibid.:244). Women preferred not to maintain terraces because of the additional labor burden. Also, the shift from cassava to higher yielding peanut crops for terraced fields resulted in an insufficient supply of cassava due to decreasing yields and area in production. Since cassava production and processing are female activities, "*Female producers lamented the reduction in income previously earned from cassava-based cottage industry which they managed. They were the ones who controlled the income earned from selling cassava products*" and they spent this money on household needs (Ibid.:246). The income from peanut production, on the other hand, was largely used to purchase inputs for peanut production. This resulted in a net decrease in family welfare.

Part of the solution is to provide labor-saving technologies. However, in low-wage countries, the rural poor who most need these technologies to reduce their workloads and to increase their productivity and incomes have so far gotten little access to it. There is no reason to suppose that this problem will be overcome with the promotion of SA. Since technology is introduced in part to reduce the cost of labor, there is little incentive to introduce technology for unpaid work, such as in those agricultural activities performed mainly by women and children (i.e., weeding, crop processing) and in domestic activities. The technology employed tends to be constant over time since those who have the greatest incentive to increase productivity are the unpaid workers themselves, who have little access to income. The incentives to make investments required to develop technology are lacking. Therefore, the appropriate technologies are often not "on the shelf". Given the lack of technology and consequent low productivity of unpaid labor, the need for unpaid labor remains high. The net result is that women and other unpaid workers are over-employed in terms of time worked and under employed in terms of income received. Lacking access to labor-saving technology, the demand for labor in rural areas in low-wage countries is high and even increasing, but this does not translate into more income. The problem is compounded by a lack of infrastructure development (i.e., roads, electricity, water, energy, health, etc.) which increases the demand for unpaid labor, reinforcing the over-under employment dilemma. The implications are that, in low-wage countries, to the extent that labor-saving technologies exist that are compatible with or promote SA (and the list of such technologies seems to be short in comparison with those that are apparently not compatible), they will not reach poor farmers and women unless they are free of charge (i.e., made by the farmers themselves or donated) or accompanied by credit and the income increases required to repay credit.

Quality of Labor

Several studies on SA make positive assertions about changes in the quality of labor from conversion. It is argued that the conversion to sustainable farming may lead to healthier work conditions as fewer hazardous agro-chemicals are used. Farmers in high-wage countries do report better working conditions due to the lack of chemicals and work that is "closer to nature". However, labor in SA may also be more physically demanding and of lower overall quality. Hand weeding, for example, is characterized as monotonous group work involving much drudgery in bending, kneeling, and lifting. Physical effort is reported to be substantially greater on organic farms in the Netherlands (Bueren et al., 1999) and Sweden (Lundqvist, 1999), the only countries where studies were found that included this type of research. There may also be a higher diversity of farm tasks as well as changes in the timing and intensity of labor that may decrease rather than enhance the quality of labor. There are implications both for the rural wage labor force and for farm family members. No research has been identified which analyzes the division of tasks, intensity of labor, distribution of rewards to labor, labor relations, the quality of labor in terms of intensity, environment, safety, and physical effort, or their relations to the quality of life in SA. However, these may be considered as the most important factors to consider when analyzing equity and social justice within the sector. The problems of hiring and retaining wage labor in both high and low wage countries have also not been examined. In addition, nearly all studies to date equate satisfaction of the farm head with satisfaction of the all farm household members whose labor and livelihoods are also connected to SA. Preliminary research in high wage countries, however, indicates that the values, incentives, benefits and satisfactions do indeed differ by sex and position within SA farm households (Chiappe and Butler Flora, 1998; Meares, 1997). Furthermore, to date there are few visible claims that wage labor in the SA sector affords wage workers greater satisfaction or better working conditions, although this may will determine the sectors' ability to compete for labor supply with conventional or traditional agricultural. It will also affect the overall social benefits and equity to be derived from conversion and eventually may affect consumers' willingness to pay more for sustainable agriculture products.

The Farm Households and Labor in Sustainable Agriculture

Most studies that analyze SA exclude consideration of the farm household. Farm labor use is considered only in relation to optimizing farm productivity. However, in most countries, family labor is critical to SA, which tends to be concentrated on family farms. Within this sector, farm households are frequently supported by multiple sources of income including transfer payments, off-farm wages, and para- and pluri-agricultural activities that are necessary to maintain household and farm enterprises. Non-market-oriented production also contributes to farm household reproduction (e.g., home gardening, fuelwood collection). Much family labor is destined for household reproduction (domestic tasks). Frequently, domestic and farm production tasks are indivisible (performed simultaneously), carried out by women and children, and are invisible or under-counted. For example, production of vegetables for family consumption often occurs in garden plots managed by women that are not considered to be part of the farm. Therefore, their output is not considered as part of total production, even though these gardens are likely to use fewer chemical inputs compared with the farm (Engel-di Mauro, 1999). Farm managerial tasks are also often carried out by women but are not considered in economic analysis. The same is true of post-harvest activities such as processing, transport and marketing if these are carried out by women and children or if the focus is on crops and on primary production.

If conversion entails greater use of family labor, then the total costs of conversion cannot be calculated without considering the opportunity costs of the additional family labor involved – that is, the cost of activities foregone to take on additional work managing the SA farm household. Research on benefits and equity should also consider shifts in livelihoods and benefits and costs foregone for different family members as these assume different tasks and responsibilities. This is particularly critical in SA not only due to the increase in labor involved, but also because new spheres of activity appear such as processing, direct marketing, converting household wastes into plant nutrients, participating in SA movements and new training activities.

It is also currently hypothesized that SA offers greater equality of opportunities to farm women and that farm women figure in the forefront of the SA movement. Women's participation in farming may increase, and there appears to be relatively higher women's participation in decision making bodies in SA in high wage countries in comparison with conventional agriculture. If women's participation in SA differs from conventional farming, it can be postulated that the gender divisions in access to resources and knowledge will probably be different in both sectors. There is also evidence to suggest that conventional patterns of male decision making persist over much of the SA sector. Women's and children's unpaid labor contributions in particular may be invisible. Their labor may be increasingly important to the success of farm operations, but they may not experience a concomitant increase in their rewards, labor satisfaction, or decision making capacities. Women's and children's satisfaction within SA may be critical not only to ensuring their labor inputs today, but also to the future sustainability of the sector as the issue of farm succession begins to arise.

Social Differentiation within Sustainable Agriculture

"Social differentiation" refers to the question of who can participate in specific development processes (in this case, either conventional or SA), and who cannot (is socially excluded). The analysis of social differentiation entails an exploration of the conditions under which specific farms do or do not convert to SA, enter into SA, succeed with SA, or dissolve. It explores factors that differentiate farms such as size, location, and resource endowments (e.g., land and labor) and relates these to the formation, survival, prosperity and dissolution of farms. Farm size differences have impacts on labor intensity and costs. Labor use in SA (in fte[3] per ha) depends on farm size. The probable reason is that smaller farms specialize in more labor-intensive activities (e.g. horticulture). However, it may also be an effect of capitalization, scale advantages, and differences in mechanization. On the other hand, conversion to SA might be difficult for some farm families or in certain regions because of the costs or availability of labor. Thus, social differentiation between farms and labor use influences the type of farms that convert to SA and probably also their degree of success.

To date, no theory of social differentiation exists in the literature on SA. This is surprising since Rural Sociology and Rural Development Studies long ago demonstrated the importance of social differentiation studies for understanding the impacts of technology development and diffusion. It can be concluded that further

---

[3] fte: full-time equivalent.

study of social differentiation, seen as the outcome of the competition over resources and the regulation of labor processes on farms, should be carried out along two lines. First, there is a process of external differentiation between conventional and SA farms and also between SA farms. Second, there is also internal differentiation within the SA sector that has implications for the overall social benefits to be derived from its expansion. This internal differentiation within farming households occurs by sex and age. It also exists between households and hired labor, and among types of hired laborers. The claims about increased quality of labor within the SA sector means that these internal relations of social differentiation should be compared with similar relations prevailing within the conventional sector.

Existing Literature and Research Needs

With regard to the results of the literature review on organic farming in Western Europe, the results show major lacunae in research published up through 1999:

1. Little exists on changes in labor demand, supply and quality occurring when farms convert to organic production. Existing data are incomplete and often incomparable, making it difficult to draw conclusions about factors that create variations between farms, farming systems and regions
2. None was found on key factors influencing labor demand, organization and quality or that interrelates these aspects.
3. Little exists on the social organization of labor in organic versus conventional agriculture, e.g., the division of family labor, the participation of women in organic agriculture and changing gender relations, the ratios between family and hired labor, the use of volunteer labor and labor exchange systems between farms.
4. Little exists on the actual quality of work and life of producers and workers in the organic sector in comparison with the conventional sector. Reported perceptions of lifestyle satisfaction focus exclusively on farm heads and have not been related to work satisfaction nor with the actual quality of work (conditions, relations), life and livelihoods of farm heads, household members or hired labor.
5. None was found on social differentiation in organic farming. Nowhere are these internal relations of social differentiation investigated or compared with relations prevailing in the conventional sector; nor is there literature addressing the dynamics of differentiation between organic farms.
6. None was found on how policies, legislation, regional economies and agro-environmental conditions influence labor use.
7. None was found that evaluates the implications of labor costs, efficiencies and returns in different organic farming systems for current and future competitiveness of high-wage versus low-wage country organic agriculture.

8. None was found that evaluates current overall employment or social effects of the conversion to organic agriculture at local, regional or national scales, much less provide recommendations for improvement.

Given the above, research needs have been identified by a consortium of organic agriculture research institutes in Europe[4] as:

- To improve knowledge of labor dynamics in different SA farm household systems and regions, considering whole farm-household efforts (e.g., processing, direct marketing and new plant protection and nutrient management tasks) and assess precise causes and conditions of changes in labor with conversion. This includes developing methods and measuring changes in labor requirements, quality, relations, supply and costs in different farm-household systems during and after conversion to permit the structural factors that induce such change or differences to be identified, and produce and make accessible data that allow changes to be better anticipated.
- To identify and analyze social dynamics and well-being (including socio-economic differentiation) in the conversion to SA for farm-household members and non-family workers (hired, volunteer) and make recommendations to improve employment, well-being, quality of work and work satisfaction. This includes developing methods and indicators to assess factors such as workload and labor conditions as well as the division of opportunities, costs and benefits among all people participating directly in production. It requires research into the qualitative experiences of welfare and work satisfaction as work environments, social networks and livelihoods are reoriented.
- To develop a quantitative analysis of labor demands in SA at the whole farm-household level including new processing and marketing enterprises and to identify existing and new solutions to labor constraints in SA farm-household systems that are common as well as specific to farm types (e.g., labor-saving and quality-enhancing technologies, management practices, cooperation strategies and policies) through in-depth research on selected farm-household types and modeling.
- To assess the implications of regional labor markets and employment policies for labor dynamics, farm and labor incomes and the competitiveness of SA under different policy and labor market scenarios, particularly with respect to their effects on the costs and efficiency of labor. The macroeconomic analysis of labor in SA is a critical component of any attempt to predict the future viability of the sector and formulate policies that stimulate its growth as low- and high-wage regions enter into increasing competition.

---

[4] This includes the University of Wageningen, the Agricultural Economics Research Institute (LEI), and the Institute of Agricultural and Environmental Engineering (IMAG) in the Netherlands, the Danish Institute of Agricultural Sciences and the Danish Institute of Agricultural and Fisheries Economics in Denmark, the University of Perugia in Italy, the Welsh Institute of Rural Studies at the University of Aberystwyth in the UK, and the Research Institute of Organic Agriculture (FiBL) in Switzerland, and the Federal Agricultural Research Center (FAL-BAL) in Germany.

- To forecast the implications and potential contributions that more widespread conversion to SA can have for rural employment generation and well-being in specific regions, relating these to the effects of current support programs and making policy recommendations for their improvement. Rigorous analysis of multiplier effects must replace simple assertions about overall employment and social effects (well-being, equity, labor quality, work satisfaction) of the conversion to SA, particularly if the growth of the sector is expected to promote rural development.
- To produce integral, multidisciplinary conceptual and methodological approaches to diagnose and forecast labor and social dynamics, employment generation and returns to labor in SA applicable at farm, regional and national levels that are fully documented and accessible to the research community.

In summary, a series of more global questions should be posed. Will resource-poor farmers, women and farm children be able to participate fully in SA? Will SA be able to achieve all that its promoters promise? Will rural people in low-wage countries be able to protect soils, plants, watersheds, forests, deserts and biodiversity, without credit, farm subsidies, diversified markets, off-farm employment opportunities, labor-saving technologies, agrarian reform and other changes which have been the subject of the agrarian development policies and debates for so many decades? Western European governments do not generally expect farmers to accept such individual and social burdens without state support and compensation. This is why governments such as those of Germany and Denmark provide specific support for conversion. Many others are contemplating it, and farmers' organizations are demanding it (Lampkin, 1994c; Clunies-Ross and Cox, 1994). Yet, in most low-wage countries, such forms of support are not only "unthinkable" in terms of current economic policies but are also critiqued by many as unsustainable.

It would be committing an equally serious error to suppose that SA is incompatible with smallholder household/farm systems or can bring few overall positive benefits. The emphasis in this article is on the need to assess the prospects and promises far more critically, taking into account a larger range of empirical and conceptual issues. There is simply no easy alternative.

References

Altieri, M. and S. Hecht (1990): Agroecology and Small Farm Development. CRC Press, Boston.

Belsky, J. (1991): Food Self-Sufficiency and Land Use in the Kerinci Uplands of Kerinci: Implications for Conservation Farming. Ph.D. Dissertation, Cornell University. University Microfilms International, Ann Arbor, Michigan.

Birnthaler, J. and M. Hagen (1989): Frauen in alternativ bewirtschafteten landwirtschaftlichen Betrieben. ASG-Kleine Reihe Nr. 37, Göttingen.

Bouwman, V. (1996): Biologische groente: hoge prijzen, veel werk. Agri-Monitor 2(6), pp. 9-10.

Bueren, T. and R. Beekman (eds) (1999): Proceedings of Kennismarkt biologische akkerbouw en vollegrondsgroenteteelt 1998. Louis Bolk Instituut, Driebergen, the Netherlands.
Burgess, J. (1991): Land Management and Soil Conservation in Malawi: A Case Study. In: Buchner et al. (eds): Gender, Environmental Degradation and Development: the Extent of the Problem. LEEC Paper DP 91-04. IIED/UCL London Environmental Economics Centre, pp. 25-36.
Chiappe, M. and C. Butler Flora (1998): Gendered Elements of the Alternative Agriculture Paradigm. Rural Sociology 63(3), pp. 372-393.
Clunies-Ross, T. and G. Cox (1994): Challenging the Productivist Paradigm: Organic Farming and the Politics of Agricultural Change. In: Lowe, P., T. Marsden and S. Whatmore (eds): Regulating Agriculture. David Fulton Publishers, London, pp. 53-74.
Conway, G. and E. Barbier (1990): After the Green Revolution: Sustainable Agriculture for Development. Earthscan Publications, London.
Dubgaard, A. (1994): Economics of Organic Farming in Denmark. In: Lampkin, N.H. and S. Padel (eds): The Economics of Organic Farming. CAB International, Wallingford, Oxon, UK, pp. 119-129.
Ellul, J. (1964): The Technological Society. Vintage Books, New York.
Engel-Di Mauro, S. (1999): Gendering the Political Ecology of Soil Use: The Case of the Ormánság, SW Hungary. Paper presented at the Conference on Gender and Rural Transformations in Europe, 14-17 Oct., 1999, Wageningen, the Netherlands.
FAO (1984): Improving Weed Management. FAO Plant Production and Protection Paper 44. FAO, Rome.
Fischler, F. (1999): Organic Farming and the New Common Agricultural Policy. SPEECH/98/0527bio, Conference on Organic Farming in the European Union – the Prospects for the 21st Century, Baden/Austria, 27-28 May.
Howard-Borjas, P. and K. Jansen (1999): Gender, Labour and Livelihoods in Sustainable Agriculture: Lessons not learned from the Green Revolution. Paper presented at the conference Women Farmers: Enhancing Rights and Productivity. Bonn, Germany 26-27 August.
IFOAM (1998): Basic Standards for Organic Production and Processing. International Federation of Organic Agriculture Movements. Tholey-Theley.
Jansen, K. (2000): Labour, Livelihoods and the Quality of Life in Organic Agriculture in Europe. In: Biological Agriculture and Horticulture, 17/2000, pp. 247-278.
Lampkin, N. (1994a): Researching Organic Farming Systems. In: N. Lampkin and S. Padel (eds): The Economics of Organic Farming. CAB International, Wallingford, Oxon, UK.
Lampkin, N. (1994b): Organic Farming: Sustainable Agriculture in Practice. In: N. Lampkin and S. Padel (eds): The Economics of Organic Farming: An International Perspective. CAB International, Wallingford, Oxon, UK, pp.3-9.
Lampkin, N. (1994c): Organic Farming and Agricultural Policy in Western Europe. In: Avalon Foundation: Proceedings of the International Seminar for Policy Makers on: The Contribution of Organic Agriculture to Sustainable Rural Development in Central and Eastern Europe. Czech Republic, 14-17 October 1993. n.p.: Avalon Foundation, pp. 161-179.
Lampkin. N. and S. Padel (eds) (1994): The Economics of Organic Farming: An International Perspective. CAB International, Wallingford, Oxon, UK.

Lundqvist, P. (1999): Working Perspective of Organic Farming. Draft paper presented to the XVIII Congress of ESRS, Lund, Sweden 24-28 August.

Marino, D, F. Santucci, R. Zanoli and S. Fiorani (1997): Labour Intensity in Conventional and Organic Farming. In: Isart, J. and J.J. Llerena (eds): Resource Use in Organic Farming. Proceedings of the Third ENOF Workshop, Ancona, 5-6 June 1997. LEAAM-Agroecologia, Barcelona, Spain, pp. 311-320.

Meares, A. (1997): Making the Transition from Conventional to Sustainable Agriculture: Gender, Social Movement Participation, and Quality of Life on the Family Farm. Rural Sociology 62(1), pp. 21-47.

Muhlebach, I. and J. Muhlebach (1994): Economics of Organic Farming in Switzerland. In: Lampkin, N.H. and S. Padel (eds): The Economics of Organic Farming. CAB International, Wallingford, Oxon, UK, pp. 131-142.

Nieberg, H. and L. Schulze Pals (1998): Economic Impacts of Conversion to Organic Farming - Empirical Results of 107 Farms in Germany. In: El Bassam, N. et al. (eds): Sustainable Agriculture for Food, Energy and Industry: Strategies towards Achievement. Proceedings of the International Conference held in Braunschweig, Germany, June 1997, vol I, 1998, James and James, London, pp. 609-614.

Offermann, F. and H. Nieberg (1999): Economic Performance of Organic Farms in Europe. Organic Farming in Europe: Economics and Policy. vol 5. Universität Hohenheim, Hohenheim.

Padel, S. and N. Lampkin (1994): Farm-Level Performance of Organic Farming Systems: An Overview. In: N. Lampkin and S. Padel (eds): The Economics of Organic Farming. CAB International, Wallingford, Oxon, UK, pp. 201-219.

Padel, S. and U. Zerger (1994): Economics of Organic Farming in Germany. In: N. Lampkin and S. Padel (eds): The Economics of Organic Farming. CAB International, Wallingford, Oxon, UK, pp. 295-313.

Pretty, J. (1998): The Living Land. Earthscan Publications Ltd., London.

Rapp, S. (1998a): Veränderung der betrieblichen Parameter (insbesondere Arbeitskräfte) bei der Umstellung auf ökologischen Landbau am Beispiel von Bioland. Diplomarbeit, Fachhochschule Nürtingen.

Rapp, S. (1998b): Öko-Landbau schafft Arbeitsplätze! Bioland 1998(2), pp. 34-35.

UNDP (1992): Environment and Natural Resources Group. Benefits of Diversity. An Incentive towards Sustainable Agriculture. UNDP, New York.

Zanoli, R. and C. Micheloni (1998): The State-of-the-art of Research on Organic Farming in Mediterranean EU Countries. IBIAGA-Faculty of Agriculture, University of Ancona. CATA-BIO, Udine, Italy.

44 Dresden Declaration: Towards a Global System for Agricultural Research for Development

To address these challenges, the GFAR[1] stakeholders gathered in Dresden, Germany, from 21 to 23 May 2000, have adopted the following Global Vision for "*Agricultural Research for Development*" which builds on the diversity and complementarity of the different GFAR stakeholders. The following is an excerpt[2]:

... Advances in agricultural research and development, including major breakthroughs in the new areas of science, have significantly contributed to meeting the challenge of food and nutrition security, agricultural sustainability, production and productivity. However, the world still faces an increasingly complex challenge of feeding its growing population and of eradicating poverty, while assuring an equitable and sustainable use of its natural resources. We, the GFAR stakeholders, believe that:

- food security, nutritional quality and safety, poverty alleviation and sustainable natural resources management are not only of concern to developing countries but are critical global issues with major impact on the well-being of the society;
- addressing these issues is a prerequisite for assuring peaceful coexistence, the attainment of human rights and basic human development in the new century;
- tackling these challenges is a matter of urgency, considering the rapid process of environmental deterioration and increasing inequalities, with long-term, pervasive impacts taking place in many parts of the world;
- agriculture, rural development and the management of natural resources are not only economic activities, but strategic dimensions of contemporary societies that have important economic, social and environmental functions. It also includes the access to resources by farmers such as land, water and genetic resources.

We share a vision for the future encompassing: (a) the appreciation of the role knowledge plays in the development of agriculture; (b) the conviction that knowledge generation and utilization is increasingly based on global research systems and networks and on farmers-led experiments and innovations; and (c) the belief that new developments in areas of natural resource management, information and communication technologies (ICT) and modern biotechnology generate new opportunities. These new developments represent an enormous potential but, at the same time, could lead to serious negative effects, widening of technology gaps and social exclusion processes. As a consequence, their socio-

---

[1] GFAR: Global Forum on Agricultural Research.
[2] The full text can be read at: http://www.egfar.org/gfar2000/index.htm.

economic, human health and environmental impacts have to be monitored, risks and benefits evaluated and then regulated as appropriate.

The GFAR stakeholders envision the development of an agriculture including crops, livestock, fisheries and forestry, which is:

- sustainable, equitable, profitable and competitive, fulfilling its functions in the context of community-centered rural development, fully recognizing the role of women in agriculture;
- diversified and flexible in its structure to cope with heterogeneous and rapidly changing agro-ecological and socio-economic environments with an important role for the farm family;
- responsive to multiple sources of knowledge and innovation, both modern and traditional.

This vision implies a progressive shift of paradigm in Agricultural Research for Development (ARD) towards a holistic *"Knowledge Intensive Agriculture"* accessible to small and poor farmers.

In implementing this vision, the GFAR stakeholders agree to adhere to the following principles:

- Programmes should clearly be subsidiary and complementary to the on-going work and provide a clearly identifiable added value.
- Agricultural research should be demand-driven and implemented through equal partnerships among GFAR stakeholders.
- Priorities for the research agenda are set with a focus on farmers' perspectives, taking into account the multi-functionality and regional heterogeneity of farming systems.
- Research design and dissemination should involve the intended users and beneficiaries, particularly farmers.

The GFAR stakeholders commit themselves to establishing the following three building blocks of the Global System for Agricultural Research for Development as first steps to implement the Global Vision:

1. The formulation of a global strategic research agenda, which capitalizes on the comparative advantages and the strengths of the different GFAR stakeholders.
2. The promotion of innovative, participatory, cost-effective and sustainable research partnerships and strategic alliances.
3. The ICT networking among stakeholders and the establishment of specialized agricultural knowledge and information systems.

We are convinced that these concerted actions can contribute to the emergence of a global system for agricultural research for development. We are also convinced that this will not succeed without additional investments in agricultural research, which implies additional efforts from the international community and the establishment of new funding mechanisms to mobilize both the public and private sectors. Therefore, the GFAR stakeholders request the policy and decision-makers to strongly support the on-going renewal of agricultural research for development.

Bio- and Crop Technologies for Agriculture

Introductory Statement

Barry Thomas and Walter Dannigkeit

The necessity to increase the production of high quality food will be a major challenge for the next 25 to 50 years. It also will affect the roles of villages in the 21st century. Biotechnology and genetic modification of crops have the potential to improve food production but used alone cannot solve all of its problems.

Genetically-enhanced crops already are growing in many countries. They demonstrate the benefits that can be derived from the first generation technologies currently in use. However, there is still a heated debate going on about these crops. To put the situation very simplistically, a division exists between those who believe that biotechnology has immense potential to increase productivity and thus the well-being of the poor and those, especially in Western Europe, who believe that its uncertainties and risks dictate more caution in its implementation. In some extreme cases, even a ban on genetically-enhanced crops is considered.

The possibilities of this new technology need to be analyzed: On which farms can it be used, and what does this mean for villages in the 21st century? Which areas most urgently need to be addressed? What are currently the biggest difficulties for public acceptance and for support from the governmental organizations? Can we develop models for the future? The articles in this section explore the science and the facts behind this controversial debate.

Jung describes the impact of biotechnology on plant breeding, stressing that the genetic improvement of crop plants will be the most viable approach to increasing food production in the future. Plant biotechnology is divided into cell and tissue cultures and molecular techniques. The impressive successes of the cultures in developing countries in the past two decades and the potentials of genetic manipulation are shown. Applicable biotechnologies are still in their infancy and so far have only been successful in altering traits which are inherited in a simple manner. Thus, improving yields through gene modification is still a dream to be realized.

Kalm highlights the potentials and problems in the utilization of biotechnology in livestock. There is great potential both in mature technologies, such as artificial insemination and embryo transfer, as well as in the developments in cloning and the technologies of gene modification, especially for future breeding programs. But only by combining the traditional and the new methods can the goal of global food security be achieved. The incorporation of biotechnology in livestock production is, however, determined by economic factors. Thus, there is a threat of the technology being concentrated in the industrialized countries with limited cooperation from the developing ones.

Zehr describes a problem in India that also confronts many other developing countries: agriculture is in transition and needs to double its productivity in order to meet the growing food demand. Without increasing production, rural poverty in India cannot be alleviated. Biotechnology can contribute to an increase in productivity. Thus, both regular and especially small-scale farmers can benefit from its utilization. According to *Zehr*, developing countries must be active participants in utilizing biotechnologies so that their specific needs for food security, poverty alleviation and natural resource management are addressed. The primary question for developing countries such as India is "*can we afford to ignore or delay the applications of biotechnological developments?*"

To be comprehensive, a discussion of biotechnology in agriculture must include the genetic resources necessary for its application. *Engels* shows the importance of genetic resources for plant breeding, food productivity and food security. He describes the international efforts to conserve these genetic resources and pleads for the establishment of a multilateral system to facilitate the access to and benefit sharing of genetic resources for food and agriculture. This could facilitate continued increases in productivity.

45 The Impact of Biotechnology on Plant Breeding

Christian Jung

Genetic improvement of crop plants will be the most viable approach to increasing food production in the future. Biotechnology can be involved at different stages of the breeding process. In general, biotechnology can be classified as cell and tissue cultures and molecular techniques. For more than 20 years, tissue culture techniques have been used for *in vitro* propagation of elite material to accelerate the breeding process. Healthy plant material also can be provided by tissue culture, e.g., virus-free potatoes. Supplying farmers in developing countries with tissue culture-propagated plant material has helped to increase yields and to protect the environment by avoiding the application of toxic agrochemicals.

Molecular techniques can be classified into analytical techniques assisting in the rapid selection of elite plant lines and the directed manipulation of gene expression to produce genetically modified plants. In the first case, molecular markers (DNA markers) have been developed which can be used for different purposes during the breeding process. Molecular markers are small subsets of the genome of a given plant species which can be identified with molecular techniques like gel electrophoresis and PCR (polymerase chain reaction). Simple protocols have been established that avoid the use of expensive laboratory equipment.

Molecular markers are invaluable for studying genetic diversity and for establishing genetic relationships between breeding materials. Genetic distance analyses have been used for selecting parental lines for hybrid breeding. Molecular markers are also applied for backcross breeding. They assist in selecting superior genotypes in early generations which can accelerate the breeding process.

Molecular markers can be used for marker-assisted selection. There are numerous examples of markers being applied to select characteristics that are controlled by only one gene, thus following a Mendelian inheritance. If the corresponding phenotype is difficult to determine, as is the case in many disease resistances, marker-assisted selection can accelerate the selection process. Markers have been applied for practical breeding, for resistance to disease (virus, bacteria, fungi), insect and nematode resistance, stress tolerance, restorer genes, self incompatibility, vernalization requirement, and growth type. The marker phenotype can be determined even at the seedling stage with only very small amounts of DNA, thus enabling an early selection of favorable genotypes.

In the past years, powerful techniques have been developed to analyze the structure of even complex genomes such as mammals and higher plants. The first plant genome to be completely sequenced is *Arabidopsis thaliana*, which has the smallest genome of all the plants with ca. 125 Mb. The complete genome sequence was finished by the end of the year 2000. Among the agriculturally important plants, rice will be the first to be sequenced. Two draft sequences have already been announced by NOVARTIS and MONSANTO, however, they are not free available to the public. Rice has a relatively small genome and is a staple food in many countries of the world. It contains about 3.5 times as much DNA as *Arabidopsis* but only about 20% as much DNA as maize and about 3% as much as wheat.

The manipulation of genes by genetic modification has attracted the most public attention. In principle, each gene of a crop plant can be altered in its expression. New genes not available from the gene pool of a given crop species can be introduced from any organism in the world. This offers the possibility of breeding plants with novel characteristics which are lacking in sexually compatible species. Today, almost every crop species can be transformed with three major techniques: *Agrobacterium*-mediated gene transfer, microprojectile bombardment and protoplast transformation.

In most cases, transgenic plants can be integrated into conventional farming systems as they differ from conventionally bred plants only in the trait that has been genetically modified. Thus, they can also be used on small farms. Assuming that a plant has between 20,000 and 40,000 genes, the alteration of only one gene seems to be negligible; however, a careful risk assessment has to be made in a step-by-step procedure. Although many transformations have been made in the last 15 years, only a few of these plants are commercially grown in the field. They belong to the class of plants with altered response to insect attack and with tolerance or resistance to herbicides. While the second characteristic seems to be of little interest for developing countries, insect tolerance is a major breeding aim for almost all crop species in warmer climates. It can be expected that insect resistant varieties will help to secure yields in the future. Since the application of toxic insecticides can be avoided, these plants also contribute to an environmentally safe production system. A major obstacle to insect resistance today is that in most cases it relies on only one gene, i.e. the *bt*-gene from *Bacillus thuringiensis*. Alternate ways to produce insect resistant plants have shown only limited success so far.

Virus diseases often cause heavy losses in crop production, especially as they cannot be controlled by pesticides. Therefore, the breeding of resistant varieties is a priority. The nucleic acid of a plant virus is coated by a protein. If the coat protein gene is transferred to plants, they will be resistant to further virus attacks. Coat protein-mediated resistance was extremely successful in different species like potatoes, tobacco, sugar beets, and tomatoes. The resistance is active against the virus from which the gene was derived and against related viruses.

To improve the quality of seed oils and starches, the manipulation of genes for enzymes involved in fatty acid and starch synthesis has been used. Also, genetic modification has changed the protein content and quality and the composition of free amino acids. Recently, the Fe-content in rice was improved by introducing three different genes: a gene encoding a heat-tolerant phytase which breaks up phytinic acid, a ferritin gene resulting in higher Fe-accumulation in the seeds, and a gene for a methallothionein-like protein improving Fe-absorption. In addition, the same transgenic rice contains a gene increasing the β-carotene content. β-

carotene is a precursor of vitamin A. Rice with higher vitamin A content will benefit the more than 400 million people suffering from vitamin A deficiency which leads to higher susceptibility to infections and to blindness. The transgenic prototypes have been introduced into the IRRI breeding program to produce varieties with superior nutritional value.

Delayed fruit ripening is essential for the postharvest storing and shipping of fruits. In tomato, genes coding for polygalacturonase have been inactivated by antisense technology. In this case, a gene construct in antisense orientation was introduced and stably expressed in plants. As a result, the mRNA of the native gene binds to the antisense mRNA resulting in double stranded RNA which cannot be translated. Thus, the enzyme will not be produced, and degradation of fruit tissue is delayed. An alternative approach relies on blocking the ethylene metabolism. Ethylene is essential to fruit ripening. If ethylene production is reduced, the same effects as with antisense technology have been observed in tomatoes and melons.

Genetic manipulation has been successful in altering traits which are inherited in a simple manner. However, improving yields and other complex characteristics was unsuccessful, because the many genes involved in the expression of these characteristics are largely unknown. Parallel manipulation of so many genes is still impossible. Therefore, the impact of genetically modified plants will be primarily on stress tolerance, food quality and disease tolerance. A recent development has attracted worldwide attention. Plants have been produced with human therapeutic proteins and edible vaccines. The production of oral vaccines in plants has numerous advantages, including the high-level recombinant protein accumulation needed for oral vaccine delivery and the simplicity of production materials. Therefore, edible vaccines may be suitable for cost-efficient and continuous vaccine production in the rural areas of developing countries.

Agenda 21 §16.1 Earth Summit, Rio 1992

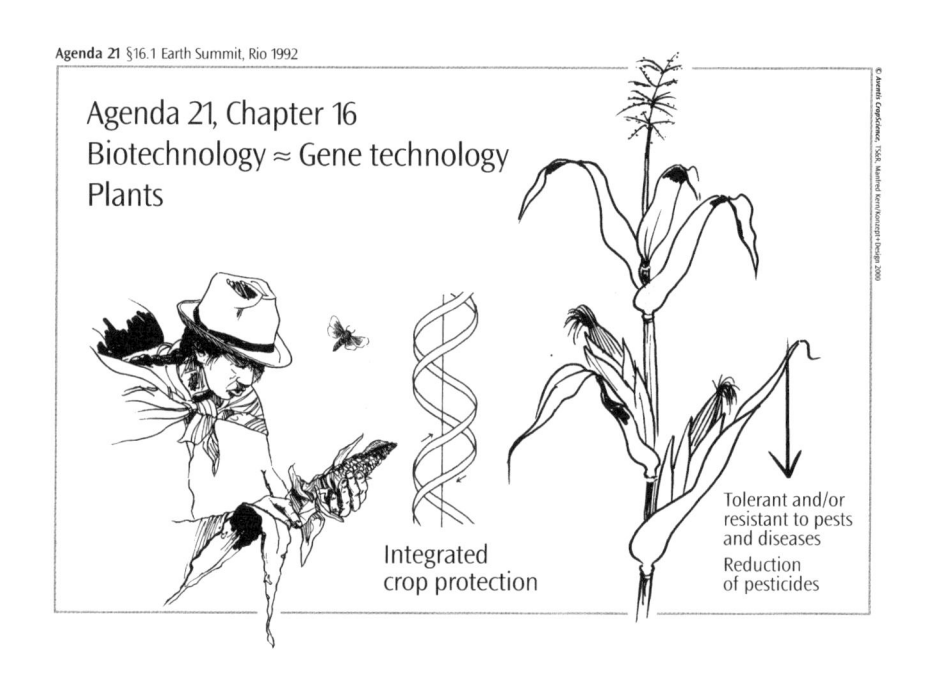

Agenda 21, Chapter 16
Biotechnology ≈ Gene technology
Plants

Integrated
crop protection

Tolerant and/or
resistant to pests
and diseases
Reduction
of pesticides

46 Potential of Biotechnologies to Enhance Animal Production

Ernst Kalm

The livestock sector accounts for over half of the agricultural output in developed countries and more than a quarter in developing countries. In response to the expanding populations and the shifts in consumption patterns resulting from increasing incomes, the livestock output globally has been growing faster than that of other sectors of agriculture. Urbanization along with population and income growth are fueling a massive increase in the demand for food from animals in developing countries. This process has been termed *"Livestock Production"* in a recent study from Delgado et al. (1999). The predicted increases in meat and milk consumption for the period from 1993 to 2020 are substantial (Table 46.1), though showing great regional variability.

Consumer demand for food of animal origin starts at a low base in most developing countries; it is equivalent to approximately one quarter of the per caput consumption in developed countries.

In view of the dramatic rise in the world population and of severe nutrition deficiencies in many parts of the world, efforts have to be made to improve the efficiency of animal production and to enhance the quantity as well as quality of animal-derived food. Improvements in efficiency arise from the development, spread and adoption of improved technologies for breeding, feeding, management, and health care of animals. There has been a substantial development of new technologies in the field of biotechnology, offering new prospects in the efficiency of livestock production. Because of the expense, most of the investment and basic research in the field of biotechnology has taken place in developed countries. The potential benefits, however, are universal. Especially in science, vision is needed to develop promising perspectives for future fields of application for scientific progress. This paper concentrates on the future prospects of biotechnological procedures and their applications in animal production.

Table 46.1. Actual and Projected Meat and Milk Consumption by Region (as per caput/kg/year)

Region	Meat		Milk	
	1993	2020	1993	2020
China	33	60	7	12
Other East Asia	44	67	16	20
India	4	6	58	125
Other South Asia	7	10	58	82
South East Asia	15	24	11	16
Latin America	46	59	100	117
West Asia/North Africa	20	24	62	80
sub-Saharan Africa	9	11	23	30
Developing World	21	30	40	62
Developed World	76	83	192	189
World	34	39	75	85

Source: Delgado et al., 1999.

Reproduction Technologies

Artificial Insemination and Embryo Transfer

The introduction of artificial insemination (AI) has changed the structures and methods of animal breeding dramatically. AI is used in 70% of cattle and in 40% of pigs. Comparatively, embryo transfer (ET) is less important. In Germany, about 20,000 embryos are transferred per year (ADR, 1998). One single AI-bull replaces 1,000 natural-service bulls, and embryo transfer increases the number of progeny per cow by a factor of 5 to 10. The cost of an embryo transfer is between DM 1,000 and DM 2,000 per calf born if on average five embryos are gained. Because of these high costs and the variable success rates, applications are limited. Currently, methods for embryo production without hormone treatment, the so-called in vitro methods, are under investigation. Collection of oocytes, in vitro maturation, in vitro fertilization (IVF) and culturing up to a transferable stage are possible without hormone treatment of the donor. Different methods are available:

1. Collection from slaughtered females: 10 – 15 oocytes, four transferable embryos;
2. Collection from the ovaries of living cows using ultrasonic or endoscopic imaging of the follicles: 8 - 10 oocytres per session.

These techniques are still costly and difficult, but their practicability has repeatedly been demonstrated (Nohner, 1997; Görlach, 1997; Besenfelder, 1997).

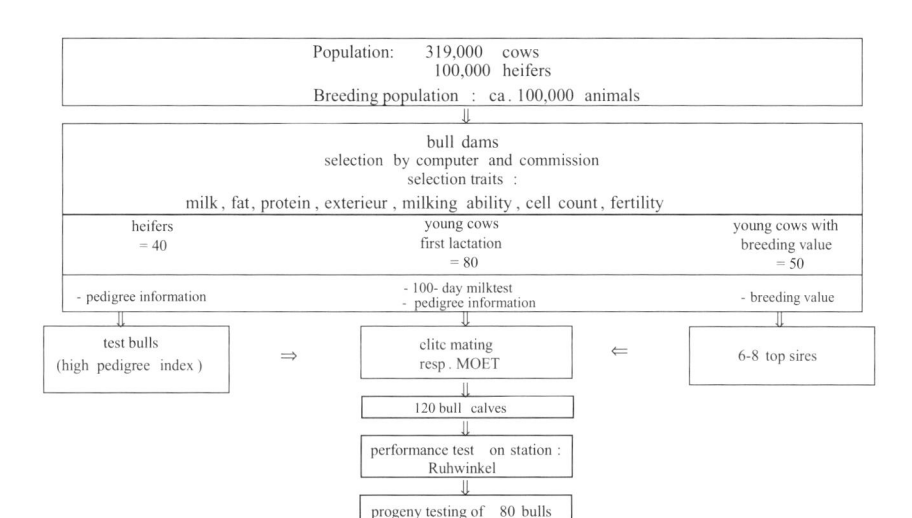

Fig. 46.1. Dairy Cattle Breeding Program in Schleswig-Holstein (1997)

```
┌─────────────────────────────────────────────────────────────────────────────┐
│                   Population:    319,000  cows                              │
│                                  100,000  heifers                           │
│                   Breeding population : ca. 100,000 animals                 │
└─────────────────────────────────────────────────────────────────────────────┘
                                      ⇓
┌─────────────────────────────────────────────────────────────────────────────┐
│                                 bull dams                                   │
│                    selection by computer and commission                     │
│                              selection traits :                             │
│        milk, fat, protein, exterieur, milking ability, cell count, fertility│
├──────────────────┬──────────────────────────────┬───────────────────────────┤
│     heifers      │        young cows            │   young cows with         │
│      = 40        │       first lactation        │    breeding value         │
│                  │           = 80               │        = 50               │
├──────────────────┼──────────────────────────────┼───────────────────────────┤
│ - pedigree       │ - 100- day milktest          │   - breeding value        │
│   information    │ - pedigree information       │                           │
└──────────────────┴──────────────────────────────┴───────────────────────────┘
          ⇓                      ⇓                           ⇓
┌──────────────────┐     ┌────────────────────┐     ┌───────────────────────┐
│   test bulls     │  ⇒  │   elite mating     │  ⇐  │    6-8 top sires      │
│ (high pedigree   │     │   resp. MOET       │     │                       │
│    index )       │     │                    │     │                       │
└──────────────────┘     └────────────────────┘     └───────────────────────┘
                                  ⇓
                         ┌────────────────────┐
                         │   120 bull calves  │
                         └────────────────────┘
                                  ⇓
                         ┌────────────────────┐
                         │performance test on │
                         │ station : Ruhwinkel│
                         └────────────────────┘
                                  ⇓
                         ┌────────────────────┐
                         │progeny testing of  │
                         │     80 bulls       │
                         └────────────────────┘
```

Embryo transfer is not extensively used. A replacement of artificial insemination with deep-frozen semen by transfer of cryoconserved embryos will only take place if prices for semen and embryos become equal for the owner of the animals. Embryo transfer has enabled the possibility of a new generation of breeding programs in cattle, known as MOET (Multiple Ovulation and Embryo Transfer) schemes. The aim is the reduction of the generation interval. Figure 46.1 shows the selection program for cattle in Schleswig-Holstein. Young heifers (15 months old) are used as bull dams, among others, and serve as donors for oocyte production. It is expected that the collection of oocytes from calves will be possible within a few years, leading to a further reduction of the generation interval for bull dams.

The introduction of MOET includes new testing procedures for bull dams, e.g., performance testing in tied housing at the test station in Osnabrück. Station testing will become less important once daily milk recording and robot milking become general practice. Instead, performance testing in large scale dairy plants offers prospects for better data on fertility and health traits.

The control of the sex of progeny from artificial inseminations will accelerate genetic progress. Cow owners then could inseminate the best third of their animals with X-sperm of elite sires in order to improve milk production, and two thirds of the cows could be used as recipients for male embryos from a beef breed.

In the past years, various attempts have been made to separate sperm populations based on their sex-determining characteristics. Through the use of flow cytometry, sperm can be sorted based on the minimal difference in DNA between X- and Y-chromosome-bearing spermatozoa. This technique has been improved substantially (Johnson, 1997; Rath et al., 1997; Wolf, 2000) and is now applicable for special use in in-vitro fertilization for elite mating.

Further advances of the technology will lead to greater use of sexed spermatozoa, depending on the species involved. This would solve the problems associated with inseminating small numbers of sexed spermatozoa and would greatly benefit AI and livestock production.

Cloning

Plant breeders use vegetative reproduction of elite genotypes in order to maintain beneficial gene combinations and to use them over a long time period. However, sexual reproduction cannot be circumvented in animal breeding.

Cattle breeding programs are characterized by:

1. low reproduction rates of cows (even with ET, only 5 to 10 female calves can be produced) and
2. large genetic variability within full sibs, i.e., the breeding value of full sib bulls may be very different:

breeding value = ½ bv sire + ½ bv dam + Mendelian sampling.

The term "*clone*" means the generation of genetically identical offspring. In plant-breeding cloning, selfing and cross breeding with cloned and selfed lines is used. Animal breeders today must be satisfied with selection within outbred populations and cross breeding between outbred lines. Therefore, cross-bred sows or cross-bred fattening pigs show a considerable amount of genetic variation as opposed to, e.g., potatoes (a single clone). The steps for embryo cloning are:

1. The cells of an embryo originating from an elite sire and an elite dam are separated;
2. Recipient oocytes from slaughtered cows are prepared;
3. The nucleus is removed from these oocytes;
4. A new nucleus from the donor embryo is inserted;
5. Cell division is induced;
6. A new embryo develops and can be used for transfer.

At this time, the success rates of cloning are still too low for use in animal breeding. An additional difficulty is the occurrence of calves which are significantly heavier than normal.

As opposed to cloning from embryonic cells, a diploid cell from the udder of a 6 year old Finn-Dorset ewe was the source of the genes for the cloned sheep, Dolly (Figure 46.2). This udder cell was fused with a denucleated, unfertilized oocyte from a Blackface recipient. Dolly is the first lamb generated in this way from 227 fusions and 29 transferable embryos implanted into 13 recipients (Campell, 1997). It remains to be seen whether or not cloning from somatic cells becomes routine. This technology could render significant benefits for animal production, e.g., in the determination of the genetic value of individuals and the multiplication of animals with special performances or traits. Cloning is interesting for the investigation of genotype-environment interactions and the subsequent usage of the results in animal production. Cloning may be applied in future cattle breeding in order to make use of heterosis effects. It is currently unclear if and when cloning will become a practicable tool in animal breeding.

Fig. 46.2. Example of Cell Multiplication (Embyro Cloning)

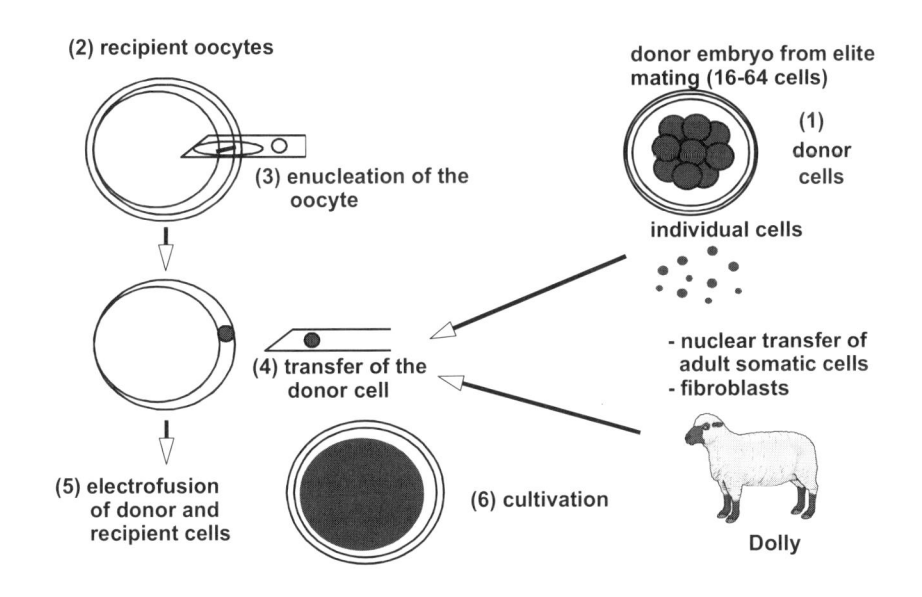

Molecular Biological Techniques

An old idea of breeders is to identify the genes for important traits on the chromosomes of potential breeding animals and to use direct genomic information in addition to phenotypic information. The purpose of genome analysis is to describe all genes of a species and to determine their location and order on the chromosomes by gene mapping. These are the prerequisites for the development of DNA tests. Research on human beings and mice has progressed farther in the sequencing of the nucleotides of all genes. The aim is to understand the biology of the organisms and the basis of genetic variability and to detect genetically induced malfunctions. Within the EU framework of the BoVMaP and PiGMaP projects, animal breeders in Germany have started work on gene mapping in pigs and cattle. Today, molecular genetic techniques are applied to identify unfavorable mutations. Table 46.2 shows some point mutations that are important for animal breeding, either because of their impact on product quality (Kappa casein in milk) or because they are causative for inherited diseases (BLAD) or unfavorable conditions (MHS). The application of DNA tests for BLAD in cattle and MHS in pigs is standard today.

Genes that directly effect production traits are of special interest to animal breeders. Thus, the Federation of German Cattle breeders (ADR) and the Federal Ministry for Education and Research (BMBF) have launched a common project on genomics in cattle. The universities in Giessen, Kiel and Munich and the research institute in Dummerstorf are collaborating on the identification of chromosome segments with importance for breeding programs.

The main interest is in traits like fertility, udder health, and quality. Twenty families with 1,200 progeny are under investigation in order to make marker-assisted selection possible. The first results are markers for somatic cell count (chromosome 18, Table 46.3). Closely linked markers may help to predict the breeding value of an animal at an early stage and to perform a preselection of newborn animals or even of embryos. Another hope is to find ways for minimizing undesired side effects of selection.

Table 46.2. Example of Gene Diagnosis in Animal Breeding

	Type 1	**Type 2**
Kappa casein (cattle)	A	B
Aminoacid substitution	Asp	Ala
DNA-sequence	GAT	GCT
	↑	↑
BLAD (cattle)	healthy	BLAD
Bovine Leucocyte Adhesion Deficiency		
Aminoacid substitution	Asp	Gly
DNA-sequence	GAC	GGC
	↑	↑
MHS (pig)	stress sensitive	stress resistent
Malignant Hyperthermia		
Aminoacid substitution	Cys	Arg
DNA-sequence	TGC	CGC
	↑	↑

Table 46.3. Results of QTL Analysis from the German Genome Scan

Chromosome	Somatic cell count [%]	Protein [%]	Fat [%]	Milk [%]
1			2	
2	10	3	5	1
5			1	1
6			6	6
7	5			
10		5		3
14		1^1	1^1	1^2
16		3	8	6
17	11	5		1
18	1	5	10	5
19		6		
22				11
23		4		5
27	8			
29			11	
XY				7

Note: [1] 1 % experimentwise, [2] 10 % experimentwise, or else chromosomenwise

Gene Transfer

In the laboratory, gene transfers in mice and also in livestock species are in use (Wall, 1996). The transgene must be integrated into germ-line cells, otherwise it is not transferred to the progeny. The development of a transgenic line needs roughly four years in pigs and eleven years in cattle; the costs can be enormous (Table 46.4).

High costs result mainly from the inefficiency of gene transfer in farm animals, with only 1 - 4% of microinjected zygotes leading to a transgenic offspring.

Transgenic dairy cows could be used to produce recombinant proteins, e.g., human serum albumin or human lactoferrin, as ingredients of their milk instead of using microorganisms.

Up until now, breeding for disease resistance was relatively ineffective in animals. Intensive research is being carried out on resistance against influenza in pigs (Müller, 1996) and against mastitis in cattle. Gene transfer could increase the genetic variability of disease resistance and also of quality traits. Health costs can comprise 10% of the total production costs. Therefore, breeding for disease resistance is of great importance.

Gene transfer could contribute greatly to minimizing both the incidence of diseases and the usage of drugs. Of course, undesired side effects in recombinant animals must be eliminated by extensive testing and selection.

Table 46.4. Costs of Producing Transgenic Animals

		Mice	Pigs	Sheep	Cattle
Costs/animal	[DM]	13	320	120	1,040
Costs/day/animal	[DM]	0.40	14	4	24
Labor time (number animals x day)		204	2,412	20,400	30,135
Costs for transgenic animal	[DM]	155	40,000	96,000	873,000
	[US $]	78	20,000	48,000	436,500

Source: Wall et al., 1992.

Summary and Conclusion

Research in biotechnology has been progressing rapidly during the last decades. There are, however, still problems to be solved to enable the successful exploitation of biotechnological advances in livestock for the benefit of animals, humankind and the environment.

Artificial insemination and embryo transfer are mature technologies, and their use has stabilized globally. Most of this activity is in developed countries and increased use in developing countries will depend on their economic situations. The cheaper methods of embryo production in vitro may extend the use of ET. For the long-term conservation of genetic resources, embryo cryopreservation is well developed in cattle but is less so in sheep, buffalos, horses and pigs. The developments in cloning have the potential for great positive impact, particularly in dairy and beef cattle. The rapid expansion of knowledge about the molecular structure of genes as well as about their regulation and expression is very important for future breeding programs. The direct identification of genes affecting important traits for economic performance or disease resistance together with marker-assisted selection would accelerate genetic improvement. Gene transfer has been used successfully for the enhancement of growth, for the improvement of disease resistance and for generating special traits in animal products, for instance to produce pharmaceutical products in animal tissues.

The incorporation of biotechnology into livestock production is determined by economic factors. It also generally requires sophisticated technological support. For these reasons, most developments in biotechnology (reproduction and genome scan) are likely to become established in the most economically advanced countries, with limited incorporation in developing countries.

The positive developments in biotechnology and genetic engineering must soon be recognized to solve nutritional deficiencies in many parts of the world. We must work together so that all humankind has enough to eat. Mother Earth has enough room and food for us all, but we have to revise our attitudes and recognize our responsibilities for the common good. Traditional and new methods must be combined so that we can move towards our goals. Of course, a single answer to all the problems is not possible. Only the intelligent utilization of all accepted methods, as well as the new methods from biotechnology, can create the driving force to establish the basic right of food for all humankind.

References

ADR (1998): Rinderproduktion in der Bundesrepublik Deutschland, 1998. Arbeitsgemeinschaft Deutscher Rinderzüchter e.V., Bonn.

Besenfelder, U. (1997): Endoscopically Guided Embryo Transfer Techniques in Livestock. Satellite Symposium II - Innovative Reproduction Techniques in Animal Production. 48. EAAP, Wien.

Campell, K.H.S. (1997): Embryo Cloning in Livestock. Satellite Symposium II - Innovative Reproduction Techniques in Animal Production. 48. EAAP, Wien.

Delgado, C., M. Rosegrant, H. Steinfeld, S. Ehni and C. Courbois (1999): Livestock to 2020: The next Food Revolution. Food, Agriculture, and the Environment Discussion Paper. 28. IFPRI/ FAO/ ILRI/ IFPRI, Washington, D.C.

Görlach, A. (1997): Embryotransfer. Berichte aus der Praxis. Arbeitsausschuß Biotechnologie der ADR 14.04.97, Saerbeck Westlandbergen.

Johnson, L.A. (1997): Current Technologies: Sexing of Sperm and Embryos. Satellite Symposium II - Innovative Reproduction Techniques in Animal Production. 48. EAAP, Wien.

Müller, M., U.H. Weidle and G. Brem (1996): Congenital Immunisation of Farm Animals. In: Houdbebine, L.M. (ed): Transgenic Animals.

Nohner, H.P. (1997): Der Embryotransfer beim Besamungsverein Neustadt a.d. Aisch. Arbeitsausschuß Biotechnologie der ADR 14.04.97, Saerbeck Westlandbergen.

Rath, D., L.A. Johnson, J.R. Dobrinsky, G.R. Welch and H. Niemann (1997): Production of Piglets: Preselection for Sex Following In Vitro Fertilization with x and y Chromosome-Bearing Spermatozoa Sorted by Flow Cytometry. Theriogenology 47, pp. 795-800.

Wall, R.J. (1996): Transgenic Livestock: Progress and Prospects for the Future. Theriogenology 45, pp. 57-68.

Wall, R.J., H.W. Hawk and N. Nel (1992): Making Transgenic Livestock: Genetic Engineering on a Large Scale. Journal of Cellular Biochemistry 49, pp. 113-120.

Wolf, E. (2000): Embryo-Sexen und Spermatrennung. 18. Hülsenberger Gespräche, 21.-23. Juni 2000, Weimar.

Agenda 21 §16.1 Earth Summit, Rio 1992

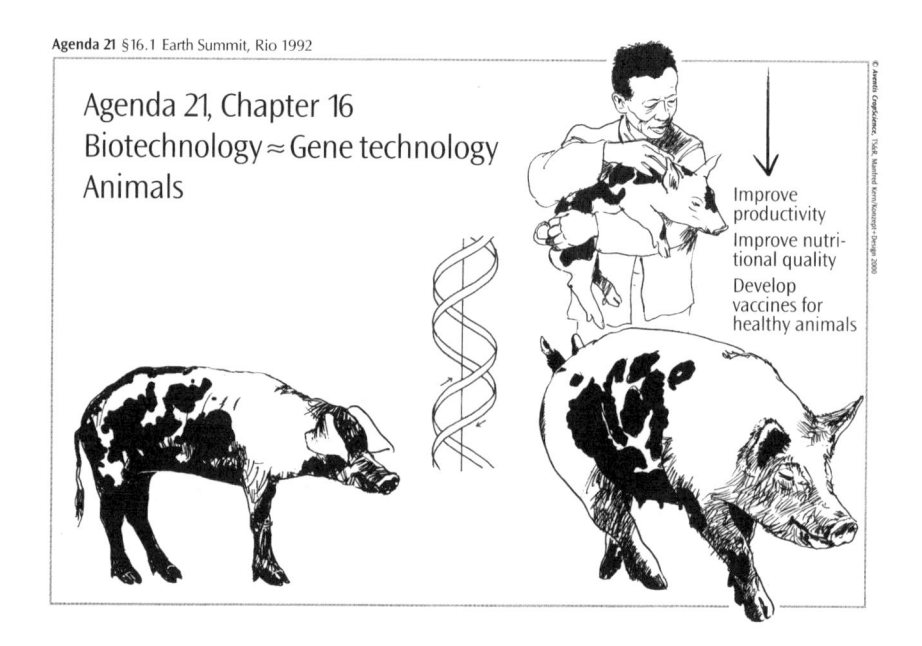

Agenda 21, Chapter 16
Biotechnology ≈ Gene technology
Animals

Improve productivity

Improve nutritional quality

Develop vaccines for healthy animals

47 Can India Afford to Ignore Biotechnological Applications?

Usha Barwale Zehr

Technology has always played a critical role in development and has made contributions in the fields of medicine, industry, agriculture (e.g., the introduction of high-yielding varieties) and, more recently, information and communication. Plant biotechnology is a new entrant in the arena of technology and has received much attention from all interest groups. Scientific innovation leads to the application of new discoveries, enhancing existing knowledge in the different fields and using new tools to come to better end products. However, plant biotechnology has tended to focus more on theory and work in the laboratory than on actual applications in agriculture. Biotechnology needs to have its agricultural potential utilized in the fields.

When defining a village, one thinks of agricultural communities, usually rural and underdeveloped. Along with the growing number of people, the number of villages in India also is growing. In 1991, there were 621,000 villages as compared to 576,000 in 1971. According to the Indian Census Bureau, in all places with a minimum of 5,000 people, at least 75% of the male working population is engaged in non-agricultural occupations. A population density of at least 400 persons per sq.km. defines an urban setting. It is estimated that 65 - 70% of the Indian population resides in rural areas and thus will be the inhabitants of the villages of the 21st century. This urban/rural definition is likely to be different in the 21st century if the reform processes in place continue to move ahead at the current pace. Today, the rural poor spend as much as 70% of their income on food. Thus, any improvement in agricultural productivity would have a substantial impact on reducing rural poverty (Deaton, 1997).

Why Biotechnology?

The population increases projected both globally and specifically for India indicate that the food requirements of the nation must be met by increasing productivity on at least the existing land area if not less. Of India's population of 1 billion, it is estimated that 300 million people are undernourished (FAO, 1999). The population in India is projected to reach 1.3 billion by 2020. The increased demand for certain commodities will come not only from the population increase but also because of improved buying power, urbanization and changes in food preferences. With the current stagnation in the productivity of major crops, all areas of agriculture will have to be reviewed. A combination of agronomy, improved germplasm, crop management and natural resource management could increase productivity in a sustainable manner and promote technologies that would help the poor.

The onset of the Green Revolution saw the introduction of high-yielding varieties of wheat and rice. The new varieties increased production using substantial complementary inputs of chemical fertilizer and other plant nutrients. They also required assured supplies of water at specific intervals. Future gains in agricultural production will be achieved with much greater difficulty. The lessons learned from the past require that, along with increased food production, the preservation of natural resources and the protection of the environment must be considered. An increase in productivity on marginal lands will be essential for meeting the overall food needs. Gains using conventional technology are now more difficult, and a combination of new and conventional technologies must be applied to address the biotic and abiotic stresses and to enhance food production on fertile as well as marginal lands.

The applications of agricultural biotechnology have the promise of bringing about a much needed new revolution in agricultural production. Biotechnological innovations to date have provided new plants that:

- give higher yields with the same or lower inputs,
- carry resistance and tolerance to abiotic stresses such as drought or salt,
- provide options for better rotation to conserve natural resources,
- maintain quality in storage thus keeping costs lower, and
- enhance natural resource management.

Of course, biotechnology cannot solve all agricultural problems. However, for the first time in the history of plant sciences, tools are available to allow a much deeper understanding of the mysteries of plants than ever before. Today's technologies allow us to be precise, fast, and specific, thus reducing farmers' risks. An improved fundamental understanding of plant physiology is also facilitating better management of our natural resources and environment.

Which Biotechnologies?

The appropriateness of technologies is necessary for their successful adoption. They must provide recognizable benefits, such as monetary, to the farmers, users and ultimately the consumers. The following classes of biotechnology are extensively used today:

- gene transfer technologies, which provide transgenic plants;
- non-transgenic biotechnological approaches for improving the efficiencies and effectiveness of conventional plant breeding methods. These also enable us to understand and identify genes that control complex traits (e.g., functional genomics, marker-assisted selection, micropropagation for disease-free plantlet generation); and
- technologies for better monitoring of our natural resources and environment.

The first generation biotechnologies provide solutions for problems that are controlled by single genes, such as insect resistance and herbicide resistance. With the progress in science and technology since the introduction of the first biotechnological products, more complex traits such as improved nutritional quality and salinity and drought tolerance are being considered.

The application of biotechnology can improve the productivity of sustainable farming and be pro-poor. Productivity can be enhanced by providing plants with built-in tolerance to insects and diseases, by engineering plants for better tolerance to salinity and drought thus improving productivity on marginal lands, and reducing the necessity of inputs thus lowering agricultural costs. These would all result in increased yields in limiting environments.

Pre- and postharvest losses due to biotic and abiotic stresses and the risks involved in marketing can be reduced by some of the technologies mentioned above. In developing countries such as India, pre and post harvest losses comprise 14 - 30%, depending on the crop. Lacking the infrastructure for transporting and storing fresh market produce (fruits and vegetables), the small farmer is often forced to sell at low prices. Improving the keeping quality and shelf life of a product through genetic engineering can provide the farmer with greater flexibility in marketing. In addition, losses due to insects in stored grain can be minimized by better storage conditions. Thus, better storage conditions and technology can alleviate harvest losses.

Edible vaccines are a newly emerging area of plant biotechnology which directly relate to human health. This innovation has the potential to eliminate the use of needles and, in some cases, may eliminate the requirement for cool storage conditions. Thus, the vaccines could be made available in remote places where facilities for proper storage may not exist due to unreliable electricity, etc.

Products from biotechnology are already on the market in developed countries, and efforts are underway in developing countries to provide the crops that are needed by small farmers and that can be produced on marginal lands. Despite the urgent need, farmers in India often do not have the opportunity to use biotechnological products, particularly transgenic plants.

Technology Transfer

Farmers have had access to technologies in India through a number of sources such as:

- national public research institutes,
- international agricultural research centers,
- international technology transfers, and
- private research.

In the evolving global agricultural market, strong partnerships between different groups will provide faster dissemination of technology. With the commercialization of genetically enhanced products, guidelines for genetic engineering and regulations for biosafety are necessary. In India, the Department of Biotechnology has such guidelines in place to monitor the safety of products before they reach the farmers. Guidelines are especially necessary for the field of technology transfer.

The most efficient way to provide a given technology to the most farmers possible is as packaged seeds. This allows both poor and rich farmers to benefit from the technology. Also, planting seeds usually requires no changes in cultural practices in order to realize the technological benefits. Farmers often enjoy higher yields with reduced inputs (e.g., insect resistant transgenic seeds would require reduced or no insecticide usage and reduce labor requirements).

Technological tools such as micropropogation or marker-assisted breeding provide benefits without altering the genetic composition of a plant. Thus, those technologies do not need the regulations required for transgenic plants (genetically enhanced plants) and can be applied immediately.

Biotechnological Potential: Sorghum as an Example

It is a common misperception that little technological work is being done for the crops of resource-poor farmers and that the potentials of biotechnological products will pass them by. Research has begun on sorghum, one of top five cereals grown worldwide. With its unique ability to grow in stress conditions, it is the crop of choice for many marginal farmers. In India, sorghum is mostly used for human consumption in the form of bread. In 1970 - 71, India grew 17.4 million ha. of sorghum. By 1998 - 99, the production had decreased to 9.9 million ha., a reduction of over 40%. A few reasons for the decline are:

- changing diets,
- production of cash crops if resources are adequate,
- increasing problems from diseases and pests, and
- stagnation in yield.

Not being a crop of the high-income countries, sorghum has received comparatively little attention in research. However, this crop has tremendous potential for good productivity on marginal land. Biotechnology can address a number of the farmers' problems with this crop.

Shoot flies and stem borers are two major pests damaging the sorghum in India. Although the losses caused by these two pests are well known, little effort has been expended to combat them. Technology is finally being applied in an effort to minimize the damage from these pests. In particular, shootfly damage to sorghum fields can result in substantial yield losses. To avoid heavy infestation, the correct timing of the sowing of the crop is essential. Shootfly resistant sorghum cultivar would give the farmers more flexibility in their sowing options and immediately increase their yield stability. With the identification of the specific Bt protein for Shootfly resistance, biotechnology should be able to accomplish this. Other resistance strategies are also being developed.

As sorghum is an important food crop in India, its nutritional improvement could have a major impact on the people's health, especially for those who live almost exclusively from sorghum bread. As compared to other cereals, sorghum is nutritionally inferior in terms of protein and starch digestibility. It is also low in the relative content of lysine, an essential amino acid. Sorghum lines have been identified with more easily digestible starches and higher lysine contents, which make the protein more easily digestible than in normal varieties. Using marker-assisted breeding and immunological screening assays, this digestibility trait is being incorporated into Indian cultivar. For the rural populations in India who rely on sorghum as their primary source of protein, raising its lysine content and improving the digestibility of the protein would increase the utilizable protein in their diets, thus improving their nutrition.

In India, sorghum is produced in two seasons. In the Kharif season (rainfed–monsoon), 55% of the farmland is cultivated, and in the Rabi season (crop grown on residual ground moisture), it is 45%. Hybrid breeding of sorghum has been targeted toward the Kharif season. Thus, during the Kharif over 90% of the sorghum acreage is planted with hybrid seeds. However, sorghum grown during the Rabi season must be drought tolerant in order to produce the desired results. As of yet, using conventional breeding methods, successful hybrids for the Rabi season that are appropriate for large acreage have not been developed. Sorghum grown in the Kharif season has an average of two times the grain yield of sorghum from the Rabi season. Technological intervention in the area of cereal genomics research could help to address this discrepancy, either by selecting for drought resistant heterotic groups or by identifying yield enhancement factors which could be transferred into Rabi season sorghum.

These improvements in sorghum would benefit all farmers and would help to achieve sustainable food security for India. Sorghum is able to survive with limited water; it is appropriate for marginal farmers and as a model for sustainable agriculture. Sorghum is also being used for poultry feed, and poultry farming is growing rapidly in India. Thus, there is an alternate use for this crop.

The above discussion of sorghum illustrates the vast potential of applications of biotechnology in different approaches: transgenic (insect resistance), marker-assisted breeding (improving nutritional quality), and functional genomics (drought tolerance, a complex trait).

Genetic Resources and Biotechnology

The collection, identification, and conservation of germplasm, which is already underway in gene banks around the world, is essential for biotechnology. Plant gene pools that are not already in gene banks must be researched and collected. Characterization facilitates the identification of unique traits, such as genes for unique proteins, pest resistance, etc. By studying genetic evolutionary linkages between plant species, a better knowledge base can be built for plants and agriculture. Gene bank managers must face the daunting task of the characterization of entire collections. Numerous national programs and CGIAR centers have begun this work.

The Indian subcontinent holds an estimated 8% of the earth's total biodiversity, with two "hot spots" of biodiversity and over 47,000 species of plants. Managing this resource requires precise tools for identification and characterization. Although costly, molecular characterization is the most effective of these tools. It is used in crops that are strategic and prioritized in relation to both biodiversity and sustainable agriculture, ensuring that these genetic resources are available for the future.

Conclusion

Indian agriculture is in transition, and it needs to double productivity although less land is available. Under stressed conditions, productivity must be enhanced and imbalances corrected. The disparity of different ecological regions must be addressed. According to a 1989 survey, 58% of the Indian farmers are marginal small landholders with an average of 0.39 ha; 18% of the small farmers have an average of 1.4 ha, representing nearly 28% of the cultivated area. Unless productivity on these farms is maximized, rural poverty cannot be alleviated. With advances in science and the judicious use of biotechnology, these challenges to Indian agriculture can be met. Farmers can benefit from improved productivity through biotechnology. The main question which must be addressed in developing countries such as India is: *"Can we afford to ignore or delay the applications of biotechnological developments?"*

References

Deaton, A. (1997): The Analysis of Household Surveys: A Microeconomic Approach to Development Policy. The John Hopkins University Press for the World Bank, Baltimore and London.
FAO (1999): The State of Food Insecurity in the World. FAO, Rome.

48 The Conservation of Genetic Resources: A Global Effort

J.M.M. Engels

Some examples illustrate the importance of plant genetic resources for food and agriculture (PGRFA) to food security and production. They demonstrate how genetic resources have been used and have sometimes contributed to the world economy. An impressive figure is reported by Pimentel et al. (1992) showing that, depending on the crop, world crop yields have increased by two to four times. The contribution of genetic resources, by the introduction of new genes and by genetic modifications through crossing with wild relatives, is estimated to be approximately US $ 115 billion per year worldwide in crop yield increases (Pimentel et al., 1997).

In several African countries and in India, yield increases of up to 18 times the previous amount were reported after disease-resistant genes from a wild cassava species from Brazil were incorporated into local varieties (Prescott-Allen and Prescott-Allen, 1982). Similarly, the fact that a grassy stunt virus-resistant gene, found in the wild rice species of *Oryza nivara* and collected in India, now can be found in almost all modern rice varieties worldwide demonstrates the importance of genetic resources for breeding purposes. Another example, based on data from CIMMYT regarding the use of wheat genetic resources, indicates that the number of landraces in the pedigrees of wheat cultivars released in developing countries increased from 12 in 1961 to 64 in 1992 (FAO, 1998). The variety "*Sonalika*" was first released in 1966, and by 1990, it was grown on over 6 million hectares. Its pedigree includes 39 landraces introduced from over 20 countries and every continent (Smale and McBride, 1996). Another study on the use of wheat genetic resources also demonstrated the importance of both landraces and wild relatives in wheat breeding. A global survey showed that over 8% of the material in wheat breeders' crossing blocks in 1994 consisted of landraces and wild relatives (Rejasus et al., 1996).

The above mentioned examples were chosen to demonstrate the importance of new and novel genetic variance for breeding purposes to make further genetic advances. Therefore, efforts to keep genetic resources available to plant breeders are essential in order to ensure the required steady increase in food crop productivity and thus to cope with the ever increasing demand for food (FAO, 1996).

Current Conservation Efforts and the Policy Framework

In response to increasing awareness of the threats to PGRFA, the number of gene banks and of germplasm collections dramatically increased during the 1970s and

1980s. In 1995, it was estimated that approximately 6 million accessions were maintained in gene banks around the world (FAO, 1998). Considering the considerable number of duplications between and even within collections, the number of genetically unique accessions is estimated to be between 1 and 2 million. Some 600,000 accessions are maintained by the Centers of the Consultative Group on Agricultural Research (CGIAR), largely consisting of germplasm from the major food crops and their wild relatives. Most of these 600,000 accessions have been placed under the auspices of the FAO as part of the International Network of *Ex Situ* Collections through agreements between the CGIAR Centers and the FAO. These so-called designated germplasm accessions are formally placed in the public domain, conserved for the long-term and freely available to all *bona fide* users. The CGIAR Centers managing this germplasm, as well as its recipients, are not allowed to claim ownership or any form of intellectual property right over the accessions themselves or over the related information (Hawtin et al., 1996).

The conclusion of the Convention on Biological Diversity (CBD) in 1993 provided a legal framework for conserving and sustainably using biological diversity and for sharing the benefits from such use. Cornerstones for any legal arrangement regarding biodiversity are the recognition of the sovereign rights of states over the biodiversity within their borders, the obligations of nations to conserve and share these resources under agreed conditions of access, and arrangements for benefit sharing on mutually agreed terms. Furthermore, Decision II/15 of the Conference of the Parties to the CBD recognizes the special nature and distinctive features of agricultural biodiversity. PGRFA are an essential subset of agricultural biodiversity.

The International Undertaking on Plant Genetic Resources, adopted in 1983, is currently being revised as a binding international instrument through inter-governmental negotiations within the FAO Commission on Genetic Resources for Food and Agriculture. This revised undertaking aims to conserve and use genetic resources in a legally binding way that is in agreement with the provisions of the CBD.

Another important instrument for the implementation of the CBD, Agenda 21 and the International Undertaking which was created in 1996, is the Global Plan of Action (GPA). This plan was adopted by 150 countries during the International Technical Conference on Plant Genetic Resources in Leipzig, Germany. It provides a basic and comprehensive technical framework for the effective conservation, development and sustainable use of PGRFA.

A Multilateral System

In addition to the importance of PGRFA for crop improvement and the current status of *ex situ* conservation and of the prevailing legal framework treated above, other important aspects and developments should be mentioned. These are essential to the development of effective and efficient global conservation efforts:

- The strong interdependence of countries regarding genetic diversity of major food crops is well documented. No country is self-sufficient, and all will have to rely on the genetic resources of other countries, frequently from other continents (de Miranda Santos, 2000).
- No country is capable of ensuring, by its own means, the conservation of sufficient genetic resources to meet its present and future needs (de Miranda Santos, 2000).
- With the advent of modern plant breeding, the rate of introgression of new traits has dramatically increased; thus, the continued access to an ever increasing range of genetic diversity needs to be ensured. Continued exchange of genetic resources is essential to provide the "raw material" for breeding programs to ensure the required steady increase in productivity (GFAR, 2000).
- There is a concern that broad forms of protection, such as plant and utility patents, may restrict research and germplasm flow and even "lock up" important sets of genetic diversity (Riley, 2000).
- Many countries are currently developing national access legislation to comply with article 15 of the CBD. It has been argued that "dedicated" legislation which sets out conditions for bilateral exchange may not be appropriate for PGRFA. It is suggested that countries consider an *exception clause* which would give them the flexibility to participate in a multilateral system (Correa, 2000).
- The expected high transaction costs for the bilateral exchange of PGRFA could diminish the exchange of germplasm and would probably impede the desired capturing of benefits (Visser et al., 2000).
- It is likely that adherence to the definition of "*country of origin,*" as per the CBD, will lead to reductions in access and use and to high transaction costs without significant additional benefits (Fowler, 2000).

Conclusion

Taking the above points into consideration, it can be concluded that only a multilateral system for facilitated access and benefit sharing of an as wide as possible subset of PGRFA will provide a win-win situation for all concerned. Genetic resources placed within such a multilateral system will be available to all *bona fide* users under globally defined and agreed conditions. This will result in continued increases in productivity, thereby benefiting those who generated the genetic diversity as well as those who use this diversity in breeding programs. Thus, an effective and cost-efficient system for its conservation will be facilitated.

References

Correa, C.M. (2000): Policy Options for IPR Legislation on Plant Varieties and Impact of Patenting. A Background Paper prepared for the GFAR 2000 Meeting in Dresden, Germany, 21-23 May 2000. http://www.egfar.org Access date: 27 July 2000.

De Miranda Santos, M. (2000): How can Multilateral Systems for Plant Genetic Resources for Food and Agriculture Benefit National Agricultural Research Systems? A Background Paper prepared for the GFAR 2000 Meeting in Dresden, Germany, 21-23 May 2000. http://www.egfar.org, access date 27 July 2000.

FAO (1996): Global Plan of Action for the Conservation and Sustainable Utilization of Plant Genetic Resources for Food and Agriculture. FAO, Rome.

FAO (1998): The State of the World's Plant Genetic Resources for Food and Agricultural. FAO, Rome.

Fowler, C. (2000): Implementing Access and Benefit-Sharing Procedures under the Convention on Biological Diversity: The Dilemma of Crop Genetic Resources and their Origins. A Background Paper prepared for the GFAR 2000 Meeting in Dresden, Germany, 21-23 May 2000. http://www.egfar.org, access date 27 July 2000.

Hawtin, G., J. Engels and W. Siebeck (1996): International Plant Germ Plasm Collections under the Convention on Biological Diversity - Options for a Continued Multilateral Exchange of Genetic Resources for Food and Agriculture. In: Proceedings of the Beltsville Symposium XXI, Global Genetic Resources. Access, Ownership, and Intellectual Property Rights, Beltsville Maryland, USA, 19-22 May, 1996.

Pimentel, D., U. Stachow, D. Takacs et al. (1992): Conserving Biological Diversity in Agricultural/Forestry Systems. BioScience 1992, vol 42, p. 360.

Pimentel, D., C.Wilson, C. McCullum et al. (1997): Economic and Environmental Benefits of Biodiversity. BioScience 1997, vol 47 (11), p. 750.

Prescott-Allen, R. and C. Prescott-Allen (1982): Genes from the Wild: Using Genetic Resources for Food and Raw Materials. International Institute for Environment and Development, London.

Rejasus, R.M., M. Smale and van Ginkel (1996): Wheat Breeders' Perspectives on Genetic Diversity and Germplasm Use: Findings from an International Survey. Plant Varieties and Seeds 9, pp. 129-147.

Riley K. (2000): Key Strategic Issues on Genetic Resources Management: A Concept Paper. A Background Paper prepared for the GFAR 2000 Meeting in Dresden, Germany, 21-23 May 2000. http://www.egfar.org, access date 27 July 2000.

Smale, M. and T. McBride (1996): Understanding Global Trends in the Use of Wheat Diversity and International Flows of Wheat Genetic Resources. Part 1 of CIMMYT 1995/96 World Wheat Facts and Trends. CIMMYT, Mexico.

Visser, B., D. Eaton, N. Louwaars and J. Engels (2000): Transaction Costs of Germplasm Exchange under Bilateral Agreements. A Background Paper prepared for the GFAR 2000 Meeting in Dresden, Germany, 21-23 May 2000. http://www.egfar.org, access date 27 July 2000.

IV. URBAN – RURAL LINKAGES

Urban – Rural Linkages

Introductory Statement

Mathias Hundsalz

The latter half of the twentieth century has seen the continuous transformation of the word's population from rural to urban, and this trend is likely to continue well into the new century. However, the rural areas still accommodate the majority of people in many developing countries, particularly in Africa and Asia. This massive and multidimensional population trend is not only the result of an ongoing migration from rural to urban areas as people search for employment, but it also represents a fundamental transformation of our societies. This transformation is characterized by changes in employment patterns, advances in transport and communication, and a globalization of national economies, with far-reaching impacts on social patterns and cultural values. Increasingly, the quality, magnitude and frequency of linkages between urban and rural areas will determine the future of population movements and the diversity of employment opportunities. It can be safely argued, therefore, that the future of rural development largely will depend on the linkages between urban and rural areas. Where these linkages are not well established and do not expand, the rural areas are likely to face further impoverishment, isolation and marginalization in a globalizing world. This will have serious impacts on further educing already limited development opportunities.

Well-managed and efficient infrastructures and services are fundamental to urban-rural and regional development with regard to the functioning of markets for the exchange of goods and services between urban and rural areas. To a large extent, the efficiency of infrastructure provision and operation determines the success or failure of productive and close relationships between cities, towns and their hinterlands. This is true for the management of production patterns in a given region (both for agricultural and non-agricultural goods), the growth of commerce and trade, the response to population growth and movements, access to education facilities and health services, and protection of the environment. In principle, well-functioning infrastructure and transport services raise the productivity of rural areas, lower production and marketing costs and increase equal access to social services. However, rural areas in poor countries such as Africa and Asia continue to be isolated from national and international markets on account of the serious constraints regarding transportation facilities. These constraints mainly refer to a lack of well-developed road systems, a severe shortage of private and public means of transport, as well as relatively high operational costs, particularly if compared to the low market value of most agricultural produce.

Examples from Africa and Asia underline the fact that rural development is lagging behind in areas where past investments in the transport and communication systems were inadequate and hampered development. *Wanmali* discusses the difficulties presented by the multisectorial and multilevel, as well as the spatial, natures of rural development and the problems created by the uneven application of the process of decentralized rural development. He emphasizes the significance of analyzing rural infrastructural services and urban-rural linkages for the future of villages. The findings can lead to the strengthening of the existing rural-urban linkages and to a *"deplugging of bottlenecks,"* thus intensifying rural economic growth and development in the region.

Emphasizing the importance of the urban-rural linkages, there is an excerpt from the *"Berlin Declaration on the urban future"*, adopted by representatives of 1,000 cities during the Global Conference on the Urban Future (URBAN 21) in Berlin in July 2000. Although the declaration is city-oriented, it confirms the relationship between urban and rural areas, *"... recognising the interdependence between the city and the region, and between the urban, rural and wilderness areas."* Hence, the step is taken from an "urban-rural divide" to an urban-rural cooperation for the sake of regional development and the world's population.

The problems affecting the infrastructure and transport sectors of rural areas in poor developing countries has been sufficiently researched. However, the area of public and private investment needs to be more closely examined as it is essential if rural areas are to be more productive in their interactions with urban centers and play a role in the global economy. Financial resources will continue to be the determining factor in designing and constructing transportation systems for regional development. Diverse functional needs, therefore, rather than predetermined conventional standards should be the basis of policy frameworks for future infrastructure investments, and options for different means of transportation should be considered.

The success of most economic activities conducted in the rural areas seems to depend upon the availability, use, and accessibility of infrastructure and other public goods. Furthermore, increased linkages appear to reduce the traditional distinctions between urban and rural areas. As villages are taking on urban attributes, so expanding urban areas are assuming rural features. The spread of cities into semi-rural areas and the increased significance of urban agriculture bear witness to this trend. *Yap and Savant Mohit* show that within a globalized world with improved infrastructure and decentralized power, a rural-urban divide is no longer relevant. Instead, *"regional agglomerations"* are becoming more and more significant. Therefore, policies should promote these regional units and their potential for the future. However, according to *Yap and Savant Mohit*, many rural areas may not be prepared for globalization and thus enforce further migration to urban centers, leaving behind those in the rural areas who have no roles to play in the global economy: the elderly, the children and those lacking skills for employment in the urban industry and service sectors.

The factors that trigger urban-rural migration are much more diverse than is often assumed and depend on complexities in the political, environmental, economic and social conditions of each country and region. Promoting small-scale enterprises, *Oucho* presents selected examples from African countries which show the increasing importance of non-agricultural employment in rural areas and the related appropriate policy interventions. He analyzes the opportunities and constraints for income generation and employment in carefully designed urban-rural linkages which create balanced regional development. He thereby criticizes former migration policies.

Besides the importance of small-scale enterprises for sustainable rural livelihood, the potential of agriculture and agribusiness as engines of economic growth in rural areas should not be forgotten. In many rural areas, the production of food, the supply of renewable energy and other agricultural products will continue to be the sources of sustainable livelihoods for years to come.

Finally, in light of the importance of urban-rural linkages in the development of both areas, the ongoing rural-versus-urban divide in international development discussions needs a comprehensive review. The categorization of "urban" and "rural" is relative and depends largely on a country's geographical and economic features. What is considered "urban" in the context of Tanzania, for instance, might well be "rural" in China. Furthermore, the conventional perception (and subsequent classification) of a region or a settlement as being either urban or rural does not necessarily correspond to the realities of rapidly transforming societies and spatial linkages in a region. In the past, perceptions of a rural-urban divide have dominated national planning and international development cooperation and have guided public investments and the management of public goods. Given the structural and functional changes in urban and rural settlements in many parts of the world, the future management of public goods must emphasize the close interrelationships between all parts of a given region. In the future, villages and cities are likely to lose even more of their traditional distinctions. As villages take on "urban elements", such as access to infrastructure, services, communications and a diversified economy, urban agglomerations of the future will be characterized by distinct communities with "rural" characteristics, such as urban agriculture. As a result, public authorities which transcend the traditional urban and rural administrative boundaries and can operate at the regional level appear to be more effective in planning investments and managing public goods such as infrastructure and services, water supply, and the environment.

Continued public intervention is needed to counter the limitations of private investments to ensure equal access and geographical coverage of infrastructure and other public goods. Furthermore, increased decentralization of administrative structures to community levels, coupled with stronger involvement of civil society institutions, is needed. This would pave the way for a more efficient and equitable management of public goods. There are several examples of regions with strong urban-rural links from countries where regional planning and administrative authorities have been set up between urban and rural areas to foster better management of the environment, infrastructure, services and communications. *Neu* analyzes the development of different rural regions in Germany, presenting a very heterogeneous picture. In some rural areas, the unemployment rate corresponds with that of the regional urban centers. However, in other regions, mainly in the structurally-weak rural areas of Eastern Germany, it is significantly higher. Thus, employment and educational programs are needed, and here, the importance of the (re-) establishment of the small- and medium-sized enterprises is stressed.

The success in diversifying the economic base in rural areas and the resulting creation of employment opportunities that are not based on farming will help to contain the ongoing exodus of rural populations to urban areas. Based on their experiences in India, *Patara and Khosla* find that the primary task and responsibility of development institutions and policies are to provide the basic infrastructure for financing, communication and transportation to small enterprises in rural areas. Having the potential to successfully compete with large corporations, small-scale production may provide a sustainable livelihood for a significant part of the rural population, including those excluded from the mainstream economy.

Although the worldwide trend is toward global urban development, *Shah* stresses the necessity to improve rural development. As long as poverty is a determining factor in rural life, sustainable urban growth is not feasible. It is argued that only by reducing rural poverty can urban living conditions be improved in a sustainable manner. *Shah's* plea underlines the fact that improved urban-rural linkages will reduce the spatial divide, challenging both – the urban as well as the rural world – to tackle poverty with joint efforts.

49 Urban-Rural Linkages in the Context of Decentralized Rural Development

Sudhir Wanmali

It is argued that most of the developing countries will remain "rural" for some time to come. This is because most of them are agricultural, most of their population stays in rural areas, and most of the constraints which hinder the realization of their potential for economic growth and development are located in rural areas. Therefore, a multisectorial, multilevel perspective is necessary for rural development if policy planners are soon to solve the problems of future growth and development. The development of both the smallholder agricultural sector and the non-agricultural sectors is central to a solution. However, the solution also depends on the methods of addressing the economic, social, political and administrative constraints that continue to hamper rural development in these countries.

One of the more recent methods adopted by several developing countries is decentralizing administrative, financial, and development planning responsibilities from the national to the local levels. However, there is a growing realization that this process has to include the removal of the above constraints so that the smallholder agricultural sector can realize its potential as an engine for local economic growth and development. This needs to be made a top policy priority. One of the major constraints needing urgent attention is the lack of relevant local infrastructure, particularly in the rural areas. Another problem is the lack of understanding about the nature and scope of urban-rural linkages, again particularly in the rural areas of these countries.

Policy researchers and policy makers are showing an increased interest in rural infrastructure and urban-rural linkages and their significance for decentralized rural development in developing countries. However, rural infrastructure and urban-rural linkages, in the context of decentralized rural development, are not easy to define. Thus, they are equally difficult to analyze. This is problematic because rural development, as emanating from the interaction of rural infrastructure and urban-rural linkages, is both multisectorial and multilevel. Rural development is also spatial in the sense that constituent sub-sectors, and elements of these sub-sectors, are located not in a spatial vacuum but in the towns and villages of a given geographical region. Further complicating the issue, the process of decentralizing such departments as administration, finance, and development planning from the national to the local levels is relatively new, and it is often unevenly carried out or not seriously put into practice in some parts of the Third World. These three difficulties are compounded by the mechanics of the decentralization of rural development itself. The emphasis, particularly in the countries of sub-Saharan Africa, appears to be more on the "what" and the "why" than on the "how to" of decentralized rural development (for the relationships between the three, see Figure 49.1).

Fig. 49.1. Framework of Rural Infrastructures, Economic Activities, and Rural-Urban Linkages

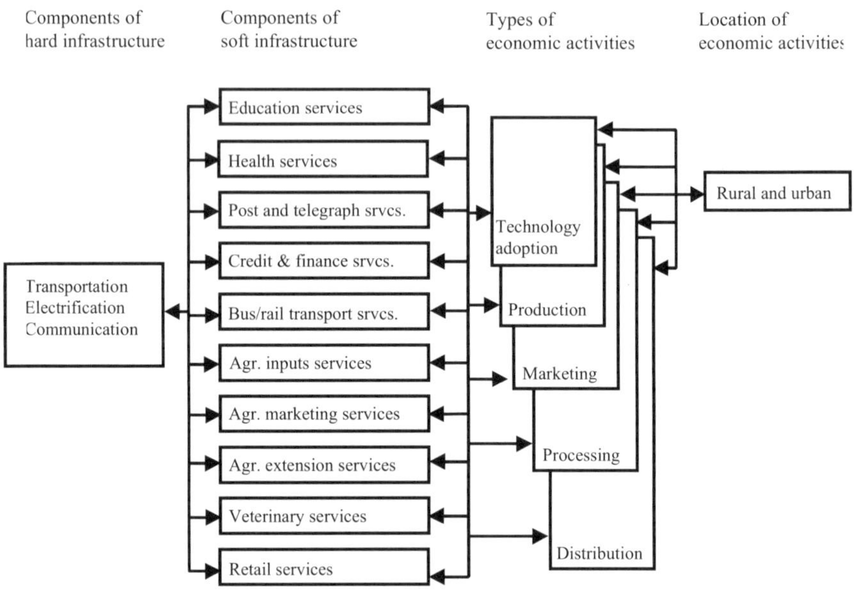

Components of hard infrastructure	Components of soft infrastructure	Types of economic activities	Location of economic activities

- Education services
- Health services
- Post and telegraph srvcs.
- Credit & finance srvcs.
- Bus/rail transport srvcs.
- Agr. inputs services
- Agr. marketing services
- Agr. extension services
- Veterinary services
- Retail services

Hard infrastructure: Transportation, Electrification, Communication

Types of economic activities: Technology adoption, Production, Marketing, Processing, Distribution

Location of economic activities: Rural and urban

Definitions of Hard, Soft, and Institutional Infrastructure

Simply defined, *infrastructures* are the foundations necessary for the building of *superstructures*. The former, therefore, can be defined depending upon the purpose, nature, and scope of the latter. For example, if the levels of human capital development are to be improved in a hypothetical developing country, one needs to provide the necessary educational infrastructure. This educational infrastructure may consist of several elements such as kindergartens, primary schools, junior high schools, senior high schools, vocational schools, colleges, universities, and research centers. The decisions about which of these elements will be used to improve the levels of human capital development will depend upon whether primary, secondary, or tertiary education is the priority in the national development plans. Furthermore, once an educational infrastructure is planned, it will have to be established in the towns and villages of that hypothetical country. Choices will also depend on where the need is most urgent. Is it in the primary education in the rural areas, or is it in the tertiary education in the urban areas? Who is using the existing educational infrastructure located in the towns and villages? What other complementary infrastructures are needed, such as roads and rail networks, bus and rail services, and links between the towns and villages? Empirical answers to such questions about educational infrastructure and its spatial context will determine the final shape of the superstructure for improved levels of human capital development. What has been noted above in connection with the example of educational infrastructure also is applicable to the infrastructures of health, credit and finance, communications, agricultural input distribution, etc. (see Figure 49.1).

Generally, when people involved in policy refer to infrastructure, they mean the physical grids of roads, telecommunications, and electrification (Lipton, 1987). These are defined as *hard infrastructure*. However, such a limited definition of infrastructure tends to provide a very one-sided picture distorting its larger role, because there are two more types of infrastructure that are equally important for economic development (Wanmali and Idachaba, 1987).

Sometimes infrastructure also refers to various types of services which facilitate economic development. These services, which are defined as *soft infrastructure*, include education, health, communications, transport, finance, input distribution, marketing, and retail services. Some of these services can be categorized as economic support services, whereas others can be categorized as social support services. Both categories are part of rural infrastructural services that are crucial particularly for the development of a smallholder agricultural sector and a non-agricultural sector in the rural areas. The absence, or inadequate provision, of such social and economic rural infrastructural services at the local level can lead to unsound economic development.

Infrastructure also can refer to governmental and non-governmental organizations, including civil society organizations, which facilitate the governance and implementation of economic development. These organizations, which are defined as *institutional infrastructure*, include local level organizations which, in collaboration with their national level counterparts, collect, collate and analyze the relevant socioeconomic data required for the governance and implementation of economic and social development at the local level. The absence of such organizations, or an inability of the existing organizations to plan and implement programs at the local level, can also hinder economic development. Furthermore, insufficient attention to the networks of soft and institutional infrastructures undermines economic development planning.

Relationships between the Different Types of Infrastructure

Previous work in South Asia and sub-Saharan Africa by IFPRI and others has shown that investment in hard infrastructure, such as roads, when coupled with simultaneous investment in soft infrastructure, such as rural infrastructural services, facilitates intensive smallholder agricultural development. It has strengthened the links to non-agricultural activities, has led to overall regional economic growth and development and, consequently, has laid the foundation for reducing poverty.

In India, for example, the location of "soft infrastructure" was planned and implemented after analyzing the demographic, economic, and functional characteristics of a regional settlement system (consisting of both towns and villages) in twenty different states in the late 1960s. A part of this research in India, conducted subsequently in three different states and nationally, has also shown that the private sector was influenced in its decision to provide complementary soft infrastructure (rural infrastructural services) after the public sector had invested in similar services (Wanmali, 1968, 1970, 1983a, 1983b, 1985, 1987, 1992; Sen, Wanmali et al., 1971; Hazell and Ramasamy, 1991; Wanmali and Khan, 1970; Wanmali and Ramasamy, 1995; Wanmali and Islam, 1995). In Bangladesh, it was noted that the construction of roads and road networks to connect rural and urban areas was responsible for increases in the productivity of agriculture through positive influences on crop yields, marketing of agricultural surplus and the overall reduction in transactions costs (Ahmed and Hossain, 1990; Ahmed and Donovan, 1992; Ahmed, Haggblade and Chowdhury, 2000).

In Zambia, Zimbabwe, Malawi and other parts of sub-Saharan Africa, improved access to rural infrastructural services facilitated the establishment, development, and reinforcement of farm/non-farm linkages which in turn resulted in regional economic growth and development (Hazell and Roell, 1983; Wanmali, 1991a, 1991b, 1997; Celis, Milimo and Wanmali, 1991; Wanmali and Zamchiya, 1992; Jha and Hojjati, 1993; Wanmali and Islam, 1997; Zeller, Schrieder, von Braun and Heidhues, 1997; Delgado, Hopkins, Kelly et al., 1998).

Several studies in South Asia show that the strengthening of old and the establishment of new institutional infrastructures at the level of districts and below was responsible for the spectacular broad-based rural development in the post-Independence period. This development in turn was responsible for the success of the Green Revolution in these countries. Studies in sub-Saharan Africa indicate that only effective decentralization of administrative, financial, and development planning responsibilities to the level of the districts and below will rejuvenate the sluggish rural sector, reduce urban-rural differences and the correct the inherent regional imbalances in economic development.

In paying adequate attention to all the above aspects of rural infrastructure, location theories, and decentralized rural development, it is important not to equate national investments in one type of infrastructure only with direct economic gains in all types of infrastructure at the local level. Long-term economic gains locally cannot be achieved without related local investment in all types of infrastructure. However, the bulk of the available analyses of various aspects of infrastructure and decentralized rural development does not consider the multisectorial and multidisciplinary aspects of the issue. The crucial aspect of the economic interdependence between urban and rural areas, as is manifested in the growth and development of towns and villages, rarely receives the analytical attention that it deserves in the current literature on regional development in the Third World.

Rural Infrastructural Services and Locational Analysis

As noted above, the elements of all infrastructure, whether hard, soft, or institutional, do not exist in a spatial vacuum but in a given geographical region consisting of towns and villages. Generally, when analyzing infrastructure, the policy researchers and policy makers have paid little attention to the aspect of the location of infrastructure in settlements (towns and villages). Thus, locational analysis of the aspects of population, economy, and functions of the distribution systems (comprising the towns and villages) of the soft infrastructure (comprising the types of rural infrastructural services noted above) is essential to understanding urban-rural linkages and their significance for decentralized rural development. Since the services noted above are available in only a few settlements but are used by a large number of settlements, economic transactions take place over a given geographical region. Depending on their distribution, urban-rural linkages can facilitate, improve or hinder the access of the rural population to the rural infrastructural services. Thus, they shape the prospects for a region's agricultural, rural, and economic development.

By looking at rural infrastructure and decentralized rural development from the perspective of location theories, which have their roots in the disciplines of economic geography and regional development and planning, useful insights into solutions to the problems of rural infrastructural services can be found (Christaller, 1933 [1966]; von Boventer, 1964). This perspective can help policy planners to comprehend the "how to" in addition to the "what" and the "why" as they tackle the problems of rural infrastructural services faced by the smallholder agricultural sector. The "how to" is placed within the framework of the decentralization of rural development which is under way in many developing countries.

When based on principles of location theory, analysis can help in understanding the distribution of rural infrastructural services and population over space and time and particularly in identifying their availability, use, and accessibility as related to the distribution of towns and villages over a given geographical region (Christaller, 1933 [1966]; von Boventer, 1964). Such analysis can facilitate a suitable categorization of the infrastructure as hard, soft, or institutional and of the population as urban or rural, agricultural or non-agricultural, skilled or unskilled, rich or poor, male or female, adult or children, and literate or illiterate. Distance, which separates the population from infrastructure, is categorized not only by geographical distance (in kilometers) but also by the time and cost it involves for users.

As was noted above, economic activities, infrastructures, and population are located within a geographical region in specific towns and villages that together constitute a settlement system. The accessibility and use of infrastructure can vary within a region. This depends on where the specific settlement providing such infrastructure is located within the wider settlement system. Locating rural infrastructural services judiciously throughout a region is crucial to improving its accessibility to the farming population. Although it is presumed that towns provide better access to services than villages, this is not always the case. Also, services which are available in distant towns cannot be of much help to village populations.

The location of rural infrastructural services in the towns and villages of a region corresponds to their capacity to sustain them. Once established, the services tend to be used not only by the local population but also by the rural population of the surrounding settlements (which forms a territory dependent on the settlement offering the service). Thus, the analysis of the relationship between a center offering a particular service and its dependent territory can provide useful information on the demographic, functional, and spatial characteristics of the settlement system. This relationship has a variety of practical implications emanating from the patterns of availability, use, and accessibility of services at both the settlement and household levels.

Availability, Use, and Accessibility of Services

Other properties of the availability, use, and accessibility of rural infrastructural services located in towns and villages that are fundamentally related to distance are worth noting. In locational studies of development, it is recognized that within a region (i) the availability of services varies directly with the size of the population; (ii) the use of services varies with the distance to the centers providing these services; (iii) the use is also affected by the income and social status of the population; and (iv) the accessibility of services declines with the distance away from towns and cities. Although there is a tendency to consider urban areas as "accessible" and rural areas as "inaccessible", this is not always true. There are degrees of "accessibility" in the provision and use of rural infrastructural services, and an identification of these degrees facilitates a better understanding of the patterns of – and the prospects for - economic development in a region (Wanmali, 1968, 1970, 1981, 1983a, 1983b, 1985, 1987, 1991a, 1991b, 1992, 1997; Sen, Wanmali et al., 1971; Wanmali and Khan, 1970; Chapman and Wanmali, 1981; Celis, Milimo and Wanmali, 1991; Wanmali and Zamchiya, 1992; Wanmali and Ramasamy, 1995; Wanmali and Islam, 1995, 1997).

In any region, therefore, there is a "*distance decay*" as one moves away from the geographical center toward the periphery. A comprehensive analysis of the relationship between such a center and its periphery, particularly involving the distribution of rural infrastructural services, is increasingly important in the current literature on regional development and planning in the rural areas of the Third World. Christaller's central place theory continues to be the underpinning of such analysis and is useful in carrying out regional development and planning (Christaller (1933), 1966; von Boventer, 1964; Grove and Huszar, 1964; Rao, 1964; Friedmann and Alonso, 1964; Friedmann, 1966; Berry, 1967; Wanmali, 1968, 1970, 1981, 1983a, 1985, 1987, 1991a, 1991b, 1992; Wanmali and Khan 1970; Sen, Wanmali et al., 1971; Alam and Khan, 1972; Misra, Sundaram and Rao, 1974; Bhat, 1976; Beavon, 1977; Heath, 1979; Henkel, 1979; Friedmann and Weaver, 1979; Mabogunje, 1980; Berry and Parr, 1988; Obudho, 1995;

Gooneratne and Obudho, 1997; Wanmali and Ramasamy, 1995; Wanmali and Islam, 1995, 1997).

Concluding Remarks and Future Course of Action

The significance of the analysis of rural infrastructural services and urban-rural linkages within the framework of location theories and of decentralized rural development has to be emphasized. This kind of analysis identifies settlement and service center hierarchies in the study regions based on the provision and use of rural infrastructural services. In doing so, it ranks the services based on the size of the population when they first were introduced in the settlements of the study region. It also ranks the settlements by their centrality scores. These are based on the type, number, and level of provision of rural infrastructural services. Since service availability is not uniform throughout all settlements, it is categorized either as service center or as dependent settlement. The analysis groups the service centers and their dependent settlements into sub-regions. The levels of service provision within the sub-regions are identified as being adequate or inadequate. Based on this, the analysis indicates the degrees of incidence of rural infrastructural services in the sub-regions of the study area. The degrees of incidence further indicate the levels of development of the rural infrastructural services. Based on these, a list of locational priorities can be established to enable the district level policy makers' decisions about the future pattern of investment in rural infrastructural services in the study region.

The practical implications of such "deplugging of bottlenecks" in the distribution of rural infrastructural services, as observed in the settlement system of the study region, are also immense for the planning of the transportation network and for the location of agro-based industries (for details, see Figure 49.2).

The results of the analysis can assist the district administration in planning for the multisectorial development of the district within the context of the budgetary allocations made by the national government. The results can further help the district administration in preparing more accurate district development plans based on local and more relevant data. Thus prepared, the district development plans will have a better chance of being successfully coordinated by the district administration, the sub-national coordinating unit, and the national development planning commission and of being included in the national development plan. Also, the district development plans will have a better chance to attract higher levels of financial resources from the government as well as from bilateral and multilateral donor agencies.

The observations noted in this paper provide a conceptual and practical framework for a plan of action based on the analysis of the location of rural infrastructural services in a settlement system which can strengthen the existing rural-urban linkages in a given developing country. These strengthened rural-urban linkages in turn can reinforce rural economic growth and development. As the plan is implemented, it can indicate the nature and scope of the hard and institutional infrastructure that is necessary to complement it. Thus, the role of transport, electrification, and communication systems in strengthening rural-urban linkages and the role of local institutions, both government and private, in implementing and managing the plan can be decided based on the requirements indicated by the plan itself. The emphasis of the action plan, however, is to define the aspect of the "how to" more than the "what" and the "why" in the planning of the development of rural infrastructural services and in the strengthening of rural-urban linkages at the local level.

Fig. 49.2. The Levels of Development and Transport in a Hypothetical Settlement System

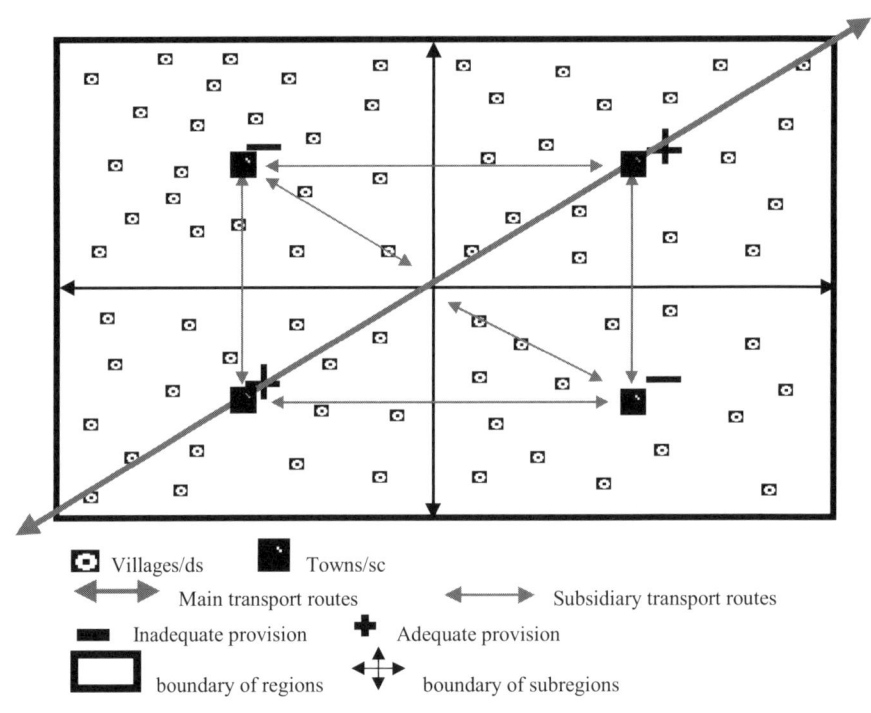

Source: Christaller, 1933, English version 1996.

References

Ahmed, R. and C. Donovan (1992): Issues of Infrastructural Development: A Synthesis of the Literature. IFPRI, Washington, D.C.

Ahmed, R. and M. Hossain (1990): Developmental Impact of Rural Infrastructure in Bangladesh. Research Report 83. IFPRI, Washington, D.C.

Ahmed, R., S. Haggblade and T. Chowdhury (eds) (2000): Out of the Shadow of Famine: Evolving Food Markets and Food Policy in Bangladesh. Johns Hopkins University Press for IFPRI, Baltimore.

Alam, S.M. and W. Khan (1972): Metropolitan Hyderabad and its Region: A Strategy for Development. Asia Publishing House, Bombay.
Beavon, K.S.O. (1977): Central Place Theory: A Reinterpretation. Longman, New York.
Berry, B.J.L. (1967): Geography of Market Centers and Retail Distribution. Prentice Hall, Englewood Cliffs, N.J., USA.
Berry, B.J.L. and J.B. Parr (1988): Market Centers and Retail Location: Theory and Applications. Prentice Hall, Englewood Cliffs, N.J., USA.
Bhat, L.S. (1976): Micro-Level Planning: A Study of Karnal Area, Haryana, India. K.B. Publications, New Delhi.
Celis, R., J.T. Milimo and S. Wanmali (eds) (1991): Adopting Improved Farm Technology: A Study of Smallholder Farmers in Eastern Province, Zambia. IFPRI, Washington, D.C.
Chapman, G.P. and S. Wanmali (1981): Urban-Rural Relationships in India: A Macro-Scale Approach Using Population Potentials. Geoforum 12, pp. 19-44.
Christaller, W. [1933] (1966): Central Places in Southern Germany. Translated from the German by C.W. Baskin. Prentice Hall, Englewood Cliffs, N.J., USA.
Delgado, C.L., J. Hopkins, V.A. Kelly et al. (1998): Agricultural Growth Linkages in sub-Saharan Africa. Research Report 107. IFPRI, Washington, D.C.
Friedmann, J.R.P. (1966): Regional Development Policy: A Case Study of Venezuela. MIT Press, Cambridge, Mass.
Friedmann, J.R.P. and C. Weaver (1979): Territory and Function. Arnold, London.
Friedmann, J.R.P. and W. Alonso (eds) (1964): Regional Development and Planning: A Reader. MIT Press, Cambridge, Mass.
Gooneratne, W. and R. Obudho (eds) (1997): Contemporary Issues in Regional Development: Perspectives from Eastern and Southern Africa. Avebury Press, London.
Grove, D. and L. Huszar (1964): The Towns of Ghana. University of Ghana Press, Accra.
Hazell, P.B.R. and A. Röell (1983): Rural Growth Linkages: Household Expenditure Patterns in Malaysia and Nigeria. Research Report 41. IFPRI, Washington, D.C.
Hazell, P.B.R. and C. Ramasamy (1991): Green Revolution Reconsidered: The Impact of High-Yielding Rice Varieties in South India. Johns Hopkins University Press for IFPRI, Baltimore.
Heath, R. (1979): Rural Service Centres in Rhodesia (Zimbabwe). Salisbury (Harare): Department of Geography, University of Rhodesia (Zimbabwe). Mimeograph.
Henkel, R. (1979): Central Places in Western Kenya. Institute of Geography, University of Heidelberg, Heidelberg.
Jha, D. and B. Hojjati (1993): Fertiliser Use on Smallholder Farms in Eastern Province, Zambia. Research Report 94. IFPRI, Washington, D.C.
Lipton, M. (1987): Agriculture and Central Physical Grid Infrastructure. In: Mellor, J.W., C.L. Delgado and M.J. Blackie (eds): Accelerating Food Production in sub Saharan Africa. Johns Hopkins University Press for IFPRI, Baltimore, pp. 210-227.
Mabogunje, A. (1980): The Development Process: A Spatial Perspective. Hutchinson, London.
Misra, R.P., K.V. Sundaram and V.L.S.P. Rao (1974): Regional Development Planning in India: A New Strategy. Vikas, New Delhi.
Obudho, R.A. (1995): The Role of Small Urban Centres in Economic Recovery and Regional Development in Western Kenya. Centre for Urban Research, Nairobi.
Rao, V.L.S.P. (1964): Towns of Mysore State. Asia Publishing House, London.

Sen, L.K., S. Wanmali, S. Bose, G.K. Misra and K.S. Ramesh (1971): Planning Rural Growth Centers for Integrated Area Development. A Study in Miryalguda Taluka. National Institute of Community (Rural) Development, Hyderabad.

von Boventer, E. (1964): Spatial Organisation Theory as a Basis for Regional Planning. Journal of the American Institute of Planners 30 (2), pp. 90-99.

Wanmali, S. (1968): Hierarchy of Towns in Vidarbha, India, and its Significance for Regional Planning. Discussion Papers No. 23 and 24, Graduate School of Geography, London School of Economics and Political Science, London.

Wanmali, S. (1970): Regional Planning for Social Facilities: An Examination of Central Place Concepts and their Application. A Case Study of Eastern Maharashtra. National Institute of Community Development, Hyderabad.

Wanmali, S. (1981): Periodic Markets and Rural Development in India. B.R. Publishing, New Delhi.

Wanmali, S. (1983a): Service Provision and Rural Development in India: A Study of Miryalguda Taluka, Andhra Pradesh. Research Report 37. IFPRI, Washington, D.C.

Wanmali, S. (1983b): Service Provision, Spatial Intervention and Settlement Systems: The Case of Nagpur Metropolitan Region, India. Annals of the National Association of Geographers of India 3 (2), pp. 27-65.

Wanmali, S. (1985): Rural Household Use of Services: A Study of Miryalguda Taluka, India. Research Report 48. IFPRI, Washington, D.C.

Wanmali, S. (1987): Geography of a Rural Service System in India. B.R. Publishing, New Delhi.

Wanmali, S. (1991a): Market Towns and Service Linkages in sub-Saharan Africa: A Case Study of Chipata, Zambia; Salima, Malawi; and Chipinge, Zimbabwe. African Urban Affairs Quarterly 6(3,4), pp. 267-277.

Wanmali, S. (1991b): Determinants of Rural Service Use amongst Households in Gazaland District, Zimbabwe. Economic Geography 67 (4), pp. 346-360.

Wanmali, S. (1992): Rural Infrastructure, Settlement System and Development of the Regional Economy in Southern India. Research Report 91. IFPRI, Washington, D.C.

Wanmali, S. (1997): Access to Rural Services and Economic development in Communal Areas of Zimbabwe. In: Gooneratne, W. and R.A. Obudho (eds): Contemporary Issues in Regional Development: Perspectives from Eastern and Southern Africa. Avebury Press, London, pp. 203-218.

Wanmali, S. and C. Ramasamy (1995): Developing Rural Infrastructure: Studies from North Arcot, Tamil Nadu, India. MacMillan (India) for Indian Council of Agricultural Research and IFPRI, New Delhi.

Wanmali, S. and F.S. Idachaba (1987): Commentaries on Infrastructure. In: Mellor, J.W., C.L. Delgado and M.J. Blackie (eds): Accelerating Food Production in sub-Saharan Africa. Johns Hopkins University Press for IFPRI, Baltimore, pp. 227-238.

Wanmali, S. and J. Zamchiya (eds) (1992): Service Provision and its Impact on Agricultural and Rural Development in Zimbabwe. IFPRI, Washington, D.C.

Wanmali, S. and W. Khan (1970): Role of Location in Regional Planning with Particular Reference to the Provision of Social Facilities. Behavioral Sciences and Community Development 4, pp. 65-91.

Wanmali, S. and Y. Islam (1995): Rural Services, Rural Infrastructure, and Regional Development in India. The Geographical Journal 161 (2), pp. 149-166.

Wanmali, S. and Y. Islam (1997): Rural Infrastructure and Agricultural Development in Southern Africa. The Geographical Journal 163 (3), pp. 259-269.

Zeller, M., G. Schrieder, J. von Braun and F. Heidhues (1997): Rural Finance for the Food Security for the Poor: Implications for Research and Policy. IFPRI, Washington, D.C.

50 Berlin Declaration on the Urban Future

Citizens and representatives of 1,000 cities, governments and civil society organizations from over 100 countries met in Berlin, from the 4th to the 6th of July 2000, at the *"Global Conference on the Urban Future – "URBAN 21"*. They adopted the *"Berlin Declaration on the Urban Future"* of which the following is an excerpt[1]:

… We, citizens and representatives of 1,000 cities, governments and civil society organisations from over 100 countries from all regions of the world, met in Berlin, from the 4th to the 6th of July 2000, at the Global Conference on the Urban Future (URBAN 21). We commend this declaration to the public and as a contribution to the Special Session of the UN General Assembly (Istanbul+5). We took into consideration the following realities:

- For the first time in human history, a majority of the world's six billion people will live in cities.
- The world is facing explosive growth of urban population …
- Urban poverty, affecting especially women and children, is on the increase, with one in four of the world's urban population living below the poverty line.
- In many countries, social conditions continue to deteriorate and the health and well-being of their citizens are threatened by the HIV-epidemic and the reappearance of major infectious diseases.
- We live in a world of great diversity, in which there is no simple answer and no single solution to the problems and challenges facing our cities.
- Many cities, confronted with hypergrowth, are failing to cope with the challenges of generating employment, providing adequate housing and meeting the basic needs of their citizens.
- Some highly dynamic cities have achieved development with equity, with poverty substantially reduced, illiteracy eliminated, the women educated and empowered, and birth rates falling.
- Other cities face an ageing population, urban decay, unsustainable use of resources and the need to adapt and change.
- No city in any part of the world is free of problems; in particular, none is truly sustainable.

We also took into account the following trends, fully aware of their positive and negative implications:

- Globalisation and the information technology revolution will increasingly create a borderless world with a new role for cities.
- …
- The world is becoming not only a community of nation states, but also a galaxy of interconnected cities.

---

[1] The full text can be read at: http://www.urban21.de/english/index.htm.

- Power is being shared more evenly between national governments, regions and cities.

We therefore recommend the following actions as the most urgent:

- Cities and other levels of government should adopt effective urban policies and planning processes, which integrate the social, economic, environmental and spatial aspects of development, recognising the interdependence between the city and the region, and between the urban, rural and wilderness areas.
- Cities should strive to alleviate poverty and meet the basic needs of their citizens by promoting economic opportunity and enabling community action.
- Cities should embrace information and communication technologies and promote the life-long education of all their citizens to become learning cities and to achieve global competitiveness.
- Cities should promote the use of environmentally friendly technologies and materials, including renewable sources of energy and higher efficiency in the use of natural resources.
- Cities should, where appropriate, consider accepting and integrating informal settlements into the existing urban structure and social life.
- Cities should promote the development of an appropriate integrated public transport system.
- Cities should attempt to achieve a good balance between the natural and built environment and should take action to reduce air, water, land and noise pollution, thereby enhancing the citizens' quality of life.
- The private sector, local, national and international, should bring to bear financial instruments and investments in a manner that promotes sustainable urban development.
- National governments should give high priority to their urban development policies in the framework of national and regional policies.
- National and regional governments should ensure that cities have sufficient power and resources to carry out their functions and responsibilities.
- The World Bank, the UN Development Programme, the UN Centre for Human Settlements, other international agencies and bilateral donors should intensify their co-operation with cities, non-governmental organisations and community-based organisations in the fields of housing, urban development and poverty alleviation.

We conclude on an optimistic note.

We are entering the urban millennium. Cities, always the engines of economic growth and incubators of civilisation, today are beset by tremendous challenges. Millions of men, women and children face a daily struggle for survival. Can we turn this around? Can we give our people hope for a brighter future?

We believe that if we harness the positive forces of education and sustainable development, globalisation and information technology, democracy and good governance, the empowerment of women and civil society, we shall truly build cities of beauty, ecology, economy and social justice.

51 Employment and Migration in the Urban Future of Southeast Asia

Yap Kioe Sheng and Radhika Savant Mohit

Urbanization has been going on for a long time, but it has now reached a stage where soon more than half of the world population will live in urban areas. In the past, urbanization was seen as detrimental to development. However, today's free-market paradigm encourages the means of production to move about freely. In this view, urbanization is a positive process, as labor is moving to places where it can be most productive, i.e., the urban areas.

As education and information penetrate the countryside, expectations for a better quality of life will rise in the rural areas. Cheap means of transportation will facilitate people's movements between urban and rural areas. Under the influence of urban culture, the lifestyle of the rural population will change, becoming more urban in character. Growing wealth in the urban areas will increase the demand for high value-added agricultural products, and this will result in an industrialization of agriculture and a reduction in agricultural employment

At the same time, there is an increasing awareness that while urban areas have been exploiting and depleting natural resources, rural agriculture and living are not completely harmless to the environment either. There is simply not enough fertile land to accommodate the ever-growing rural population which is, therefore, moving into marginal, less fertile and environmentally sensitive areas. Modern agriculture, on the other hand, often pollutes the environment.

The more positive view of urbanization also does not mean that all is well in the urban areas or that the rural areas can be abandoned. Cities are still causing major damage to the environment, and urban problems abound. Large sections of the population live in rural areas, and many of them cannot take advantage of the fruits of economic development. They are and remain among the poorest of the poor. Therefore, the development of rural areas continues to be urgently needed.

The following questions can be raised: (a) If urbanization is an irreversible and positive process, what is the implication for rural development? (b) With urban culture penetrating the rural areas and people moving between urban and rural areas, is the distinction between urban and rural areas still relevant? (c) As the world is becoming one global market place, should rural development be planned within a national development framework, or should one rather think of regional development (covering urban and rural areas) within a global framework?

The Traditional Rural-Urban Divide

Southeast Asia encompasses several countries with a variety of cultures, economic regimes and political systems. However, many of these countries seem to be affected by similar trends - free market economic policies, industrialization, and urbanization - resulting in a convergence of conditions. It may, therefore, make more and more sense to speak of Southeast Asia as a region, despite the different situations in the countries.

Among the continents, Asia is not the most urbanized nor does it have the highest rate of urbanization. Latin America and Europe are far more urbanized, and Africa is urbanizing much more rapidly. However, in absolute numbers, urbanization in Asia affects more people than in any other continent in the world. Asia is changing from a predominantly rural to an increasingly urban continent. In Southeast Asia, more than half the population of Brunei, Malaysia, the Philippines and Singapore lives in the urban areas, while Indonesia will reach this level of urbanization around 2010 - 2015, according to UN statistics.

Asian governments have long denounced rural-urban migration and urbanization as the main obstacles to national development, claiming that they generate poverty, unemployment, crime and social disorder, slums and squatter settlements and degradation of the urban environment. Governments adopted policies that focused on rural development in the expectation that improved living conditions in the rural areas would stop the population from migrating. They tried (unsuccessfully) to reduce urbanization or redirect rural-urban migration to secondary towns and rural areas through transmigration, migration controls, de-urbanization and industrial dispersal policies, often at the expense of human rights or the environment.

Indonesia's transmigration program aimed at stopping people on the densely populated island of Java from moving to the urban areas by shifting them to less populated islands. An evaluation found that most trans-migrants had benefited but that there had been inadequate protection of indigenous people and the environment (World Bank, 1994). Trans-migrants faced the hostility of the local population who had different cultures and religions and who saw their land being taken away for the settlement of newcomers. Because of the difficulty in obtaining land suitable for agriculture, trans-migrants were often given plots of converted rainforest.

In 1970, the governor of Jakarta tried to limit migration by decree. He ordered that migrants could only enter the city if they had a Jakarta residence permit, guaranteed accommodation and regular employment. Most migrants could not meet these requirements. The short-lived policy increased corruption and the cost of living for the poor, but it only partially interrupted the flow of migrants. The authorities also tried to reduce the employment opportunities of poor migrants by limiting the number of rickshaws in the city, which involved a quarter of a million people in 1972. Employment was permanently lost, but urbanization continued (Castles, 1989:251; Gilbert and Gugler, 1992:251).

After its victory in 1975, the Khmer Rouge evacuated the urban areas of Cambodia in order to build a new egalitarian agrarian society. During the war, refugees from the countryside had increased the population of Phnom Penh from 0.4 to 2 million, with US aid forming the only economic basis. This situation was clearly unsustainable after 1975, but the main factor for the de-urbanization was probably the anti-urban ideology of the Khmer Rouge: cities were dangerous, immoral and full of inequality. Many people died during the evacuation and from the harsh life in the rural areas. After the fall of the Khmer Rouge, rural migrants invaded Phnom Penh and other towns in search of security and employment.

Since the 1950s, successive governments in Thailand have allowed the rural population to clear virgin forests for agriculture. This provided a growing rural population with subsistence and reduced the need to move to the urban areas to make a living. In the 1960s, this policy also aimed at denying territory to communist insurgents. Between 1961 and 1988, the area covered by forest declined from 171 million to 90 million *rai* (1 *rai* = 1,600 m^2). Because of a lack of organization, 10 - 12 million people, around a quarter of the rural population, are now squatters on untitled land in forest reserves (Phongpaichit and Baker, 1995:48-67).

The Board of Investment of Thailand divides the country into three zones. The first covers Bangkok and six surrounding provinces; the second covers ten provinces surrounding Zone 1; the rest of the country forms the third zone. Privileges for investors increase in zones further away from Bangkok. Over the years, investments and employment have increased in Zones 2 and 3, but this is due to market forces rather than privileges (Poapongsakorn, 1995:139). The privileges are not sufficient to outweigh the advantages of Bangkok: its social and physical infrastructure, the presence of suppliers and a large market, its large skilled labor force, and the proximity to the government and to transport hubs.

While the expressed policies of governments aimed at rural development and at a reduction in rural-urban migration, actual decisions often promoted urban industrialization and widened the gap between rural and urban incomes. Governments invested in urban areas disproportionately to the size of the urban population, unmoved by the magnitude of rural poverty. They made or promoted massive investments in urban infrastructure and in health and educational services in urban areas. Substantial subsidies were often given to urban services, while urban firms were rarely taxed for the negative externalities they caused such as pollution and congestion.

All this happened despite the expressed policies, primarily because cities are the places where the elite live and the government is located. The capital city is the showcase of the country. More importantly, however, the higher economic rate of return on investments in urban areas was considered more important than the social impact of investments in rural development to alleviate poverty and reduce disparities. Urban areas are contributing an ever growing percentage to the national economic development, and urban investments were expected to benefit the rural areas at least indirectly.

The insincerity concerning rural development is gradually disappearing. Now, it is acceptable to invest "disproportionately" in urban areas. With the spread of economic policies that give the lead role for resource allocation to the market rather than the government, the positive aspects of urbanization and rural-urban migration are being highlighted. Market mechanisms and competition are expected to bring about a better allocation of production factors than a more regulated economy with administrative control and central planning (Martinussen,

1997:261). Capital and labor are supposed to be free to go where they can get the best return.

Economies of scale allow governments to deliver public services and utilities, education, health care and family planning at much lower costs in urban areas. Cities grow not only horizontally but also vertically and therefore use less land. Multi-unit housing requires less energy for cooling than single housing units. The negative environmental effects of urban activities are easier to control through regulatory and economic measures than those of rural activities. Distances within the city are relatively short, and mass transit systems can carry large numbers of passengers at relatively low costs. Recycling of solid waste is worthwhile, because large quantities of materials are available.

Increasing Convergence and Disparities

Better and cheaper transportation and communication links have decreased the significance of geographical distances, and people interact and move more easily between rural and urban areas. Temporary, seasonal and circular migrations are common phenomena. Migration is not limited to movements between urban and rural areas. It also includes movements between countries and even regions. Remittances and returning migrants do or at least can play important roles in the development of rural areas and small towns.

Rural-urban migration and international labor migration have contributed to a spread of urban and foreign cultures. Education and information technologies have further enhanced the spread of an urban lifestyle. The economic, socio-cultural and environmental influence of cities - what McGee (1994) called *desakota* - extends far into their hinterlands with export-processing factories in the middle of the rice fields and villagers commuting by bus, train or motorbike to urban areas for employment. At the same time, interest in urban agriculture as a source of income for the urban population and as a means to improve the environment is growing (Tacoli, 1998:158). In other words, the geographical and the socio-cultural gap between urban and rural areas will be narrowed.

The increased demand for high value-added food products will result in a decline of small-scale farming, a reduction in agricultural employment and the industrialization of agriculture, agro-industry and non-farm employment. This will further urbanize rural areas, particularly those that can meet the demands of the urban population. In this respect, the distinction between urban and rural areas may lose its relevance. While population and building densities in rural areas will remain low, the culture, lifestyle and occupations of the rural population and their environmental impacts will become urban in character.

The industrial sector is increasing its demand for skilled labor, and a high value-added industry and service sector will increase the demand for better educated people. When a true knowledge-based economy becomes reality, this trend will be further reinforced. Those lacking the required education will not be able to participate in such an economy. The transition to the new economy risks leaving a portion of the population behind, particularly among a rural population that does not have full access to education.

In many countries, there is not enough unclaimed fertile land left to accommodate an ever-growing rural population. Farmers move into marginal, less fertile and environmentally sensitive areas, causing deforestation, erosion and floods as well as reducing the habitat of wild animals and jeopardizing biodiversity. Governments are becoming aware of the environmental problems caused by the expansion of farmland. However, what are the alternatives: the creation of more off-farm employment or the promotion of more rural-urban migration?

An ultimate consequence of free-market policies, of the urban economies of scale and agglomeration and of the environmental concerns may be the movement of a large majority of the rural population to the urban areas. Will those remaining in the rural areas be the ones who have no role to play in the urban economy: the elderly, the children and those lacking the knowledge and skills to be employable in the urban industry and service sectors?

Betting on the educated, the strong and the entrepreneurial and uninterested in developing safety nets for the disadvantaged, the urban economies may not be too keen to see those people move to the cities as well. Since rural-urban migration cannot be regulated within a country and should not be regulated, because free movement within one's country is a human right, some may see autonomy for the city and its immediate hinterland.

There are a number of references in the literature about the desirability of a return of the city-state with the advent of the global economy. Economists look at the examples of Hong Kong and Singapore. These cities did not need to take care of large rural populations, but at the same time they could draw on an unlimited labor pool from across their borders. They could also invest in nearby countries for the labor-intensive parts of the production process. Their example has raised the question about the need for rural areas. In a global market, the proximity of raw materials loses its importance, while trade and intensive agriculture in peri-urban areas may be able to secure at least part of the food supply.

In his book, "*The End of the Nation-State: The Rise of Regional Economies*," Ohmae (1995:80-81) suggests the rise of regional states, i.e., sub-national regions that have all that is needed to operate in the global economy. In his view, regional agglomerations are the meaningful units of urban life, because they have all the essential ingredients for successful participation in the global economy. They can tap into global investment capital to develop their industries using information technology and have a substantial market of consumers (Ohmae, 1995:100). Kunio (1999:74-78) comes to a similar conclusion for the regions in Southeast Asia.

With trade liberalization, there is no reason for cities to have to buy agricultural products from the rural areas in the same country. Economic globalization will turn the world into one market place, and the cities will be the first to benefit. Rural areas could in turn sell their products anywhere. Where the rural areas are prepared for this new situation, they will prosper. However, many rural areas may not be prepared, because they are too dependent on the national bureaucracies and on the physical and economic infrastructure that is directed towards the country's major cities.

Regional Autonomy

Development needs to be inclusive and equitable. It should not only benefit the rich but also the poor, not only the urban but also the rural population. It should benefit men as well as women and not only people of the working age but also the elderly and children. However, the emphasis on market forces is widening income disparities between the urban and rural areas and between the rich and poor, with the widest gap between the urban rich and the rural poor. Free-market policies bet on the strong, the skilled, the entrepreneurial and aim consciously at the survival of the fittest. Who will take care of the weaker sections of society?

Governments are increasingly promoting a decentralization of responsibilities to lower levels of government. Up until now, this has benefited mainly the urban areas, because cities have more political clout, and rural areas are considered backward. There may be a case for regions within countries to make use of the opportunities of decentralization to profile themselves and exploit their comparative advantages, not so much within a national but rather within a global economy.

Greater autonomy for regions (i.e., rural areas and the nearby towns) within countries can create possibilities for economic development less dependent on capital cities. Greater autonomy would allow regions to seek economic and infrastructure links with the best urban and regional partners and would allow for regional economic development in a global economy. This will obviously require capacity building and infrastructure development that links regions to the global market rather than only to the capital city.

It would also allow regions of countries to link up economically with cities and regions of other countries. Growth triangles such as the Singapore-Johor-Riau triangle and the Hong Kong, Taiwan-Guangdong-Fujian triangle are examples of urban and rural areas in different, adjacent countries that have joined forces for development based on comparative advantages (Lee, 1995).

Greater regional autonomy may also ensure the preservation of local cultures and possibly more effective environmental policies than those that emanate from the center. Globalization of the economy and decentralization of decision making powers are trends in today's world which, as many observers have pointed out, can lead to an increase in regionalism and possibly the rise of the region state. With differences in development levels between regions within a country, regionalism highlights the identity of the region. Regional autonomy enables its population to decide its own policies to meet its needs and deal with the global economy. The regions are not cities but are different parts of a country with divergent interests.

However, a central government or nation state is necessary to play a role in facilitating and promoting the balancing of development between different regions and population groups. Within a globalizing world, the rural-urban divide is no longer relevant. The unit of development should be the region, and policies should aim at maximizing their potential within a global rather than a national economic framework. The role of the central government would be to facilitate the entry of regions in the global economy. This will require institutional change and capacity building at a regional level and the development of political institutions that can ensure good governance.

References

Castles, L. (1989): Jakarta: The Growing Centre. in: Hill, Hal (1989): Unity and Diversity: Regional Economic development in Indonesia since 1970. Oxford University Press, pp.233-253.
Gilbert, A. and J. Gugler (1992): Cities, Poverty and Development: Urbanization in the Third World. Second Edition, Oxford University Press, Oxford.
Kunio, Y. (1999): Building a Prosperous Southeast Asia: From Ersatz to Echt Capitalism. Curzon, Richmond.
Lee, T.Y. (1995): The Johor-Singapore-Riau Growth Triangle: The Effect of Economic Integration. In: McGee, T.G. and Ira Robinson: The Mega-Urban Regions of Southeast Asia. UBC Press, Vancouver.
Martinussen, J. (1997): Society, State and Market: A Guide to Competing Theories of Development. Zed Books, London.
McGee, T.G. (1994): Labor Force Change and Mobility in the Extended Metropolitan Regions of Asia. In: Fucks, R., E. Brennan, J. Chamie, F. Lo and J. Uitto (eds): Mega-City Growth and the Future. United Nations University Press, Tokyo.
Ohmae, K. (1995): The End of the Nation State: The Rise of Regional Economies. Harper Collins Publishers, London.
Phongpaichit, P. and C. Baker (1995): Thailand: Economy and Politics. Oxford University Press, Kuala Lumpur.
Poapongsakorn, N. (1995): Rural Industrialization. In: Krongkaew, Mehdi (ed): Thailand's Industrialization and its Consequences. St.Martin's Press, New York, pp. 116-140.
Tacoli, C. (1998): Rural-Urban Interactions: A Guide to the Literature. In: Environment and Urbanization, vol 10, No. 1, April 1998, pp. 147-166.
World Bank (1994): Indonesia Transmigration Process: A Review of Five Bank-Supported Projects. Report No. 12988. World Bank, Washington, D.C.

Agenda 21 Principle 3, Earth Summit, Rio 1992

Principle 3 The right to development must be fulfilled so as to equitably meet developmental and environmental needs of present and future generations.

1992 2000 2100 2500 2750 3000

Sustainable Development

52 Employment and Migration Perspectives in Africa

John O. Oucho

A clear dichotonomy exists between urban and rural areas, but literature shows that links between these areas are an important feature in developing countries. Any index of the application of development points to the hackneyed fact that, whereas urban areas remain centers of employment opportunities, rural areas are still suffering from migrations to the cities. The synergy between these two areas sustains urban-rural linkages. In the urbanizing world (UNCHS, 1996) in which economic reforms, changes in governance and a vibrant civil society are dominant, the nature, scope and implications of urban-rural links deserve to be analyzed from the perspectives of employment and migration which are part of the dynamic calculus of urban-rural linkages. However, since both employment and migration are studied from various disciplinary stances, it is difficult for scholars to form a consensus on the issues, even within a given discipline.

Non-agricultural Employment for Rural Development: Retrospect and Prospect

For over two decades, many African countries have made conscious efforts to stem rural-urban migration and the concomitant unemployment by trying to create opportunities for non-agricultural employment at the places where out-migration begins. Unfortunately, many of these experiments failed. This was mainly because the chasm, real or imagined, between rural and urban areas has widened over time. Against this dismal background, it is imperative to attempt a new set of policy interventions.

There are several instances where the creation of non-agricultural employment failed to arrest the flow of rural-urban migration in Africa. In Kenya, for instance, rural transformation came in the wake of the post-independence land settlement program. In this program, new farmers were able to raise their incomes. The sugar industry augmented rural incomes, and smallholder agriculture assumed growing importance. However, since the mid-eighties, the structural adjustment program with its attendant economic reforms has allegedly mismanaged the finances and crippled commercial rural agriculture, which has had to lay off many employees. Corrupt practices, including importing sugar in a country which had become self-sufficient in the product, has crushed the sugar industry. From 1991 to 1993, ethnic clashes in several settlement areas in the country caused farmers to be evicted from their lands (Oucho, 1999). In neighboring Tanzania, the *Ujamaa* (socialism) policy of the country's socialist government, which existed before capitalism took over in the early nineties, was intended to create, among other things, organized agricultural employment to stem rural-urban migration. The Tanzanian experiment was an obvious failure, and in its wake, increasing rural-urban migration has exacerbated the grave economic difficulties in the country. Finally, in several West African countries – Cote d'Ivoire, Ghana, Nigeria and Sierra Leone – conscious efforts to create rural employment faltered as a result of economic crises and political problems that are still not completely resolved. As a response to structural adjustment programs, the former rural-urban migrants who often had sustained links with their rural homes have begun an urban-rural return migration.

What policy interventions need to be employed to create diversified employment in rural areas of Africa? It should be kept in mind that the continent's fate has been complicated by economic adversities in both urban areas, where hope has faded, and rural areas, where it is lost.

Non-agricultural employment is essentially non-farm employment. In a detailed survey, Lanjouw and Lanjouw (1995) provide useful evidence covering different geographical areas of the developing world. They define the non-farm sector as including "...*all economic activities except agriculture, livestock, fishing and hunting...and involves activities undertaken by farm households as independent producers in their homes, the sub-contracting of work to farm families by urban-based firms, non-farm activity in village and rural town enterprises and commuting between rural residences and urban non-farm jobs.*" Indeed, much of non-farm activity, especially that engaging female labor, remains unremunerated and therefore an unquantified aspect of non-agricultural employment. Some findings in different African countries affirm the importance of non-farm employment as a source of income. In Kutus Town in central Kenya, Evans and Ngau (1991, cited in Lanjouw and Lanjouw, 1995) found that the wealthiest quarter of households received 52% of their income from non-farm sources compared to 13% for the lowest quarter. In Burkina Faso, Reardon et al. (1992, cited in Lanjouw and Lanjouw, 1995) found a similar pattern. Among African women, several non-farm activities generated income: beer brewing in Botswana, Malawi and Zambia; fish processing in Senegal and Ghana; pottery in Malawi; rice husking in Tanzania; and retailing and vending (Bagachwa and Stewart, 1992; cited in Lanjouw and Lanjouw 1995).

The importance of non-agricultural income, which sustains agricultural productivity and the income accruing from it, is clearly a factor in many African countries. However, in the absence of policies and programs having to do with non-agricultural activity, the place of non-agricultural employment remains unclear.

Income Generation and Employment Opportunities in Urban-Rural Linkages

Economists who have studied rural-urban migration have propounded extremely useful theories and models to underpin determinants of the process. In their conceptual frameworks and using empirical evidence, they identify rural-urban income and employment opportunities differentials as crucial to urban-rural links. Although this is not the place to reiterate economic theories and models, a brief examination of selected ones is appropriate. However, economists, like other social scientists, do recognize non-economic factors.

Urban-rural remittances play a crucial role in generating and sustaining employment opportunities through urban-rural linkages. In rural settings, they sustain domestic consumption and survival strategies, social networks and links between urbanites and their rural family and relatives. They constitute wages for hired farm labor which replaces the youthful out-migrants from farming households, and, in a pooled form, they provide and substitute the government provision of social infrastructure and services (Oucho, 1985; Oucho, 1996). Conversely, in urban areas, rural-urban remittances reciprocate with the provision of foodstuffs and money for new, settling and even long-term urban in-migrants. Moreover, they are the source of agricultural produce sold in the informal sector of urban areas, e.g., open-air markets. Varied forms of urban-rural linkages exist and give meaning to the interdependence of the two settings in many African and indeed also in other Third World countries.

Regrettably, it is increasingly evident that these forms of urban-rural linkage are being eroded by economic crises, economic reforms and poverty in many African countries.

The last point implies that employment opportunities are diminishing as a viable part of urban-rural linkages. The African world still languishes in the "urban bias" of development which, as Lipton (1988) argues, explains "*why poor people stay poor.*" The need for research on this issue cannot be overemphasized.

Future Prospects of Urban Agriculture

Both urbanization and urban agriculture in Africa can be examined in two historical epochs: the colonial and the post-colonial periods. In the earlier period, European colonial powers prohibited urban agriculture, finding that it interfered with the grandeur and cleanliness of cities and wanting to distinguish them from the "bush". In the independence era, civic authorities initially frustrated spontaneous efforts in urban agriculture but later finally relented (UNDP, 1996:38-41). Meanwhile, urban agriculture has become an integral part of urban economies in African countries. Not limited to the so-called urban poor, also the middle class participates in urban agriculture and thus can augment their incomes (Mlozi et al., 1992:287).

For the future, strategies are needed to sustain urban agriculture in an environmentally friendly manner and to the satisfaction of both urban authorities and residents. The overarching strategy should be urban planning, where particular areas are designated for urban agriculture in much the same way as other conventional land uses – commercial, industrial, residential and recreational. This is necessary as the important sector of urban agriculture has both formal and informal characteristics. As related by Mlozi et al. (1992:293) in their study of urban agriculture in six Tanzanian towns of varying sizes, urban authorities should protect and improve presently unusable land so that it can be cultivated by the urban poor. They should be assisted in forming small groups or cooperatives in order to benefit from bank credit, and national governments should help to improve agriculture on the basis of an inventory of the urban poor and the available land. Thus, the future strategies should be the concern of the state as well as the civic authorities in the interest of the urban farmers. But encouragement of urban agriculture must be based on sound policies that protect the fragile urban environment, enhance the survival of urban populations and create a symbiotic relationship between urban agriculture and other land uses. For these to be realized, careful monitoring and evaluation of urban agriculture is mandatory for all urban areas.

Past Migration Policies versus Future Migration Strategies: Prospects for Spatial Population Distribution

This section first examines past policies as a basis for considering future strategies for addressing rural-urban migration and urban-rural links. Against that background, it speculates on prospects for a viable spatial distribution of population which is likely to stimulate regional development. Using the available literature, we revisit past policies, which broadly fall into three categories: (i) government policy responses aimed at stemming rural-urban migration and fostering sustainable urban-rural linkages, (ii) international agency policies targeting urban growth, and (iii) urban-rural linkages policy within the framework or in the wake of structural adjustment programs (Becker et al., 1994:120-134).

Broadly speaking, *"policies either aim to reduce the stock of potential migrants, decrease the flow of migrants to the city by inducing migrants to stay in their rural homes, create attractive alternate (non-primate city) destinations for the migrants, or successfully absorb the flow of migrants into existing urban centres"* (p. 120). For the sake of brevity and focused analysis, Table 52.1 presents the past policies and suggests future strategies for influencing rural-urban migration from the perspective of spatial distribution and regional development.

The first category of policy options consists of government policy responses. Rural development policy has been meant either to stem or to stop rural-urban migration through the "growth pole" or "growth center" models, such as in Kenya in the decade from 1968 - 1978. This was later undertaken by the government's Rural Trade and Production Centers program (Becker et al., 1994:121-2). Yet this approach has had little if any impact. An important objective of the growth center model was to stimulate urban and industrial development to enable the centers to absorb increasing numbers of in-migrants. Apart from inadvertently triggering a step-wise migration process to the cities (Riddell, 1978:255 cited in Becker et al. 1994:123), the decentralization model is unfeasible in Africa where sparsely settled populations increase its costs (p.123). Other policy prescriptions included redirection of migrants elsewhere, for example, by land colonization projects sending people to settle in agricultural areas, by resettlement of populations displaced by the construction of dams, or by forced resettlement as under Tanzania's *Ujamaa* (socialism) of 1967 - 77.

Table 52.1. Past Rural-Urban Migration Policies and Strategies Affecting Future Spatial Population Distribution and Regional Development

Past Policies*	Future Strategies
Government policy responses:	*Governmental and non-governmental strategies:*
Rural development to stem/discourage out-migration.	Revitalization of rural development programs/projects in the face of economic adversities.
Urban and industrial development to absorb in-migrants	Strengthening informal sector urban and industrial opportunities.
Targeting migrants: decreasing rural-urban flow	Assistance to potential rural-urban migrants based on needs assessment.
Urban migrants' return migration and redirection elsewhere	Monitoring and evaluation of returnees' conditions for improvements.
Creation of growth poles for decentralized development	Critical assessment of growth poles in order to enhance their roles
Compulsory resettlement schemes (dams-displaced persons, Ujamaa, etc.)	Redesigning failed schemes and adjusting to globalization
Relocation of the national capital	Retrospective and prospective interpretation of relocated capitals.
International agency policies facilitating urban growth:	*Partnership between international agencies and national governments & NGOs:*
Loans and grants: project-oriented assistance, short-term loans and relief aid in response to civil strife and agricultural disasters.	Critical appraisal of international agencies' loans and grants in order to recast them. Stocktaking of effects of SAP strategies for recasting and new humane strategies.
Structural adjustment programs (SAP) favoring rural development: shrinking public sector and parastatal employment, reduced tariff protection, exchange rate devaluation and de-emphasis of import substitution industrialization.	Strategies for eliminating the pangs of SAP Institutionally sustainable rather than "limited success" programs for long-term (e.g., second and third generation urban migrants, rural returnees and stayers.
Post-SAP policies Reactionary programs to SAP (Ghana, Nigeria, Guinea, Madagascar)	Replication of success stories in particular countries Employment (including self-employment) opportunities for urban-rural and intra-rural migrants.

Source: * Based on text from C.M. Becker et al., 1994:120-134.

Unfortunately, all these have experienced disheartening failures, fomented ethnic strife and precipitated as well as deepened the chasm between the minority landed gentry and the landless majority in many African countries. Another government policy response has been the relocation of the national capital – from Mafeking in South Africa to Gaborone in Botswana, from St. Louis in Senegal to Nouakchott in Mauritania. This has also taken place within one country as from Zomba to Lilongwe in Malawi, from Dar es Salaam to Dodoma in Tanzania and from Lagos to Abuja in Nigeria (p.124). Except for the first two international transfers of capitals, the rest involved unjustifiable costs and has not met the expected objectives.

The second category of policy options falls within the framework of international agencies' policies which have incessantly underscored loans and grants and, since the eighties, emphasized structural adjustment programs. Becker et al. (1994:130) document several examples, citing success stories such as the World Bank-funded "site and service" schemes in many African countries. These have also included loans for transport sector investment, for infrastructure and utilities (energy, water and telecommunications) and for the expansion of secondary and university education. Then came the Structural Adjustment Programs (SAP) which involved a variety of reforms that affect the public sector, give prominence to private enterprise and inject changes in urban-rural links. The results of these programs had significant "demonstration effects" for replication, but economic deterioration in the countries which had adopted them has made them unsustainable. This leads us to the post-SAP policies in which governments have reacted to the changing economic, and indeed political, climates through retrospection and contemplation of future strategies. In Ghana, Nigeria, Guinea and Madagascar, agricultural reforms have spurred the economy. Retrenchment of the civil servants has reduced the wage bill, and several positive results have been reported. On the negative side, food prices have escalated as the prices for agricultural commodities have plummeted, and massive devaluations of national currencies have dealt heavy blows to the people's purchasing power (pp.131-3). The balance seems to be skewed to the left in most African countries.

What then are the appropriate strategies for the future? Some of these are presented alongside the past policies in Table 52.1. The idea here is to consider the policies using both hindsight and foresight. Against the background of past government policies, they must be reviewed critically in order to prescribe appropriate strategies which recognize the changes that have taken place since the policies were adopted, the *in situ* and post-SAP economic reforms, the wave of globalization and the tenets of sustainable development. In a world where governmental and non-governmental institutional partnerships are assuming greater importance than before, it is necessary for governments, international agencies and NGOs to take a fresh critical look at their delivery of services to the populace, the weaknesses of delivery approaches and the improvements that need to be made. The civil society is no longer passive. It is vibrant and critical to what these institutions do. Finally, the effects and experiences of SAPs demand critical appraisal in order to sustain the best practices and to discard the undesirable ones.

Conclusion

This paper has attempted to answer a number of questions concerning urban-rural linkages from the perspectives of two important factors in the process, namely employment and migration. It is becoming increasingly evident that non-agricultural employment will assume greater importance in the future as non-farm activities gain a firmer hold while agriculture diminishes. With most African countries perceiving rural industrialization as a panacea for development, they necessarily will have to develop well-conceived strategies and put into place suitable programs to realize their dreams.

The future of urban agriculture, on which the food supply of most African urbanites has depended, looks bleak. As environmental degradation increases, urban land use moves farther away from agriculture. If the momentum of industrialization gathers speed in the future, urban agriculture will inevitably become a thing of the past. It will therefore become necessary to find viable alternatives for feeding the burgeoning city populations in the face of declining urban agricultural activity. Both central government and civic authorities should endeavor to formulate policies based on the wealth of information from both the past and the present as well as the implications of urban agriculture.

In the same vein, African central and local governments should work in concert to review the past and current migration policies in order either to reformulate or newly formulate them. As migration is largely responsible for the spatial distribution of the population, including urbanization, it plays a catalytic role in sustaining urban-rural linkages. Unfortunately, migration research has been starved of resources by the United Nations system, international NGOs, foreign aid agencies and national governments in Africa. It remains the stepchild of demography, even though it is the "Cinderella" of development. If we pause to ponder Africa's political climate, economic situation and social decay, the importance of migration for employment creation in urban-rural linkages cannot be overemphasized.

References

Baker, J. and O. Pedersen (eds) (1992): The Rural-Urban Interface in Africa. Nordiska Afrikainstitutet (The Scandinavian Institute of African Studies in cooperation with the Centre for Development Research, Copenhagen), Uppsala.

Drakakis Smith, D. (1992): Strategies for Meeting Basic Food Needs in Harare. In: Baker and Pederson (eds), pp. 258-283.

Mlozi, M.R.S., I.J. Lupanga and Z.S.K. Mvena (1992): Urban Agriculture as a Survival Strategy in Tanzania. In: Baker and Pedersen (eds): The Rural-Urban Interface in Africa. The Scandinavian Institute of African Studies, Uppsala, Sweden, pp. 284-294.

Oucho, J.O. (1985): Co-Existence of Rural-Urban Migration and Urban Rural Links in sub-Saharan Africa. Development Policy and Administration Review XI (2), pp. 33-60.

Oucho, J.O. (1990): Migrant Linkages in Africa. Retrospect and Prospect. In: Union for African Population Studies. Conference on the Role of Migration in African Development: Issues and Policies for the 1990s. UAPS, Dakara, Senegal.

Oucho, J.O. (1996): Urban Migrants and Rural Development in Kenya. Nairobi University Press, Nairobi.

Oucho, J.O. (1999): Undercurrents of Ethnic Conflict in Kenya. Mimeo.

Riddell, J.B. (1978): The Migration to the Cities of West Africa: Some Policy Considerations. In: Journal of Modern African Studies 16.

UNCHS (United Nations Centre for Human Settlements) (1996): An Urbanizing World: Global Report on Human Settlements 1996. Oxford University Press.

UNDP (United Nations Development Programme) (1996): Urban Agriculture: Food, Jobs and Sustainable Cities. UNDP Publication Series for Habitat II, vol 1, UNDP, New York.

53 Reunified Germany: Separate Rural Developments

Claudia Neu

Agriculture has been losing importance in the creation of employment in Germany for many years now. The effects of this structural change vary widely among the different regions of Germany. In Western Germany, the release of workers from the agricultural sector has been offset by the creation of new jobs in other sectors. Particularly in Southern Germany, small- and medium-sized enterprises (SMEs) are providing employment in rural areas. These enterprises are mainly active as suppliers for the automobile or electrical engineering industries. The rural areas of Bavaria and Baden-Würtemberg, both in Southern Germany, have unemployment rates that are only slightly – if at all – higher than these states' average unemployment rates, which are about 6%. (The unemployment rate in Germany as a whole is about 10% - 11%).

The situation in the traditionally agricultural-oriented northeast of Germany is completely different. The rural areas of the new federal states of the former East Germany (*Neue Bundesländer*) are still suffering from the aftereffects of German Reunification, especially the de-industrialization and the restructuring of the formerly socialist agricultural sector. Unemployment rates of more than 20% are not rare here.

For more than 40 years, the developments of agriculture and of the rural areas in Germany were characterized by different strategies in the East and West. What both strategies shared was the payment of high subsidies, though with different political aims. In the West, small farms and SMEs remained the model for the development of rural areas after World War II. Subsidies cushioned – and thereby often hampered – structural changes in the agricultural sector. This was completely different in the former German Democratic Republic (GDR). Here, the agricultural sector was rearranged according to the socialist model. The aim was to create a collective, industrially organized agricultural sector. This aim was achieved by means of several political interventions such as land reform, collectivization, and industrialization. As was the case in agriculture, enterprises - and small- and medium-scale trades and entrepreneurial activities - were destroyed by the process of nationalization. Only a few small farms and workshops survived. Large-scale agricultural organizations characterized the rural areas of the GDR much more than the rural areas of West Germany. The GDR's agricultural cooperatives (*Landwirtschaftliche Produktionsgenossenschaften* (LPG)) and state farms (*Volkseigene Güter* (VEG)) did not solely produce agricultural goods but also ran large production-support departments such as "tradespeople's brigades" or transport units. In addition, they took over social, cultural and infrastructural functions such as child care and youth clubs as well as road and housing construction. In 1989, the year the wall came down, over 3,850 LPGs and 464 VEGs existed in the GDR.

Transformation in Rural Areas

Since the break-up of the GDR and the reunification of Germany, significant changes have been taking place, especially in Eastern Germany. In the course of the economic transition from a planned to a market economy, the whole East German agricultural sector was reorganized. Under the banner of "privatization," the nearly exclusive state and cooperative farming system (LPG and VEG) was subdivided. Today, small farms (owner/operator farms) exist alongside farms organized as legal entities such as agricultural cooperatives, limited liability companies, civil law associations and stock corporations (*Agrargenossenschaften, GmbH, GbR, Aktiengesellschaften*).

Following the collapse of the GDR and its reunification with West Germany, socialist agriculture was restructured. Small farms were reintroduced, and agricultural cooperatives had to change their legal status or be dissolved. As of 1998, about 28,000 agricultural enterprises existed in Eastern Germany. 22,000 of these were undertaken by individual entrepreneurs, 3,000 were general partnerships (*Personengesellschaften*), and 2,900 were corporations (Agrarbericht, 2000). In comparison, in Western Germany 98% of the almost 420,000 agricultural enterprises were run by individual entrepreneurs. The process of adjustment to the social market economy and the European domestic market was accompanied by a huge wave of layoffs, due to the fact that the businesses that had emerged from the former LPGs now had to concentrate on agricultural production. Support departments had been drastically reduced and the infrastructural support of the rural areas largely abandoned. During the transformation process, about 80% of the former LPG/VEG employees lost their jobs in agriculture. In 1989, about 880,000 people were working in agriculture in the former GDR. Almost ten years later, in 1998, only 140,000 people were making their living as farmers (small farms, cooperatives and personal enterprises) in Eastern Germany. However, field studies[1] in the state of Brandenburg have shown that losing jobs in agriculture generally is not to be regarded as synonymous with unemployment (Neu, 2000). In 1996, 40% of the former agricultural employees were still working: 16% of them were working in re-established agricultural cooperatives, 21% had found new jobs in sectors other than agriculture (communal administration, skilled trades, unskilled labor) and 3% were establishing their own small businesses (skilled trades, family farms, catering). In contrast, about 53% of the former farm employees were out of jobs: about a third (34%) had retired, 15% were unemployed and 4% were working in job creation programs. Information about 7% of the former employees could not be acquired.

---

[1] In 1995 and 1996 I analyzed the situation of four former LPGs (two livestock and two crop farms) in the state of Brandenburg. These four farms transformed themselves to Agrargenossenschaften (farm cooperatives). My research focused on the occupational and social mobility of the 719 former employees of the four aforementioned LPGs between 1989 and 1996, taking their personal data into account. In addition, I conducted personal interviews with 69 of these former staff members.

Newly Created Jobs Outside of Agriculture in the Rural Areas of Eastern Germany

The structural change within the East German agricultural sector in the last ten years led to a reduction in employment and, simultaneously, to a significant drop in employment in the manufacturing sector in Eastern Germany. New jobs in rural areas were created mainly in the service sector, the crafts professions and retail trade. The latter results mainly from the huge number of shopping centers – containing supermarkets, garden centers and builders' markets – that have sprung up on the outskirts of large villages, towns and smaller cities. The workers released from the agricultural sectors found new job opportunities, e.g., as salespeople or stock workers. On the other hand, these shopping centers have, in many areas, displaced small retail trade. Thus, many city centers in Eastern Germany today are increasingly empty, and this is not only because of a lack of purchasing power. There are many developments already beginning to work against this trend, but still, many stores remain empty because new businesses cannot be found to fill them or because old businesses, such as tea shops and fashion boutiques, have had to close their doors due to a lack of customers.

The new entrepreneurial businesses in rural areas, especially in the skilled trades (for example, electricians, roofers, and structural and civil engineers), were often established as the offspring of former LPGs. After the fall of the Berlin Wall, all departments of the LPGs not directly associated with agricultural production were dissolved. A few of the LPGs' former master craftsmen started their own businesses and employed some of their former colleagues and assistants. Some of the newly-established businesses in the private service industry such as bakeries and restaurants were also formed in this way. Many of the newly self-employed, as well as the new single-family farmers, have above-average qualifications, from master-craftsman upwards, and often possess management experience. Nevertheless, the impetus provided by medium-sized businesses for employment as a whole remains insignificant, as numerous new businesses are still fighting for survival and are facing the necessity of lay-offs to cut costs.

In many agricultural regions of Eastern Germany, the economic upswing did not take place as expected, and as a result, high unemployment remains a problem, particularly for women. Therefore, extensive job creation schemes and retraining measures have been initiated to provide a large number of people with a point of entry into working life. The job creation schemes with limited time periods are concentrated on efforts for town and village renewal, road construction and the dismantling of old LPG facilities. Often, such projects support such cultural establishments as youth and senior citizens' clubs. Special women's projects concentrate on locally oriented projects such as the extension of social services, for example, the care of the sick and elderly, party services and the production of alternative products (for example, hemp products, cheeses, and sea buckthorn liquor).

The hope that population growth in the areas surrounding the large cities of Germany will lead to an increase in employment has not yet been fulfilled.

Typically, those moving out of the cities are young families seeking reasonably priced land to build on and attractive, child-friendly neighborhoods in smaller communities. They tend to prefer areas with an urban quality of life and a nearly urban infrastructure. However, it is still necessary for these working people to commute to their workplaces in the cities. The best example of this kind of development is the suburban belt surrounding Berlin.

As the employment opportunities in the rural areas of Eastern Germany are still far from adequate, political measures will have to be taken in the future as well. Above all, it is crucial that educational policy measures give young people the skills and knowledge to keep pace with current developments in the new electronic media. Especially in Eastern Germany, there are still not enough openings for in-house traineeships. This is the result of the lack of medium-sized enterprises. For this reason, the establishment of new companies should be supported. Spreading a culture of entrepreneurship – in schools, universities and local communities – would help to increase the rate of business creation. Furthermore, it should be relatively easy to obtain the risk capital needed to establish new enterprises in Germany. Also, partnerships between the public and private sectors seem to be important keys to the successful operation and application of government programs designed to support small- and medium-sized enterprises, because they encourage the transmission of informal knowledge and networking. It is also helpful when these businesses have an experienced business manager as a consultant or advisor for the initial period of getting established.

Conclusions

Germany's rural development presents a very heterogeneous picture. On the one side, there are the prosperous rural areas of Southern Germany, and on the other, the structurally weak rural areas of Eastern Germany. In the new federal states, there will be no rapid increase in employment in the foreseeable future, despite a certain consolidation of the East German economy. The small and medium-sized enterprises of Eastern Germany must first be (re-) established and will only be in a position to provide a significant number of jobs and traineeships several years from now. The new electronic media could accelerate this process or provide a stimulus for entirely new developments.

Until then, employment and educational programs should provide support and counteract the high unemployment which is especially acute among women and young people. However, these programs usually offer only temporary solutions (part-time, seasonal and project-related jobs) and seldom provide "normal" jobs. A limited amount of employment has been created in tourism, in the production of local specialty products and in social services. But here, too, a stable employment situation has yet to be established. As these different part-time and seasonal jobs seldom develop into full-time employment – and because the German social security system is still based on the model of a complete full-time working life – new policy solutions (especially for the pensions, health care and welfare systems) are urgently needed to ensure that the social security of people in the structurally weak rural regions is not further endangered.

In the rural areas of Eastern Germany, but certainly not only there, this will depend above all on the active commitment of local-level policy makers and engaged community members to creating more jobs, improving and developing villages and towns and expanding activities in the social sphere. Perhaps it will even be those distressed rural areas that generate the innovative approaches and ideas for future models of living and working in which the now separate realms of paid work, household and family work and community work will be cohesively and meaningfully integrated.

References

Bundesministerium für Ernährung, Landwirtschaft und Forsten (2000): Agrarbericht der Bundesregierung. Bonn.
Neu, Claudia (2000): Die Transformation der ostdeutschen Landwirtschaft: Chancen und Risiken für (ehemalige) Genossenschaftsbauern. In: Hinrichs, W. and E. Priller: Eigenverantwortliches Handeln in der Transformation, Sigma Verlag, Berlin, pp. 125-151.

54 Large-Scale Creation of Sustainable Livelihoods

Shrashtant Patara and Ashok Khosla

Development that can take families, communities and nations to higher levels of peace and prosperity is a universal dream. However, as we enter a new millennium after years of unbridled industrialization, this dream remains unfulfilled for nearly half of the world's population.

Unquestionably, the world today is a better place to live than it was a hundred years ago. The revolutions in materials, energy and information technologies have provided a dramatic range of possibilities for satisfying basic human needs and for extending our capacities to shape our destinies in many new ways. Many contemporary economies have demonstrated how much "progress" is possible by adopting an aggressive use of technology and enterprise. But their experiences have also shown the need for careful selection of the types of technology and forms of enterprise to avoid wholesale destruction of human, social and environmental values. It has become clear that how something is produced, where it is produced and for whom it is produced are issues as important as what is produced.

We have come to realize over the last three decades that there are crucial social and environmental imperatives that must govern the development process and ensure that it is sustainable. In varying degrees across the world, there is broad acceptance of the concept of sustainable development, not only among environmental groups but even governments and the private sector.

Very little, however, is known about how to make this concept operational in a way that changes *everyone's* lives for the better without compromising the life support systems, natural and societal, of future generations. The need for basic goods and services for the poor, jobs for the unemployed and action to save the environment continues to be unmet.

Sustainable Technology

With the evolution of societal perceptions, aspirations and conditions and with recent developments in science, design, new materials and production processes, technological innovation is becoming increasingly important for solving the problems of poverty. New products and technologies, many with significant positive social and environmental spin-offs, are now available for mass distribution as a result of the application of sophisticated scientific and technological knowledge. Technology that serves the long-term goals of development can be defined as "sustainable technology".

Sustainable technology is the offspring of the marriage between modern science and traditional knowledge. It is a method, a process, a design, a device or a product to open up new possibilities and potentials for improving the quality of life. It requires frameworks for innovation and delivery very different from those that exist today, either on the global or the village levels. Throughout the Third World, there is an evident and pervasive need among both the rural and the urban poor for a whole variety of technologies ranging from cooking stoves and lamps to gasifiers and windmills. Tens, if not hundreds, of designs are available for each of these technologies. They are scattered in laboratories, workshops and archives throughout the world. However, these needs have not led to a more widespread demand, and the existing technical capacity has not led to supply.

The reasons that such changes have not taken place are complex and interlinked. A combination of economic, social, political and cultural - not to mention scientific, technical and institutional - factors have greatly inhibited the supply and demand for sustainable technology. A technology must be measured by how well it satisfies the needs of the end client and by how effectively it takes advantages of the opportunities and constraints of the production and marketing processes. Contrary to earlier thinking, sustainable technologies need to compete in the marketplace. To design technologies that can reconcile the conflicting requirements of the market, the environment and people requires systems for innovation and delivery comparable in sophistication with those of the most successful multinational corporations.

Many technologies for such enterprises already exist. So does the demand for their products. What prevents the poor from setting up such enterprises is their lack of access to these technologies and their inability to put together the financial capital required. What prevents them, once set up, from becoming profitable is the absence of entrepreneurial and management skills, infrastructure and marketing channels. Much more public investment is needed to provide these, but not nearly as much as is being made available today for the benefit of large, urban industries.

In the last decade, globalization has added to the complexity of the local and national economic models within which solutions must be found. Our concepts today of human development and of what constitutes well-being are not the same as they were at the turn of the last millennium - or even at the turn of the last century. In a world made dramatically smaller by technological innovations in transportation, communication and information processing, the opportunities available to people to find fulfillment have exploded in terms of range and variety. And their expectations have also grown, embedded in such concepts as "progress" and "economic development". Short of some unforeseen catastrophe, few societies would accept returning to the way people lived in earlier times.

Conventional wisdom maintains that large scale industry will supply the goods and services needed by people in a clean, efficient and cost-effective manner. Our conviction is that this is not possible. There has been considerable progress in manufacturing efficiency, but that is about all. In any case, this efficiency is derived from increasing degrees of automation. The modern economy would appear to be creating a world where cheap machines produce ever-cheaper products for other cheap machines to use. As a consequence, human beings have less to do. It is common to see more and more automation in the face of more and more unemployed people. With each day that passes, we have more products chasing less purchasing power. Today's labor displacing technologies and mechanistic economic structures can only lead to growing supply and stagnant demand – until, perhaps, we reach a catastrophic environmental transition as supplies collapse altogether and both human populations and their demands collapse with them. In effect, the very nature of the technology used puts a cap on the extent to which economic development can take place. Material intensities, mass movement of resources, transport energy and distribution costs associated with such scales of manufacturing and marketing will continue to be at levels that nature cannot support.

Clearly, a better mix of industries of different sizes is now needed. Given the continued failure of policies to address the needs of the small, mini- and microsectors, a proper balance will require greater encouragement and incentives to the smaller industries. There are, of course, sectors for which the economies of scale favor large, mechanized production units. These include steel making, oil refining, petrochemicals and automobile manufacture. But there are many sectors where economies of scale are not relevant. Most industries producing basic goods for rural populations are commercially viable even at quite small-scales. And because of the low capital requirements and short gestation periods, they can have high returns on investments - in some cases even double those for their larger counterparts.

Are there any alternatives? Sustainability on a global scale must be driven by a mix of clean and efficient production systems at all scales, including the micro and small, that create jobs by the millions. And these jobs must be productive to improve people's incomes. Essentially, developing societies will need large numbers of technology-based sustainable livelihoods.

Sustainable Livelihoods

Any simple solution proposed for a complex set of social and economic problems must inherently be suspect. Yet, if there is a *one point agenda for sustainable development*, it surely has to be the large-scale creation of sustainable livelihoods. Sustainable livelihoods create goods and services that are widely needed in any community. They give dignity and self-esteem to the workers. They create purchasing power, and with it greater economic and social equity, especially for women and the underprivileged. And they do not destroy the environment. In short, a sustainable livelihood is a remunerative, satisfying and meaningful job that enables each member of the community to help nurture and regenerate the resource base.

Sustainable livelihoods, and the human security they engender, underlie the one set of issues that is common to all nations and societies, at all stages of development. They provide a powerful synthesizing, unifying concept that can bring the most disparate interests together to design more viable economic systems for the future in any country, rich or poor.

Sustainable Enterprises

Sustainable livelihoods using sustainable technologies will require sustainable enterprises. Sustainable enterprises produce goods and services that are needed to better the lives of the great majority of people, including those who have been left outside the mainstream economy. At the same time, being environment-friendly, they minimize waste, use renewable and residues and generally conserve resources.

Taking the complete cycle from biomass generation to end-product use, entire jobs can be created at costs of a few hundred dollars; the environment can be enriched at no cost at all; and the basic needs of whole communities can be met through the additional purchasing power created.

If the full economic and environmental costs of the processes and resources used in manufacturing and delivering products is taken into account, and no "perverse" subsidies are allowed for energy, transportation, financial and other services, small-scale production can become quite competitive.

Our work in sustainable building materials and technology has shown results which, when multiplied on a large scale, can have astonishingly significant impact. Over 200 Micro Concrete Roofing tile enterprises have, for example, been set up in the last five years in rural areas with private investments of about Rs. 90,000 (US $ 2,000) each. Each enterprise produces 4,000 - 5,000 square meters of durable roofing per year, employs 5 - 8 persons, saves 40 - 60% in energy over its nearest competing product, the fired clay tile, and is profitable. At a national market share of only 5%, 3,000 - 4,000 such enterprises could be set up in India and serve the needs of people in areas that large industry would find impossible to reach.

Such enterprises can create several workplaces, each at a capital investment of US $ 200 to US $ 1,000. In addition, they indirectly lead to the creation of several more jobs, upstream or downstream, usually at an even lower capital cost. Such workplaces, in the village or small town, yield incomes for workers whose purchasing power is comparable to, if not better than, those created at a hundred times the cost in large urban industries. At the same time, they permit very high returns on investment, sometimes with payback periods of less than a year.

Examples of other Development Alternatives products include building materials, water pumping and water purification systems, recycling of waste materials, handmade paper, energy from renewable fuels and local infrastructure such as sanitation, communication and transportation.

The design and operation of rural enterprises is a complex, still unfamiliar business. The technology-environment-finance-marketing linkages have to be mastered while keeping the overhead costs low. This must occur without access to highly qualified engineers, management specialists, marketing experts, friendly bankers or market infrastructure, either for buying raw materials or for selling products.

An interesting solution to these seemingly insuperable obstacles lies in building franchised networks of small, private enterprises capable of growing and processing biomass to manufacture products for both the urban and local markets. To be successful, the franchise arrangements will have to provide high technological and marketing inputs and access to capital. Technology and Action for Rural Advancement (TARA), the commercial wing of Development Alternatives, actively franchises minienterprises based on appropriate and sustainable technologies.

The paradox of our economy is that there is virtually no source of funding today that can actually deliver adequate financial credit in this intermediate range (which might properly be termed "minicredit") where it has the greatest potential impact, both on the generation of employment and on the national economy.

These initiatives will have to come from the non-governmental sector and, more widely, the "independent sector", hopefully with direct encouragement and support from the government.

Technological, Financial and Marketing Linkages

All that small enterprises need to beat the large corporations at their own game is better access to technology, finance (not necessarily cheaper finance), and marketing channels. The primary role of the public sector in facilitating these is to provide the basic infrastructure for financing, communications and transportation. To the extent that the support providers need concentrations of other material inputs, energy, services and knowledge in order to be viable, they are, by and large, present in urban areas. The link between them and the rural enterprises is crucial.

Activation of these links will augment the creation of sustainable livelihoods by factors of ten, or even a hundred. It is self-evident to us that this is, therefore, the primary task and responsibility of development institutions in the immediate future.

Agenda 21 Principle 8, Earth Summit, Rio 1992

Principle 8 To achieve sustainable development and a higher quality of life for all people, States should reduce and eliminate unsustainable patterns of production and consumption...

Global Economic Disparities

year 2000

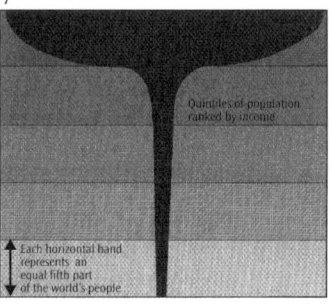

year ?

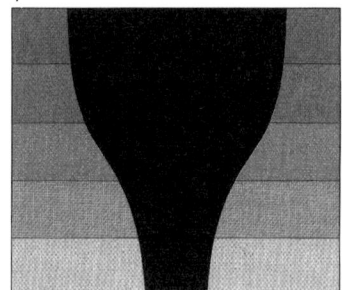

Richest fifth

Quintiles of population ranked by income

Each horizontal band represents an equal fifth part of the world's people

Poorest fifth

55 Urban World - Rural World

Kirtee Shah

While quoting statistics to establish the fact that half of the world is now urban, the tendency is to ignore that, by the same token, half of the world is still rural. In fact, it is more rural than urban in Asia and Africa, where development challenges are the most difficult. For instance, 74% of India is still rural – a staggering 740 million people! In view of the growing urbanization trends such as the increasing contribution of cities to national economic growth and the complex problems of environmental management, resource mobilization, infrastructure provision and governance, a focus on cities is timely and unavoidable. However, it needs to be recognized that rural areas are still afflicted by poverty. They suffer from a lack of basic services, unemployment, under-employment, deficient infrastructure, paucity of investment resources, a declining contribution of agriculture in the GDP and the consequent marginalization of populations dependent on it, inadequate housing, recurrent natural disasters, social and economic inequality and persisting structures of exploitation. Taken together, these factors constitute a formidable "push factor" for migration from rural to urban areas. Stable and economically viable rural settlements will ensure manageable urban growth. Therefore, sustainable and viable rural development strategies should constitute an integral part of molding a new urban future. Villages cannot remain the suppliers of food and raw materials and also be the dumping ground for rotting urban waste. They have a role to play and should be allowed and facilitated to do so.

Inevitability of Urbanization

Historically, cities are the products of many forces and influences. There is little doubt that the most dominant of these shaping current urbanization trends and modern city development is economic globalization and its accompanying development model. There are many factors, such as Structural Adjustment Programs, global integration and liberalization of national economies, businesses forming corporations, the role and power of multinational corporations and giving free reign to market forces, which on the one hand spur economic growth, technological transformation and ensure high levels of prosperity to a select few. However, on the other hand, they can deepen poverty, widen inequality, cause exclusion and marginalization, promote wasteful consumerism, undermine national sovereignty, destroy the environment, deplete natural resources, tilt the investment balance in favor of cities (especially the big cities), and cause cultural alienation. Thus, people's capacities to find solutions rooted in their culture, social norms, value systems and traditional wisdom are seriously damaged. In spite of this, many argue that the force of globalization and the resultant urbanization are irreversible and inevitable.

Questioning the inevitability of urbanization – especially the inevitability of resource depleting, polluting, exploitative and in some ways dehumanizing urbanization – is an important step in moving towards sustainable development. It is essential to recognize that the urbanization we experience and the cities we live in are products of the economic policies we pursue and the development model they promote. It is the result of conscious choices we have made, not divinely ordained. If the policies and the model change, the urbanization trends and cities will also change. There is nothing inevitable about it.

Urban versus Rural

Is the urban-rural relationship inherently adversarial? Is the relationship exploitative? Though urban experts advise against "either-or" formulations, the matter does bear some resemblance to the relationship prototypes of small-big, rich-poor, north-south, developed-underdeveloped, agricultural-industrial, etc. The following two statements on the nature of modern cities give an indication of possible areas of tension between cities and their rural hinterlands:

Cities occupy just 2 per cent of the world's land surface, yet they use three quarters of the world's resources and discharge a similar percentage of waste.
Greater London has a foot print 125 times larger than its own area. With only 12% of Britain's population, it requires an area equivalent to all the country's productive land.[1]

The well-known disparities between per capita availability of environmental services (water, power, sanitation), per capita investment in social services (education, health, family welfare) and the economy (agriculture plus irrigation versus industry) in rural and urban areas within a given country indicate a bias in favor of the urban areas. Urban productivity has its price. The question is: are villages paying that price? Undoubtedly, urban productivity also benefits the rural sector. The related question is: is the price disproportionate to the benefits?

Urban-Rural Poverty Link

Although the percentage of rural-urban migration compared to the overall urban population growth in developing countries is declining (natural increase and reclassification of settlements are other components in this equation - in low-income India, natural increase provides 41% of the urban population growth, migration 40% and reclassification 19%), poverty-induced migration still contributes a fair amount to urban population growth and increases the incidence

---

[1] Urban foot print, according to Canadian economist William Rees, is *"the area of land required to supply a city with food and timber, and the vegetarian cover needed to absorb its discharges of carbon dioxide."*

of urban poverty. While urban poverty is not just the spillover of rural poverty, it still is effected in both nature and scale by the influx of poor migrants from rural areas, especially in the South Asian countries of India, Pakistan and Bangladesh. Tackling rural poverty, therefore, is an essential part of improving urban living conditions in general and alleviating urban poverty in particular. How this is accomplished is important. The strategy should include:

(a) achieving higher growth rates in the *"indirect route"* approach.[2] 7 to 8% instead of 4%, *"to allow wealth to trickle down to the poor;"*
(b) avoiding the five "bad" patterns of growth listed in UNDP's Human Development Report: jobless growth, ruthless growth, voiceless growth, rootless growth, and futureless growth;
(c) working towards sustainable agriculture practices;
(d) improving the productivity and performance of specific target groups in focused antipoverty programs (the *"direct route"* approach to poverty alleviation);[3]
(e) higher investment in human resource development, especially in basic services like water, sanitation, education, primary health care, family planning, etc., and
(f) empowering orientation, especially for women in development planning and action.

---

[2] The *"indirect route"* approach means the trickle-down effect of rapid economic growth.
[3] India's efforts with a colossal investment of US $15 billion in twenty years has produced only a moderate success, with only 12% of the assisted families crossing the poverty line.

Index

access to
  basic services 12, 21
  credit 71, 78, 101
  drinking water 9, 190
  education 19, 43, **44**, 115, 126
  food 97
  forest resources 55
  information 5, 11, 13, 14, 110, 228, 235, **237**, **240**, 261
  infrastructure and services 22, 34, **43**, 47, 69, 71, 101, 110, 115, 186, **352**
  knowledge 315
  land 107, 219
  markets 10, 92, 204
  NGOs 69
  public goods 16
  research 8
  resources 115
  rights 18
  technologoy 19
  technology 7, 10, 25, 395
  water **195**, **215**, 219, **223**
administration 5, 39, **59**, 71, 84, 92, 101, 142, 208, 242, 260, 355, **361**
Agenda21 78, 85, 89, 288, 346
agricultural
  production 3, 11, 16, 22, 38, 40, **46**, 56, **67**, 108, 111, 155, 170, 173, **175**, **185**, **203**, **279**, **281**, 296, 307, **340**, 386, 387
  productivity 16, 21, 22, **44**, 115, 199, 279, **283**, 295, 321, **323**, 339, 345, 347, 358, 378
  research 19, **45**, 115, 173, 182, 282, **289**, **295**, 298, 305, 317, **321**, 341
  trade 123, 291, 373
assistance
  development **98**, 106, 109, 236, 242, 252, 275, 370, 381, 383
  technical 99, 209, 242, 381
authorities 86, 105, 112, 211, 370
  abuse of **59**
  administrative 23, 275, 353
  local 6, 18, 22, 116, 135, 138, 227, 229
  public 260, 353

  urban 21, 379, 380, 383
awareness
  public **228**, 253
  raising 13, 25, 28, 51, 53, 87, 111, 125, 156, 192, **228**, 260, 345, 369

biodiversity 3, 40, 83, 155, 170, 291, 295, 299, 303, 318, **344**, **346**, 373
biosafety 342
biotechnology 11, **45**, 279, **283**, 296, 321, 323, **325**, **329**, **339**
breeding 17, 165, **285**, 323, 324, **325**, **329**, **340**, **345**

capacity building 18, **57**, 109, 130, 261, 374, 375
civil society 7, 18, 22, 24, 37, 51, 91, 92, 112, 155, 194, 236, **251**, 260, 353, 357, 367, 377, 382
Common Agricultural Policy (CAP) 83
community 23, 47, **60**, 77, 78, 82, 88, 94, **100**, **109**, **111**, 131, 142, 235, 236, **240**, 251, 252, **260**, 339, 353, 367, 368, 389, 391, 393, 394
  development 35, 109, **259**
  eco 147, 148
  farming 209, 212, 339
  international 155, 189, 191, 208, 322
  local 44, 57, 61, **65**, 93, 117, 173, **198**, 388
  rural 3, 9, 18, **57**, 85, 89, **104**, **119**, 153, 198, 261, 286, 299
  rurban 148
  sustainibility **145**
  water users **215**
company 4, 34, 47, 79, 84, 125, 142, 198, 256, 386, 388
  energy supply 131, **141**
  services **283**
  telecommunication 277
concept **3**, 25, 54, 56, 77, 81, 119, 147, 180, 275, 290
  development 78, 87, 115, 391, 392
  energy 88
  participation 99
conservation

biodiversity 17, 79, 92, 295, 296, 336, 343, **345**
  energy 152, 290
  environmental 34, 55, 86, 88, 155
  soil 14, 56, 111, 155, 156, 308, 311, 312
  water 16, 56, 190, **223**, 281, 308
consumption 77, 134, 169, 379
  energy **129**, **141**, 146, 150
  food 264, 270, 287, 314, 342
  fuel 284
  mineral fertilizers 160, 186
  needs 82, 116
  over 63, 64
  patterns 21, 24, 40, 61, 94, 124, 282, 329
  resources 281
  soil 9
  water 189, **204**, **225**
Convention on Biological Diversity (CBD) **346**
cooperation 71, 88, 92, 94, 126, 173, 197, 230, 290, 317, 323
  development 22, 24, 36, 39, 353
  international 16, 189, 190, 203, 207, 208
  technical 190
  urban-rural 352
cooperative 7, 380
  agricultural 69, 217, **385**
  credit 70
  farmers 46, 123
  milk 112
  producers 249
costs
  agricultural **341**
  capital 145, 146
  environmental 81, 305, 394
  for public drinking water 302
  health 335
  investment 209, 211
  labor 74, **308**
  of embryo transfer 330
  of gene transfer 335
  of labor 309
  of production 165, **298**, 310, 335
  of village infrastructure **81**, 135, **136**, 261, 372
  of village infrastructures 102
  transaction 347, 358
  transport 268, 275, 393, 394

village infrastructure 142
village infrastructure 277
countries
  developing 3, 10, 11, 16, 21, 23, 27, 41, 44, 45, 47, 78, 115, 129, 130, 133, 149, 155, 157, 158, 160, 161, 162, 169, 190, 199, 204, 206, 210, 226, 238, 252, 256, 259, 261, 275, 280, 287, 296, 309, 321, 323, 329, 336, 341, 345, 351, 355, 359, 361, 377, 398
  high-income 11, 45, 238, 239, 275, 342
  industrialized 4, 11, 16, 79, 129, 130, 131, 141, 157, 239, 241, 275, 279, 280, 300, 302, 323
  low-income 4, 12, 14, 43, 45, 239, 275
  middle-income 18, 43, 275
credit 207, 224, 313, 318, 357, 380
  and financial services 44, 46, 47, 68, 104, 108, 115
  associations 57, 70, 93, 112, 116, 220
  lack of 162
  micro 26, 58, 71, 74, 263, 271, 395
  rural 115
crop varieties 45, 298, 308, 326, 327
  high yielding 159, 163, 164, 182, **339**
  local 345
cropping 38, 163, 170, 177, 308, 312
  patterns 206, 207
  rotations 79, 159
  systems 181, 296
culture 8, 38, 82, 108, 121, 147, 198, 248, 369, 372, 388, 397

decentralization 6, 18, 33, 39, **51**, 102, **109**, 117, 137, 353, 355, 358, 359, 374, 380
deforestation 98, 131, 146, 149, 150, 152, 159, 165, 173, 176, 203, 373
degradation
  economic 63
  environmental 59, 98, 321, 383
  of energy resources 9
  of genetic resources 8
  of natural resources 11, 14, 40, 131, 295
  of soil 9, **155**, **169**, 286, 312
  of the urban environment 370

political 63
de-industrialization 385
democracy 4, 6, 33, **54**, **59**, 86, 101, 190, **215**, 368
desertification 40, **155**, 170, **173**, 287
development
  economic 8, 22, 41, 139, 200, 236, 263, 275, 277, 279, **357**, 369, 371, 374, 393
  funds 14, 35, **93**
  integrated **125**
  of new water supplies **191**
  options 22, 61, 109, 136, 145, 171, 224, 279, 306, 380
  regional 115
  rural 3, **4**, **22**, 25, **33**, **35**, 71, 74, 78, **83**, **85**, **91**, **93**, **97**, **112**, 123, **133**, **185**, **237**, **264**, 275, 307, **321**, **351**, **355**, **369**, **377**, **385**, **391**, **397**
  socioeconomic 133, 152, 241
  sustainable 3, 7, 14, 34, 78, 83, **85**, **91**, 94, 98, **107**, **119**, 125, 129, 155, 223, **281**, 290, 295, 368, 382, **393**, 398
  technological 13, 25, 295, **323**, 329, 344
  urban 22, 354, **367**, **369**, 380, **397**
  water resources **203**, 215, **223**
development concepts
  regional 13, 22, 23
  rural 19, **36**
digital divide 26, **235**, 238, 251, 252, 253, **259**
disincentives 309
donors 36, 38, 46, 82, 99, 101, 103, 109, 135, 145, 152, **208**, 212, 229, 236, 361, 368

economy
  global 10, 23, 85, 251, 291, 345, 352, 373, 374
  information 238
  knowledge-based 259, 372
  national 260, 395
  rural 17, 19, 43, 48, 62, 68, 137, 142, 164, 204
  urban 21, **373**
education 6, **8**, 11, 18, 33, 38, 40, 44, 48, 51, **57**, 59, 64, 77, 78, 86, 92, 94, 110, 125, 130, 152, 224, **228**, 235, **237**, **252**, 256, 260, 277, 287, 295, 351, 353, 356, 368, 369, 372, 382, 388, 398, 399
efficiency
  economic 85, 223
  energy 8, **129**, **151**
  market 269
employment
  agricultural 369, 372, 378
  nonagricultural 352, **377**
  opportunities 17, 22, 28, 77, 83, 92, 135, 235, 238, 318, 351, 354, 370, **379**, 388
  patterns 23, 78, 351
  rural 23, 33, 75, 280, **306**, **369**, 378
  urban 21, **369**
empowerment 6, 16, 57, **58**, 60, 65, 100, 104, 146, 200, 236, 264, 271, 368
energy
  industrial use 151, 296
  research 130, 136, 138, 145, 149, 151
  sources 131, 138, **145**, **149**
  supply 8, 14, 88, **130**, **133**, 141, 149
  system 14, **129**, **133**, **141**, **145**, **149**
entrepreneur 9, **65**, 74, 108, 115, 237, 249, 373, **385**, 392
environmental
  accounting 23
  benefits 152, 299, 302, 391
  damage 212, 281, 295
  degradation *see degradation*
  functions 78
  impacts 23, 112, 225, 275, 296, 321, 372
  issues 65, 77, 193
  management 110, 397
  problems 21, 302, 373
  quality 81
  refugees 174
  requirements 79
  transformation 169
environmentally friendly
  agriculture 125, 307
  energy concept 88
  technologies 281, 368
environmentally sensitive areas 369, 373
environmentally sound
  agriculture 9, 86
  infrastructures 92

water utilization 10
equity 56, 61, 65, 81, 82, 93, 94, 97, 98, 103, 145, 197, 219, 221, 223, 235, 237, 305, 314, 318, 367, 393
erosion control 155, 165, **169**, 178, **181**, 206
extension 11, **45**, 108, 112, 115, 150, 182, **212**, 263, 312

farmers
　African **179**, **185**, 209
　commercial 198
　female 217, 220
　organic 123
　organizations 46, 102
　resource poor 12, 318, 342
　resource rich **282**, 342
　small-scale 4, 11, 16, 45, 46, 115, 125, 182, 192, 198, 237, 261, **286**, **312**, **313**, 318, 322, 324, 341, 342, 343, 373, 386
　urban 380
farming
　activities 38, 133, 136, 341
　industrial-style 300
　low external input 282
　organic 88, 94, **123**, 166, 282, **298**, 301, 302, 303, **305**
　pioneering 282
　practices 46, 81
　precision 279, 282, **283**
　prescription 282
　subsistence 166, 282, 287
　systems 46, 79, 104, 173, 207, 209, 212, 286, 299, 307, 309, 312, 316, 322, 326
　traditional 289
fertilizer **157**, 169, **178**, **185**, 207, 269, 281, **283**, **287**, 290, 298, 302, 303, 308, 340
fisheries 57, 322, 378
food
　availability 281
　demand 16, 33, 155, 279, 280, 290, 324
　import 205
　safety 11, 280, 295, 300, 301
　security 4, 8, 10, 35, 44, 73, 77, 108, 165, 169, 191, 279, 288, **289**, **295**, 321, 323, 343, 345

forest 8, 23, 52, 59, 61, 64, 151, 203, 205, 318, 371
　afforestation 149, 288
　land 112, 169
　management 54, **55**
　organizations 217
　rain 170, 175, 176, 243, 370
　reforestation 206, 288
　zones 97, 98, 101
forestry 79, 85, 91, 151, 322
framework
　democratic 5, 7
　economic 110, 375
　institutional 6, 14, 25, **31**, 35, 56, 57, 102, 117, 209, 212, 219, **229**, 369
　legal 18, 102, 209, 220, **229**, 259, 346
　political 5, 17, **31**, 33, 57, **83**, 102, 107, 212, 345, 352, 368
　urban-rural linkages 24
future
　of agriculture **11**, **16**, **213**, **279**, **281**, **297**, **305**
　of rural areas 4, 13, 22, 23, 25, **86**, 279, 351, 352
　of urban agriculture 383

GATT (General Agreement on Tariffs and Trade) 277, 291
gender
　equality 44, 53, 108, 120, 190, 197
　inequality 107, **216**
　perspective 29, 93, 107, 110, 218, 305
　poverty 104, 311
　water management **215**
genetic diversity 308, 325, **347**
genetically modified
　organisms 288
　organisms (GMO) 288, 290, 302
　plants 281, 284, 285, 288, 289, 297, 325, 327
GIS (Geographic Information System) 135, 283, 296
globalization 3, 12, 14, 17, 23, 27, 35, 40, 64, 85, 86, 155, 351, 352, 373, 374, 381, 382, 392, 397, 398
government 22
　agencies 56, 112, 192, 194, 199, 208
　central 6, **51**, 109, 208, 209, 374, 383
　decentralized system of 108

initiatives 108
local 6, 40, **47**, 57, **59**, 72, 108, 110, 208, 209, 383
local institutions 23, 113
local organization 39
national 6, 27, 33, **47**, **59**, 182, 259, 361, 368, 380
policy 68, 380, 382
programs 388
regional 368
regulation 34, 47, 68
rural self 69, 71
socialist 377
GPS (Global Positioning System) 283, 284
grassroots
innovators 93, 95
institutions 60, 116
level 5, 94, 199
movements 7, 14, **93**, **97**, **115**, **123**
organizations 7, 37, **97**, 116
Green Revolution 281, **305**, 340, 358
growth
agricultural 125, 133, **298**, 307, 309
economic 5, 19, 22, 26, **43**, 129, 133, 200, 203, 251, 279, 282, 295, 352, 353, 355, 357, 361, 368, 397
income 8, 329
Internet 246, 252
of the internet 259
of world food demand 290
plant 169, 181, 289
population 155, **157**, 169, 189, 203, 279, 282, 286, 351, 387, 398
rural 6, 11, 14, 33, **43**, 355, 358
urban 8, 22, 354, 358, 367, 380, 397

health 7, 21, 33, 38, 108, 110, 111, 191, 241, 252, 260, 265, 280, 287, 295, 296, 313, 321, 341, 343, 357, 367
aide 243, 247
care 6, 12, 18, 44, 48, 54, 59, 125, 235, 237, 242, 247, 248, 258, 329, 388, 399
centers 115, 138, 152, 242, 243, 245, 247
child 58
education **244**
information 242
problems 190, 243

services 43, 115, 133, 243, 269, 351, 371
tele **241**
urban 12
workers 242, 243, 244, 245
herbicides 165, **283**, 296, 326, 341
household
electrification of **134**
farm 307, 309, 311, 314, 379
female-headed 104, **311**
plots **68**
poor 10, 52, 55, 265, 273
rural 69, 104, 115, 134, 138
human
actions **8**, 171
capital 29, 80, 82
captial 253
development 7, 260, 277, 321, **356**, 392, 399
needs 8, 186, 195, 197, 281, **295**, 391

ICM (Integrated Crop Management) 279, **290**, 296
ICTs (information and communication technologies) 10, **235**, **237**, **251**, **255**, **259**, **263**, **275**, **277**, 296, 321, 322, 368
implementation
arrangement 98, 102, 103, **105**, 199
of local activities 109, 113, **137**
policy 14, 37, 38, 64, 80, **84**, 86, 110, 130, 192
process 195
program 58, 137, **208**, **228**
project 56, **97**, 112, 150, 156, 207, 223
incentives 6, 9, 19, **46**, 80, 81, 113, 130, 173, 182, 209, 211, **224**, 236, 238, 259, 313, 393
inadequate 9
income
distribution 77, 81, **115**, 374
farm 204, 210, 301, 305, 311, 312, 377
generation 45, 47, 69, 70, **100**, 112, 145, 152, 264, 277, 353, 378
non-farm 48, 378
rural 43, 55, **67**, 86, 99, 191, 210, 371, 377, 379
sources 7, 12, 13, 55, **68**, 73, 92, 108, 120, 279, 372, 378

urban 371, 379
industrialization 22, 152, 158, 280, 370, 371, 383, 391
    of agriculture 4, **297**, **369**, 385
information
    management **251**, 260, 264, 277
information society 3, 236, **238**, **251**, 260, **275**
infrastructure
    availability 352, **360**, 398
    basic 21, 238, 263, 354, 395
    development 98, 104, 141, 313, 374
    economic 374
    local 104, 355, 394
    rural 17, 72, 258, 279, 283, **355**
    social 71, 115, 252, 379
    urban 371, 388
    urban-rural 5, 12, 22
innovation
    grassroots **93**
    indigenous 3
    institutional 6, 8, 18
    rural 7, 8, 16, 38, 147
    technological 3, 6, 10, 19, 33, 236, 241, 252, **281**, 340, 391
insurance 7, 46, 69, 115, 245
inter-governmental mechanism 261, 262, 346
investment
    in infrastructure 22, 248, **255**, 351, 357, 361, 382
    in natural resources **181**
    in water supply **191**, **207**, **224**
    labor **310**
    local 93, **97**, 100, 102, 103, 106, 358, 371
    private 186, 352, 353, 394
    protection of 291
    public 45, 186, 352, 353, 392
    urban 368, 371
    urban-rural **397**
IPM (Integrated Pest Management) 11, 45
IPNS (Integrated Plant Nutrition Systems) 11, 279, **290**
IPR (Intellectual Property Rights) 45, 259, 291, 346
IRM (Integrated Resource Management) 223, **290**

irrigation 10, 45, 111, 139, 145, 152, 185, **189**, **191**, **203**, **215**, 226, 286, 290, 398

jobs
    creation 85, 88, 91, 94, 123, 146, 152, 165, 200, 204, 255, 385, 386, **387**, **391**
    farm 86, 386
    non-farm 8, 378, 387
    urban 12

knowledge **8**, 13, **95**, 119, 220, 228, 251, 272, **295**
    development 26, 39, 83, 235
    divide **251**
    indigenous 58, 156
    informal 388
    lack of 9, 109, 141, 161, 182, 213, 217
    scientific 58, 297, 391
    systems 18, 94
    technological 120, 391
    traditional 19, 29, 103, 392
    urban 19
knowledge intensive agriculture 282, 307, 322

labor
    availability 315
    demand 44, 67, 111, 307, 308
    farm 45, **305**, 379
    force 43, 104, 371
    in sustainable agriculture **307**
    intensity 280, 307, 308, 310, 315, 373
    productivity 287, **305**
    requirements 280, 287, **305**, 342
    rural 16, 23, 280, 308
land
    arable 11, 70, 157, 185, 284
    availability 157, 170, 296
    reform 44, 385
    tenure 24, 173, 182
    use 14, 69, 79, 91, 171, 173, 207, 296, 379, 383
LCP (Least-Cost Planning Program) **142**
livelihood 22, 29, **43**, 54, 107, 145, 173, 191, **255**, 259, 280, 288, 307, 311
    improvement 3, 6, 19, 34, 97

rural **6**, 8, 19, 33, **70**, 111, 123, 353
security 236
sustainable 353, **391**, **393**
livestock 45, 98, 155, 173, 185, 269, 298, 307, 322, 323, **329**, 378
low external input agriculture 16, 279, **282**

malnutrition 22, 44
market **47**
  economy 70, 386
  failures 93, **115**
  forces 130, 173, 298, 371, 374, 397
  global 14, 27, 237, 259, 369, 373
  infrastructure 47, 395
  input 47, 115, 175, 186
  integration 34, 71
  international 17, 23, 351
  local 27, 69, 209, 272, 395
  national 28, 394
  output 47, 115, 175, 186
  policies 85, 373, 374
  regional 69
  rural finance **115**
migration 4, **22**, 33, 55, 69, 119, 165, 174, 275, 312, **351**, **369**, 370, 372, **377**, 380, **381**, 398
modernization 19, 69, 108, 136
  of rural areas 131, 133

NGOs (non-governmental organizations) 28, 44, 47, 52, 60, 87, 93, **99**, 111, 126, 138, 193, 195, 252, 261, 290, 301, 357, 368, 382, 383
  rural 6, 56, 74
nitrogen **159**, 176, 177, 186, 302, 308
nutrition 44, 108
  crop **162**, **341**
  deficiencies 329, 336
  human 287, 291, 321

participation 24, 33, 37, 59, 86, 91, 308
  community **51**
  equal 150, **260**
  farmer 208, **210**
  global 236, **252**
  in decision making 93
  local 14, 117, 146, 156, **275**
  political **51**
  private sector 224, **230**
  public 226, 238, 249

regional 5
rural 67, 94, 99, **102**, 113, 134, 173
social 28
water management **193**, **215**
women 107, **315**
partnership 22, 37, 84, 97, 125, 156, 182, **251**, 260, **322**, 341, 381, 382
  administrative 57
  public-private 24, 45, 138, 388
  urban-rural **91**
pesticides 169, **283**, 296, 298, 302, 308, 326
PGRFA (plant genetic resources for food and agriculture) 345
plant protection 281, 286, 290, 317
policy
  agrarian 33, **69**, **85**, 133, 186, 280, 286, 318
  development 5, **6**, 27, **36**, 65, 79, 173, 253
  employment **317**
  energy **130**, 149
  failures 4, 190, 393
  global 26
  implications 3, 273
  makers 3, 58, 175, 192, 207, 213, 220, 263, 273, 295, 303, 305, 322, 355, 359, 389
  priorities 33, 36, 355
  public goods **77**
  rural 4, 29, **36**, 78
  rural development 14, 23, **33**, **43**, **83**, **85**, 370, 380
  urban 4
  urban-rural 23, **352**, **368**, 371, 380
pollution 19, 21, 81, 169, 290
  control 145, 229
  urban 368, 371
  water 79, 181, 196, 204, 223, **225**, 228
population
  rural 7, 8, 10, 16, 19, 25, 33, 36, 43, 46, 67, 70, **73**, 86, 93, 103, 119, 131, 133, 152, 255, 308, 343, 354, 359, 369, 372, 393
  urban 12, 17, 21, 43, 69, 275, 367, 371, 372, 380, 398
  world 3, 18, 136, 203, 275, 281, 329, 369
poverty
  increase 21, 27, 28, 64, 311, 370, 397

reduction 8, 13, 19, 21, 22, 23, 28,
    29, **43**, **91**, 100, 105, **107**, 173,
    252, 264, 273, 279, 321, 357, **367**,
    371, 391, 399
  rural 4, 13, 19, 24, 33, 48, 56, 70, 73,
    86, 97, 104, 115, 203, 242, 324,
    339, 344, 354, 371, 397
  urban 21, 24, 367, 399
  urban-rural **398**
privatization 70, 186, 230, 275, 386
production
  animal 16, 185, **329**
  crop 70, 111, **158**, 175, 179, 185,
    284, 295, 308, 326
productivity 12, 16, 129, 251, 257, 264,
    269, 275, 351, 398, 399
protection
  biodiversity 295
  climate 40, 129, 142
  cultural and social heritages 34, 86
  indigenous people 370
  IPR 291, 347
  marginal land 165
  minorities 36, 40
  security and peace 30
  water resources 225, 228
public goods 7, 34, 47, **77**, **352**

reform
  agrarian 22, 28, 83, 318, 382
  economic 162, 377, 379, 382
  educational 246
  institutional 66, 192, 193, 197, 224,
    228, 229, 260
  local level 86, 192
  Russian 33, **67**
  structure 28
  Ugandan law 107
research 8, 9, 11, 16, 93, 125, 217, 285,
    329, 336, 355, 379, 383
resource management 8, 35, 38, 65,
    111, 113, 216, 223, 287, 321, 324,
    339
resources
  energy **9**, **136**
  financial 14, 75, 77, 82, 107, 134,
    182, 296, 352, 361
  genetic 3, 8, 11, 17, 25, 295, 296,
    321, 324, 336, **343**, **345**
  human 37, 46, 58, 113, 119, 149,
    252, 255, **256**

  natural 3, **8**, **23**, 25, 28, 33, 56, 60,
    65, 80, 81, 85, 92, 111, 119, **127**,
    145, 151, 164, **165**, 169, 171, 173,
    181, **185**, 279, 290, 295, 321, 340,
    368, 369, 397
  rural 94, 100
  water 16, 25, 169, 173, 189, **191**,
    204, **223**, 228
rights 6
  human 7, 29, 33, 55, 61, 64, 189,
    321, 370, 373
  land 156, 219
  legal 64, 66
  property 28, 44, 46, 81, 107
  social 14, 30
  sovereign **59**, **63**, 346
  user **81**, 116
  water 47, 189, 200, 207, 219, 225
rural electrification 9, 131, **134**, 146,
    152
rural life 4, 13, 16, 33, 55, 70, 133, 257,
    258, 277, 354
rural organizations 7, 14, 18

sanitation 9, 21, 43, 54, 252, 257, 394,
    398
SAP (Structural Adjustment Program)
    186, 378, 380, 381, 382
sector
  agricultural 22, 33, 38, 39, 45, **67**,
    69, 86, 108, 136, 192, 261, 355,
    359, 385, 387
  energy 149
  industrial 22, 28, 30, 198, 223, 372
  informal 379, 381
  non-agricultural 13, 170
  policy 194, 213
  private 5, 6, 10, 12, 24, 37, 46, 108,
    138, 159, 186, 208, 212, 229, **230**,
    236, **251**, 275, 287, 291, 322, 358,
    368, 388
  public 46, 230, 260, 275, 287, 358,
    381, 382, 395
  rural 28, 150, 358, 398
  small-holder 11, 16, 318
  urban 145
  water 196
seed
  distribution 287
  genetically modified **284**, **342**
  hybrid **281**, 343

improvements **281**, 286, 287
industries 28, 45, 286, **287**, 291
markets 12
subsidies 207
self-help groups 57, **58**, **111**
settlements 7, **21**, 79, 92, 226, 236, 239, 275, 353, 358, **359**, 368, 370, 377, 397, 398
  resettlement 381
small-scale enterprises 12, 19, 26, 71, 100, **352**, 386, 394
small-scale irrigation schemes **195**, **209**, 217
soil
  erosion 56, 98, 159, **169**, 173, 177, 181, 287, 373
  fertility 14, 155, 156, **159**, **169**, **178**, 287, 312
  fertility management **159**, 173, 179, 185, 186
  fertility managment 47
  management 9, 14, **155**, **157**, **169**, **173**, **175**
  nutrient mining 8, **162**
  productivity 155, 159, 170, 173, 181, 287
subsidies 21, 68, 71, 179, 182, **207**, 235, 248, 280, 298, 299, 312, 318, 371, 385, 394
sustainability 40, 60, 63, 64
  agriculture 170, 186, 204, **281**, 287, **305**, 321, 343, 399
  associations 194
  ecological 36
  economic 36
  global 59, 61, 66, 129, 393
  investments **99**
  principle of 86, 88
  principles 88, 145, 147, 228
  rural 19, 78, 93, 147
  social 36
  urban 145
  water supply 223

technologies
  adapted 11, 13
  adoption 16, 153, **181**, 252, 280, 287, **305**, 329, 340
  agricultural 10, 16, **279**, **281**
  energy 14, 130, 136, 145, **149**
  erosion control 181

irrigation 198
labor-saving 280
modern 10, 16, 19, 26, 28, 236, 263, 296
new 5, 10, 11, 12, 16, 26, 94, 133, 236, 261, 288, 291, 295, 305, 323, 329
options 4, 11, 223, 280, 340, 352
sustainable 145, **391**
traditional 19
transfer 12, 130, 137, 291, 296, **341**
water 16, 47, 120, 189, 229
trade 6, 16, 40, 55, 80, 115, 173, 237, 277, 298, 301, 351, 387
training 40, 55, 92, **100**, 108, 110, 112, 115, 125, 133, 137, 194, 212, **242**, **252**, **256**, 260, 296, 308, 315
transportation 22, 270, 277, **351**, **357**, **369**, 382, 385, 392
  efficiency 205
  energy efficiency 9
  energy efficient 131, 148
  lack of 242, 341
  rural 47
  system 368
  systems 12

unemployment 27, 33, 56, **67**, 353, 370, 377, **385**, 397
urban
  centers 3, 12, 13, 19, 41, 77, 115, 133, 159, 200, 229, 261, 352
  economies 373, 379
  life 16, 21, 120, 372, 373
  population *see population*
urban footprint 8, 24
urbanization 3, 12, **21**, 43, 48, 155, 157, 203, 223, 329, 339, **369**, 379, 383, **397**
urban-rural linkages 7, **12**, **17**, **21**, **352**, **355**, **367**, **377**, **379**

village pay phones **263**

water
  availability 112, 158, **204**, 206, 210, 296
  contamination 155, 229
  erosion 79, 98, 155
  groundwater 120, 165, 181, 189, 194, 223, 229

industrial use 191, 193
management 9, 100, 152, 190, **191**, **192**, **196**, **208**, 212, **215**, **223**, 229
market 194
markets 189
productivity 159, **191**
reform 47, 192, 196
wastage 8, 189, 194, 223
wind
  energy 136, 138, 147, **149**, 392
  erosion 79, 155
WTO (World Trade Organization) 277

Druck: Strauss Offsetdruck, Mörlenbach
Verarbeitung: Schäffer, Grünstadt